Progr Colloid Polym Sci (2002) 121: 67–71

ERRATUM

S. Siegel
D. Vollhardt

Temperature-dependent domain growth in 9-hydroxypalmitic acid monolayers

Due to a communication gap between Kiel and Heidelberg, not all authors of the contribution:

"Temperature-dependent domain growth in 9-hydroxypalmitic acid monolayers"

Progr. Colloid Polym. Sci. (2002) 121: 67–71 were named. The authors are:

S. Siegel, D. Vollhardt, and D. A. Cadenhead.

The Editor and Publisher

Springer-Verlag Berlin Heidelberg GmbH

Progress in Colloid and Polymer Science
Editors: F. Kremer, Leipzig and G. Lagaly, Kiel

Volume 122 · 2003

Aqueous Polymer – Cosolute Systems

Special Issue in Honor of Dr. Shuji Saito

Volume Editor:
Dan F. Anghel

Springer

The series Progress in Colloid and Polymer Science is also available electronically (ISSN 1437-8027)

- Access to tables of contents and abstracts is *free* for everybody.
- Scientists affiliated with departments/institutes subscribing to Progress in Colloid and Polymer Science as a whole also have full access to all papers in PDF form. Point your librarian to the LINK access registration form at http://link.springer.de/series/pcps/reg-form.htm

ISBN 978-3-642-05512-6
ISBN 978-3-540-36114-5 (eBook)
DOI 10.1007/978-3-540-36114-5

http://www.springer.de

Originally published by Springer-Verlag Berlin Heidelberg New York in 2003
Softcover reprint of the hardcover 1st edition 2003

Production: PRO EDIT GmbH, Elke Beul-Göhringer, Haberstr. 7, 69126 Heidelberg, Germany, Fax: +49-6221-379-777, e-mail: elke.beul-goehringer@proedit.de

Typesetting: SPS, Madras, India

Cover Design: Estudio Calamar, F. Steinen-Broo, Pau/Girona, Spain

Cover Production: design & production, 69121 Heidelberg, Germany

Printed on acid-free paper

Progr Colloid Polym Sci (2003) 122: V

PREFACE

Dr. Shuji Saito

This issue is dedicated to Doctor Shuji Saito on the occasion of the 50th anniversary of his first paper published in the Journal of German Colloid Society. The paper entitled "Die Solubilisation von Polyvinylazetat" by N. Sata and S. Saito appeared in the *Kolloid Zeitschrift* (1952) 128: 154. According to the title, the paper dealt with the solubilization of poly(vinyl acetate) (PVAc) in aqueous sodium dodecyl sulfate (SDS) solutions. The authors observed that PVAc, a water-insoluble polymer, is completely dissolved by micellar SDS solutions, a result that mostly intrigued them. One has to recall that at that time, the micellar solubilization was the only available theory to explain the dissolution of hydrophobic compounds in micellar systems. The theory worked well in the case of low molecular weight compounds like hydrocarbons, oleophilic dyes, etc., which are solubilized in the inner core of the micelle, but how can a small micelle accommodate a giant polymer molecule? To solve this problem Sata and Saito originally proposed a model consisting of surfactant aggregates formed along the polymer backbone. The model based on simple viscometric measurements was later in the eighties confirmed by the advent of the more sophisticated neutron scattering technique and is nowadays called the "necklace model".

During his career, Dr. Saito published about 70 original and review papers, which is a very good score for a person who mainly worked in a cosmetic company and not in the academic field. The majority of his papers belong to the interaction between ionic or nonionic polymers and charged and uncharged surfactants. Dr. Saito also paid attention to the interactions of ions with polymers, of nonionic polymers with polymeric acids, as well as to the surfactant micelles and to aqueous solutions of tetraalkylammonium salts. His review chapter on "Interactions of Polymers and Surfactants" published in *Nonionic Surfactants. Physical Chemistry*, M. J. Schick (Ed.), Dekker, New York, 1987, became a classic for all those who act in this field. I would like to mention in this short introduction about the work of Dr. Saito the lines written by Dr. E. D. Goddard: "As regards the subject matter of this chapter, it is appropriate first to mention the name of S. Saito, who can properly be termed the father of this field of research" (E. D. Goddard and K. P. Ananthapadmanabhan, *Interactions of Surfactants with Polymers and Proteins*, CRC Press, Boca Raton, 1993).

My first encounter with Dr. Saito's work was more than 30 years ago, when I started a research project on polymer-surfactant interaction. The search of literature revealed as the most frequent author the name of S. Saito. His papers were a source of inspiration for our research group, and after we published the first paper in Kolloid Zeitschrift und Zeitschrift für Polymere in 1972, we were deeply impressed to receive a congratulation letter from Dr. Saito. In 1989 we had the privilege to have Dr. Saito as invited lecturer at the 3rd Romanian Symposium on Colloid and Surface Chemistry held in Timişoara. During the visit, he truly enjoyed the people and the country, and decided to come back in the future. This happened after his retirement and in 1992 he spent almost two months in our institute. On this occasion we had the opportunity not only to learn useful things from a master, but also to get a friend.

The most distinctive feature of Dr. Saito that stroked me from the very beginning was his eagerness. I remember the visits paid together to the Village Museum, the National History Museum, and the National Gallery or in the outskirts of Bucharest. His vivid eyes scrutinize everything and he asked a lot of questions about our history and customs, culture, art and traditional architecture aiming to know as much as possible about us. A proof in this respect is that he started to learn Romanian, to be able of reading in original the poems of Eminescu, the Romanian National poet. Dr. Saito is also a talented painter. His gift was revealed to us in many landscapes and portraits he did during the stay in Bucharest, and everyone who opens this volume will have the proof of his skill.

I can not conclude these lines without mentioning the unanimous enthusiasm with which both the Editor-in-Chief of Colloid and Polymer Science, Professor Gerhard Lagaly, the Springer Publishing House and the authors invited to contribute a paper responded to the initiative to publish this volume. I wish to express my gratitude to all persons who helped me in this endeavor of bringing an idea to life.

Dan Florin Anghel

Progr Colloid Polym Sci (2003) 122: VI–VII

CONTENTS

Progr Colloid Polym Sci (2003) 122: 1–7
DOI 10.1007/978-3-540-36114-5

Chaoyang Wang
Zhen Tong
Fang Zeng
Biye Ren
Xinxing Liu
Shuizhu Wu

Surfactant structure effects on binding with oppositely charged polyelectrolytes observed by fluorescence of a pyrene probe and label

Dedicated to the 50th anniversary of Dr. Shuji Saito's first paper on polymer–surfactant systems published in Colloid and Polymer Science

C. Wang · Z. Tong (✉) · F. Zeng
B. Ren · X. Liu · S. Wu
Research Institute of Materials Science, South China University of Technology, Guangzhou 510640, China
e-mail: mcztong@scut.edu.cn
Tel.: +86-20-87112886
Fax: +86-20-87112886

Abstract Poly(2-(acrylamido)-2-methylpropanesulfonic acid) (PAMPS) and three copolymers containing about 40 mol% of AMPS and *N*,*N*-dimethylacrylamide with a single label of naphthalene or pyrene and with both labels were synthesized. The binding of cationic surfactants cetyltrimethylammonium bromide (CTAB) and dodecyltrimethylammonium chloride (DTAC), and nonionic surfactant octaethylene glycol monododecyl ether ($C_{12}E_8$) on these anionic polyelectrolytes was observed in dilute aqueous solutions with the relative emission intensity of excimer to monomer, I_E/I_M, of cationic probe 1-pyrenemethylamine hydrochloride (PyMeA·HCl), the nonradiative energy transfer (NRET), I_{Py}/I_{Np}, between the labels, the emission intensity ratio, I_1/I_3, of the first peak to the third peak and the fluorescence anisotropy, *r*, of the pyrene label. The binding with cationic surfactants made the polyelectrolyte chains much curlier, leading to an increase in I_E/I_M and I_{Py}/I_{Np}. The intermolecular NRET occurred at higher CTAB concentrations owing to the hydrophobic aggregation between CTAB tails bound on different polyelectrolyte chains, which induced an increase in *r* of the pyrene label. No intermolecular aggregation was observed from the intermolecular NRET and *r* of the pyrene label for DTAC-bound polyelectrolytes owing to its weaker hydrophobicity of 12 carbon atoms in the tail, shorter than that of CTAB. As shown with constant values of the previously mentioned photophysical parameters of the labels with increasing $C_{12}E_8$ concentration, there was no obvious binding for $C_{12}E_8$ on the anionic polyelectrolytes. However, the presence of PAMPS promoted the micelle formation for $C_{12}E_8$ below its critical micelle concentration as detected by the excimer emission of PyMeA·HCl probe.

Keywords Surfactant · Stoichiometric aggregation · Fluorescence label · Fluorescence probe

Introduction

Recently, the binding process and the complex structure of surfactants with oppositely charged polyelectrolytes have been intensively studied owing to fundamental and application interests [1, 2]. The binding process in aqueous solution is considered to be composed of two steps: initial binding of surfactant ions on the charge sites of polymer chains and propagating aggregation of surfactant molecules to form micellelike aggregates in a highly cooperative manner [3, 4]. The driving force is the electrostatic attraction between the charge groups of the polyelectrolyte and the ionic heads of surfactant molecules as well as the hydrophobic interaction between the bound surfactant alkyl tails, the hydrophobic moieties of the polymer and the surfactant tails. Therefore, the surfactant structure, especially the alkyl tail length and the headgroup charge, plays a very important role in the binding process and the conformation and aggregation of bound polyelectrolyte chains.

Fluorescence techniques are powerful in investigating the polyelectrolyte–surfactant binding [5, 6]. Pyrene chromophore is frequently used for this purpose owing to its special photophysical properties and high efficiency of excimer formation [7, 8]. Chu and Thomas [9] investigated the interaction between poly(methacrylic acid) and alkyltrimethylammonium bromide (C_nTAB) through the photophysics of pyrene and its derivatives and found that the critical aggregation concentration (cac) and critical micelle concentration (cmc) of C_nTAB decreased with the increase in the surfactant tail carbon number, *n*. Choi and Kim [10] observed the same result from the binding of C_nTAB on poly(acrylic acid) (PAA) by I_1/I_3 of the free pyrene probe. Kogej and Skerjanc [11] also found similar results in the interaction of poly(styrenesulfate) (PSS) with C_nTAB but the cac remained approximately unchanged for $C_{16}TAB$ and $C_{18}TAB$. Almgren and coworkers [12, 13] studied the interaction of C_nTAB or alkyltrimethylammonium chloride with PSS and sodium polyacrylate by time-resolved fluorescence quenching of pyrene and reported that the aggregation number increased with the increase in the surfactant tail carbon number, *n*. Saito [14, 15] pointed out no obvious binding for nonionic surfactants on water-soluble polymers. Vasilescu et al. [16] observed the binding of nonionic poly(oxyethylenic) (PEO) surfactants on PAA with fluorescence of a pyrene probe. They found that the cac was lower than the cmc and the aggregation numbers of the clusters were smaller than those of free micelles, but the difference was very small.

Solid complexes of polyelectrolyte–surfactant are generally produced by precipitation from their aqueous solution [2] and one characteristic factor of the complex is the equivalent stoichiometric composition of the complex with respect to the charge [17]. Understanding the formation of a 1:1 stoichiometric polyelectrolyte–surfactant complex in dilute aqueous solution is significant in the structure design for the solid polyelectrolyte–surfactant complex [18]. Owing to binding with oppositely charged surfactants, the conformation and aggregation of polyelectrolyte chains in dilute solution obviously change with reduction of the electrostatic repulsion along the chain and enhancement of the hydrophobic interaction among surfactants and among bound polymers.

Recently, we observed the fluorescence of dansyl-labeled polyelectrolytes in presence of cetyltrimethylammonium bromide (CTAB) and dodecyltrimethylammonium chloride (DTAC) in dilute aqueous solutionas well as in the solid [19]. It is interesting that the polyelectrolyte with higher charge density forms the less polar micellelike aggregates of the polyelectrolyte–surfactant complex in dilute aqueous solution; the opposite occurs in the solid state. We used the excimer emission of a cationic pyrene probe, 1-pyrenemethylamine hydrochloride (PyMeA·HCl), to follow the binding of CTAB on poly(2-acrylamido-2-methylpropanesulphonic acid) (PAMPS) and found that the 1:1 stoichiometric aggregate was formed in the solution without precipitation [20, 21].

In this work, we attempted to observe changes in conformation and aggregation of strong polyelectrolyte chains induced by binding with cationic surfactants CTAB, DTAC, or nonionic surfactant octaethylene glycol monododecyl ether ($C_{12}E_8$) in dilute aqueous solutions by fluorescence nonradiative energy transfer (NRET). For this purpose, naphthalene- and/or pyrene-labeled sulfonate copolymers were chosen.

Materials and methods

Materials and polyelectrolyte samples

PyMeA·HCl (Aldrich) was used as received. Fluorescently labeled monomers *N*-(1-naphthylmethyl)methacrylamide (NpMAm) and *N*-(1-pyrenylmethyl)methacrylamide (PyMAm) were synthesized in our laboratory following the procedure of Refs. [22] and [23], respectively, and were confirmed with ^{1}H NMR spectra. Ionic monomer 2-acrylamido-2-methylpropanesulphonic acid (AMPS, Fluka) was used without further purification. *N,N*-Dimethylacrylamide (DMAA, Kohjin, Japan) was distilled before used. CTAB (Fluka) and DTAC (Fluka) were recrystallized from ethanol. $C_{12}E_8$ (Fluka) was used as received. Other reagents and solvents were all purchased from commercial sources and purified according to the standard procedure. Highly pure water was obtained by deionization and filtration with a Millipore purification apparatus.

Three naphthalene- or/and pyrene-labeled copolymers of AMPS and DMAA, referred to as ADN50, ADP50 and ADNP50, respectively, were polymerized in dimethylformamide solution initiated by 2,2′-azobis(isobutyronitrile) [24]. The total monomer concentration was 0.5 mol/l with 1 mol% of NpMAm or/and PyMAm. The solution was degassed three times with nitrogen bubbling and sealed in a polymerization bottle, which was maintained at 60 °C for 12 h. The mixture was poured into a large excess of ether to precipitate the copolymer. The copolymers were purified by precipitation from methanol into ether three times. Finally, the aqueous polymer solution was dialyzed against pure water for 1 week and lyophilized. The composition of the copolymers was estimated from the C, N and H content determined by elemental analysis with a Heraeus CHN-O apparatus. The chromophore content in the samples was verified by UV absorbance. Gel permeation chromatography (GPC) measurement was carried out with a Waters GPC instrument at 40 °C using a 0.1 mol/l Na_2SO_4 aqueous solution for the elution and narrowly distributed PEO as the standard. The results are listed in Table 1.

PAMPS was synthesized according to the method described previously [20]. The molecular weight of PAMPS determined with GPC was 1.9×10^6.

Fluorescence measurements

For measuring the fluorescence spectra of the cationic probe PyMeA·HCl, naphthalene and/or pyrene-labeled polyelectrolytes in aqueous solutions with surfactants, concentrated solutions of polymers, PyMeA·HCl, CTAB, DTAC and $C_{12}E_8$ were prepared. Then, desired amounts of concentrated solutions were mixed in a

Table 1 Molecular weight and composition of fluorophore-labeled polyelectrolyte samples

Sample	$M_w/10^4$	M_w/M_n	F_{AMPS}	NpMAm (mol%)	PyMAm (mol%)
ADN50	6.0	3.06	0.401	1	0
ADP50	4.7	2.28	0.371	0	1
ADNP50	7.0	2.48	0.396	1	1

volumetric flask and diluted to test concentrations. The concentration of the probe PyMeA·HCl used in this work was fixed at 3×10^{-6} mol/l. The polyelectrolyte concentration is expressed by the molar concentration of the repeat unit AMPS.

Steady-state fluorescence spectra and fluorescence anisotropy were recorded with a Hitachi F-4500 fluorescence spectrometer. All the solutions were kept for more than 12 h to ensure the binding equilibrium, then stirred prior to measurement. All measurements were performed on air-equilibrated solutions at 25 °C. The excitation wavelength was 290 or 340 nm as indicated. I_E/I_M is defined as the emission intensity ratio of excimer (480–490 nm) to monomer (395 nm) of the pyrene probe PyMeA·HCl. The energy-transfer efficiency was characterized by the emission intensity ratio I_{Py}/I_{Np} for the pyrene label at 396 nm to the naphthalene label at 338 nm. I_1/I_3 is the emission intensity ratio for the first peak (375 nm) to the third peak (386 nm) of the pyrene label. The fluorescence anisotropy, r, is defined as

$$r = \frac{I_{\parallel} - I_{\perp}}{I_{\parallel} + 2I_{\perp}} , \qquad (1)$$

where $I_{\parallel}$ and $I_{\perp}$ are emission intensities measured parallel and perpendicular to the polarizing direction of the excitation light, respectively.

Results and discussion

Surfactant binding determined by the cationic probe

In water, PyMeA·HCl dissociates into the cationic probe $PyMeAH^+$ and Cl^-. $PyMeAH^+$ can bind on anionic polyelectrolyte PAMPS by electrostatic attraction. We observed the fluorescence spectra of PyMeA·HCl before and after binding on PAMPS [20, 21]. At concentrations up to 1×10^{-4} mol/l, there is no sign of excimer formation in the spectrum of PyMeA·HCl in water owing to the electrostatic repulsion between the probe cations. The presence of PAMPS causes a reduction in the monomer intensity and induces the typical pyrene excimer emission of a broad, structureless band centered at about 490 nm. This is due to the binding of $PyMeAH^+$ on the polyelectrolyte, which results in an increase in the local concentration of the probe chromophore along the polymer chain and promotes the excimer formation.

The fluorescence spectra of PyMeA·HCl excited at 340 nm in aqueous solutions of nonionic surfactant $C_{12}E_8$ are shown in Fig. 1. When the $C_{12}E_8$ concentration is above its cmc of 8×10^{-5} mol/l in water solution [16], the monomer emission intensity decreases and the

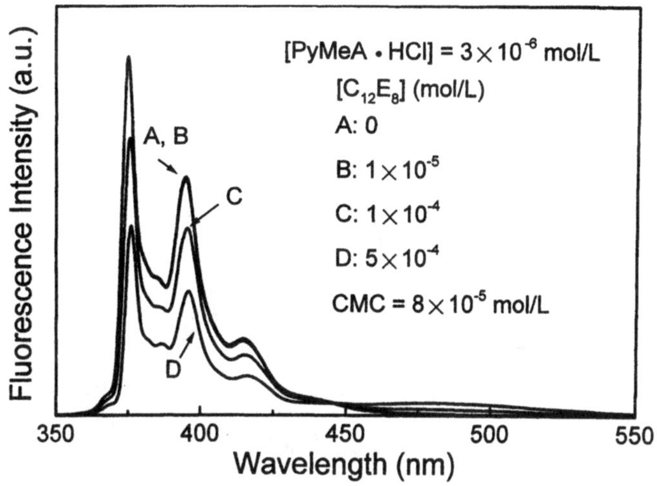

Fig. 1 The fluorescence spectra of 1-pyrenemethylamine hydrochloride (*PyMeA·HCl*) in aqueous solutions of nonionic surfactant octaethylene glycol monododecyl ether (*$C_{12}E_8$*)

excimer emission appears (curves C and D). This means that the ionic probe $PyMeAH^+$ can enter the $C_{12}E_8$ micelle and form the excimer owing to the stacking in the micelle. The fluorescence spectra of PyMeA·HCl in aqueous solutions of cationic surfactants CTAB [20] or DTAC are similar to curves A and B in Fig. 1, and are not shown here. No excimer is observed even at CTAB or DTAC concentrations higher than their cmc. These facts indicate that the excimer emission of the ionic probe $PyMeAH^+$ can only be observed either bound on PAMPS chains or encapsulated by the $C_{12}E_8$ micelle; therefore, I_E/I_M can be used to detect the competitive binding of $PyMeAH^+$ with cationic surfactants on PAMPS.

The effect of CTAB, DTAC, and $C_{12}E_8$ on the I_E/I_M of PyMeA·HCl in the presence of PAMPS is shown in Fig. 2a–c, respectively. The data on the ordinate indicate the I_E/I_M value in PAMPS solutions without surfactants. All the curves in Fig. 2a and b for the cationic surfactants are almost similar but are quite different from those in Fig. 2c for the nonionic surfactant. With increasing cationic surfactant concentration, the I_E/I_M value at a given PAMPS concentration firstly increases and then decreases, showing a maximum on every curve. Binding of the cationic surfactants on PAMPS leads to much curlier polyelectrolyte chains, which results from charge neutralization and hydrophobicity enhancement of the polyelectrolyte. This in turn enhances the excimer formation between bound probes and increases I_E/I_M. Because the pyrene probe and the surfactants are cationic species, they competitively bind on the anionic polyelectrolyte PAMPS. When the surfactant concentration is beyond the cac of about 8×10^{-6} and 1×10^{-5} mol/l for CTAB and DTAC, respectively [19, 20], the binding will be dominant for the cationic surfactants owing to the

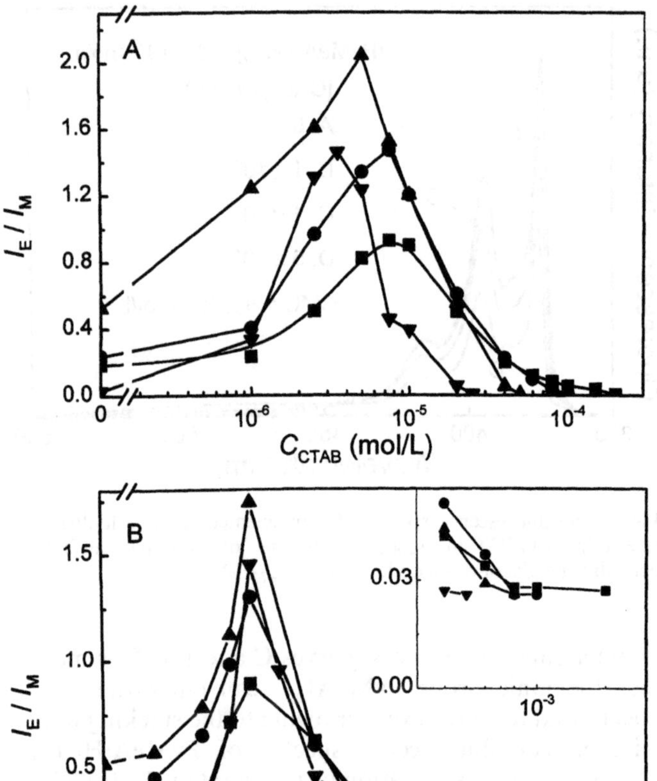

Fig. 2a–c I_E/I_M of the pyrene probe cation $PyMeAH^+$ as a function of surfactant concentration in aqueous poly(2-acrylamido-2-methylpropanesulphonic acid) (*PAMPS*) solutions with [PyMeA·HCl] = 3×10^{-6} mol/l. Three surfactants were used, **a** cetyltrimethylammonium bromide (*CTAB*), **b** dodecyltrimethylammonium chloride (*DTAC*), **c** $C_{12}E_8$. [PAMPS]: 2.4×10^{-5} mol/l (*down triangles*); 4.8×10^{-5} mol/l (*up triangles*); 9.6×10^{-5} mol/l (*circles*); 1.9×10^{-4} mol/l (*squares*). The inset in **b** is an enlargement for the higher c_{DTAC} of about 10^{-3} mol/l

cooperative interaction of both electrostatic attraction between the surfactant headgroup and the charge site of polyanion and hydrophobic aggregation among the alkyl tails of the surfactants. Consequently, the bound probe $PyMeAH^+$ is gradually replaced by bound surfactants from the charge site of PAMPS, resulting in the decrease in I_E/I_M. The dependence of I_E/I_M on PAMPS concentration observed in Fig. 2a will be discussed later.

The fluorescence spectra are different for binding DTAC compared to CTAB at high surfactant concentrations. As discussed in a previous paper [20], binding of CTAB will exclude the bound cationic probe completely and lead to $I_E/I_M = 0$, where the CTAB concentration just equals the concentration of the PAMPS repeat units. This fact manifests itself in the formation of 1:1 aggregates of CTAB and PAMPS at surfactant concentrations higher than the cac without precipitation. The binding of DTAC does not decrease I_E/I_M exactly to zero even at concentrations up to 1×10^{-3} mol/l as shown by the enlarged inset in Fig. 2b, which is about 1 order of magnitude higher than the PAMPS concentrations used. This can be attributed to the lower hydrophobicity of DTAC with the tail of 12 carbon atoms, shorter than that of 16 carbon atoms in CTAB. Thus, the hydrophobic aggregation force of bound DTAC is not strong enough to completely remove the bound $PyMeAH^+$ from the polyelectrolyte. Hayakawa et al. [25] also found that the bound cationic probe proflavine was expelled from anionic polymer pectate with the competitive binding of cationic surfactant dodecyltrimethylammonium bromide.

The situation shown in Fig. 2c for binding $C_{12}E_8$ is completely different. I_E/I_M is kept at a low value, close to that in PAMPS solution, when the $C_{12}E_8$ concentration is lower than 2.5×10^{-5} mol/l, where $C_{12}E_8$ neither binds on PAMPS nor forms micelles. At the same time, the probe cation binds to PAMPS individually and separately as if there is no surfactant in the solution. Then, I_E/I_M begins to increase at a $C_{12}E_8$ concentration of 2.5×10^{-5} mol/l and reaches a maximum at 10^{-4} mol/l. The latter is a little higher than the cmc of 8×10^{-5} mol/l for $C_{12}E_8$ [16]. As pointed out in the preceding paragraph, the PyMeA·HCl probe can enter the $C_{12}E_8$ micelle and form excimers, while PyMeA·HCl at a concentration of 3×10^{-6} mol/l cannot form excimers in the absence of micelles. Therefore, the increase in I_E/I_M with $C_{12}E_8$ concentration is an indication of micelle formation. It is worth noting that the $C_{12}E_8$ concentration at which I_E/I_M begins to increase is lower than the cmc. This means that the cac for $C_{12}E_8$ to form micelles in the presence of polyelectrolyte is somewhat lower than its cmc though $C_{12}E_8$ is a nonionic surfactant. In otherwords, the micelle formation of $C_{12}E_8$ seems to bepromoted by PAMPS. Saito [14, 15] found a conformation change for the PAA chains induced by addingnonionic PEO surfactants and suggested that hydrophobic attraction of the surfactants to PAA and hydrogen bonding between the acidic groups of the polymer and the headgroups of the surfactants produced this effect.

When the surfactant concentration is beyond 10^{-4} mol/l, there are enough $C_{12}E_8$ micelles to induce dilution of $PyMeAH^+$ and to decrease the excimer intensity. However, because the electrostatic attraction between $PyMeAH^+$ and PAMPS is stronger than the hydrophobic interaction between $PyMeAH^+$ and $C_{12}E_8$ micelles, only a slight decrease in I_E/I_M is observed in this case.

As proposed by Chandar et al. [26], the surfactant concentration at the I_E/I_M maximum in Fig. 2a and b corresponds to the cac. The dependence of the cac on PAMPS concentration is shown in Fig. 3. The cac for CTAB first increases, then levels off with increasing PAMPS concentration, in contrast to the constant cac for DTAC. We have no convincing explanation for this phenomenon at present, but the difference in the binding force of CTAB versus DTAC appears to provide some hints. When the PAMPS concentration becomes lower, a smaller amount of CTAB is sufficient to replace the bound probe $PyMeAH^+$ from the PAMPS chains owing to its stronger binding force. At higher PAMPS concentrations or for DTAC, a higher amount of surfactant, which is the equilibrium value of the cac, is required for the same purpose.

Bound surfactant effect determined by NRET of labeled polyelectrolytes

NRET between energy donor and energy acceptor covalently attached to the same or different polymer chains provides a sensitive tool to probe changes in chain conformation and aggregation [27–31]. Naphthalene and pyrene are often used as the energy donor and acceptor, respectively, because of their large spectral overlap, and naphthalene can be selectively excited at 290 nm.

I_{Py}/I_{Np} due to the NRET as a function of surfactant concentration, c_S, is plotted in Figs. 4 and 5 for the solutions of doubly labeled sample ADNP50 and for the mixtures of naphthalene-labeled ADN50 and pyrene-labeled ADP50, respectively. The former mainly stands for the intramolecular NRET, while the latter is induced merely by the intermolecular NRET. On increasing the surfactant concentration, I_{Py}/I_{Np} in Fig. 4 for CTAB binding slowly increases below the cac, rapidly increases above the cac, and finally decreases after formation of the 1:1 surfactant–polyelectrolyte complex. The trend is similar to that observed in our previous work [24]. At the same time, I_{Py}/I_{Np} in Fig. 5 for CTAB binding abruptly increases at CTAB concentrations beyond 4×10^{-5} mol/l and reaches a maximum where c_S equals the concentration of the charged sites of the polyelectrolyte samples. The observed changes in I_{Py}/I_{Np} are caused by binding of CTAB on the labeled polyelectrolytes, which makes the chain more hydrophobic and curlier, increasing the intramolecular NRET. This binding also induces the intermolecular aggregations between CTAB-bound polyelectrolyte chains after micelle formation on the chain, increasing the intermolecular NRET.

Because the alkyl tail of DTAC is shorter and its hydrophobicity is weaker than CTAB, I_{Py}/I_{Np} in Fig. 4 for intramolecular NRET caused by DTAC binding is lower, the maximum value being about half of that for CTAB. However, the c_S at the I_{Py}/I_{Np} maximum also equals the molar concentration of charged groups of the

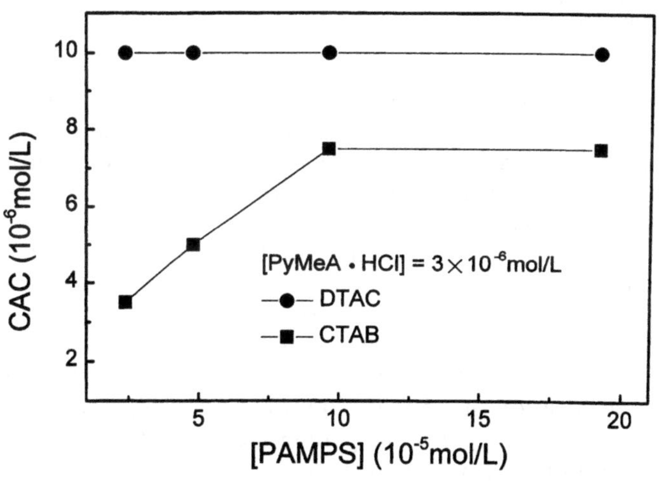

Fig. 3 The dependence of the critical aggregation concentration (*cac*) on PAMPS concentration for CTAB (*squares*) and DTAC (*circles*) from Fig. 2a and b

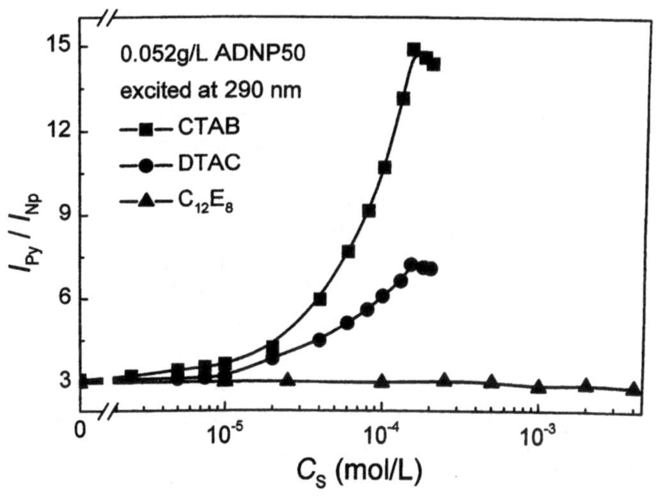

Fig. 4 I_{Py}/I_{Np} due to the nonradiative energy transfer (*NRET*) as a function of surfactant concentration, c_S, in aqueous solution of doubly labeled copolymer of 2-(acrylamido)-2-methylpropanesulfonic acid (*AMPS*) and *N,N*-dimethylacrylamide (*DMAA*) (*ADNP50*) at 0.052 g/l

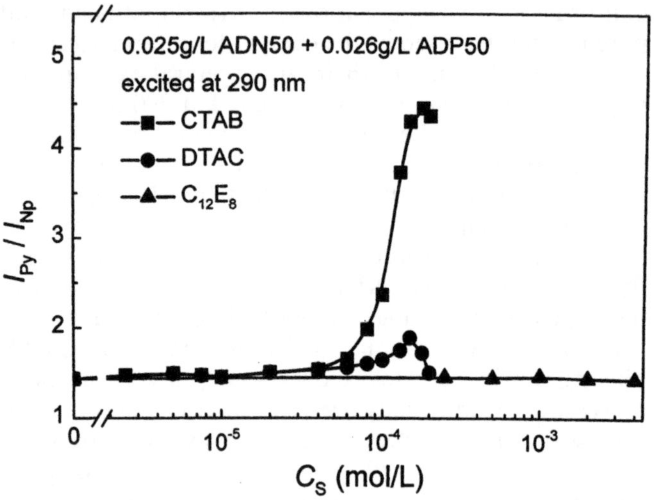

Fig. 5 I_{Py}/I_{Np} due to the NRET as a function of c_S in the mixture solution of naphthalene-labeled copolymer of AMPS and DMAA (*ADN50*) at 0.025 g/l and pyrene-labeled copolymer of AMPS and DMAA (*ADP50*) at 0.026 g/l

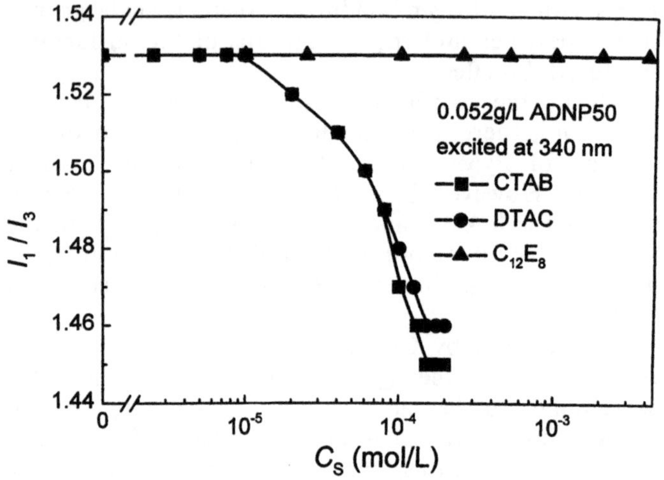

Fig. 6 I_1/I_3 of pyrene labels of doubly labeled ADNP50 in aqueous solution of with the concentrations indicated against c_S

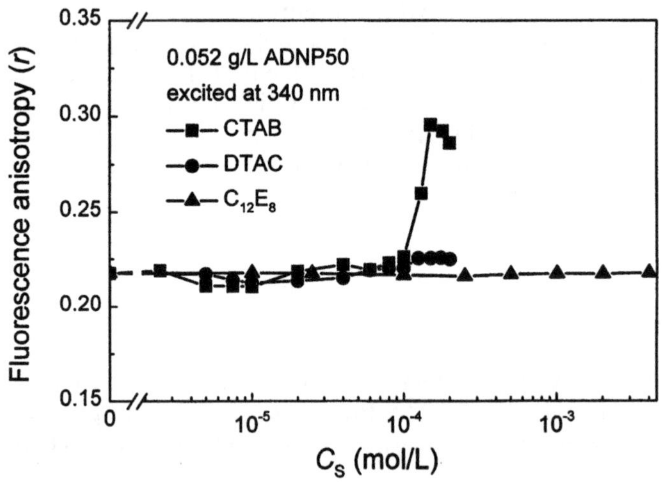

Fig. 7 Fluorescence anisotropy r of pyrene for ADNP50 in aqueous solution against c_S

polyelectrolytes. The I_{Py}/I_{Np} versus c_S curves in Fig. 5 obviously illustrate the effect of the surfactant structure on the intermolecular aggregation of surfactant-bound polyelectrolyte chains. The DTAC-bound polyelectrolyte chains cannot form intermolecular aggregates, so after the formation of micellelike aggregates, I_{Py}/I_{Np} recovers its value without DTAC owing to the isolation by the free DTAC in the solution.

The neutral surfactant $C_{12}E_8$ cannot bind on the polyelectrolytes, its presence has no obvious effect on either the intramolecular or the intermolecular NRET as seen from the Figs. 4 and 5.

Observations from pyrene label spectra

If the excitation wavelength is 340 nm, only the emission of pyrene can be observed because naphthalene cannot be excited at this wavelength. The intensity ratio I_1/I_3 of the first to the third emission peak of pyrene is known to be sensitive to the polarity of the microenvironment, becoming lower in less polar media [32]. I_1/I_3 of the pyrene label on the sample ADNP50 in aqueous solutions is plotted in Fig. 6 against the surfactant concentration c_S. The I_1/I_3 value is almost constant at about 1.53, when $c_S \leq 1 \times 10^{-5}$ mol/l for CTAB or DTAC, then decreases with increasing c_S and achieves lower values of 1.45–1.46 when $c_S \geq 1.5 \times 10^{-4}$ mol/l. This reduction of I_1/I_3 in CTAB or DTAC solutions suggests that the pyrene label does not enter the micellelike aggregate of surfactant bound on polyelectrolyte (if it is the case, I_1/I_3 should be about 1.25 [26]).The decrease of I_1/I_3 is due to the increasing hydrophobicity of the local environment surrounding the polyelectrolyte chains with binding cationic surfactants. Morishima et al. [33] considered that the pyrene label was prevented from penetrating into the hydrophobic core of the micelle by a constraint exerted by the polymer chain.

To monitor the mobility of the pyrene labels, the fluorescence anisotropy, r, was measured. Plots of r as a function of surfactant concentration are shown in Fig. 7. We can see an abrupt increase in fluorescence anisotropy for the CTAB binding and a maximum at a CTAB concentration of 1.5×10^{-4} mol/l. As the microenvironmental viscosity around pyrene increases, the depolarization of the probe is restricted, inducing an increase in fluorescence anisotropy. The I_1/I_3 data tell us that the pyrene label does not enter the CTAB micelle on the polyelectrolyte, so the increase of r is deduced by the intermolecular aggregation between CTAB-bound polyelectrolyte chains owing to the hydrophobic interaction between the CTAB tails. The fluorescence anisotropy for the DTAC binding slightly increases with surfactant concentration. This indicates no intermolecular aggregation between the DTAC-bound polyelectrolyte chains and is consistent with the NRET result in Fig. 5.

I_1/I_3 and r for the ADNP50 solutions with $C_{12}E_8$ are constant over the entire range of surfactant concentration and denote no surfactant binding on the polyelectrolyte.

Conclusion

The structure of surfactants affects their binding behavior on the polyelectrolytes and the chain conformation and aggregation of bound polyelectrolytes. Though the nonionic surfactant $C_{12}E_8$ does not bind on the polyelectrolyte, the presence of polyelectrolyte promotes its micelle formation, leading to a cac value of 2.5×10^{-5} mol/l, slightly lower than its cmc. The cationic surfactants CTAB and DTAC can bind on the oppositely charged polyelectrolytes to form micellelike aggregates beyond their cacs and 1:1 stoichiometric binding in respect of the charge is achieved for the former at about 10^{-4} mol/l without precipitation. The difference in the alkyl tail length between CTAB and DTAC induces an obviously different associative strength for the bound polyelectrolytes. The CTAB-bound polyelectrolyte chains are much curlier and can aggregate intermolecularly through the hydrophobic interaction among the bound surfactants, but the DTAC-bound polyelectrolyte chains cannot form intermolecular aggregates.

Acknowledgements Financial support for this work by the National Natural Science Foundation of China (29725411 and 90206010) and the Natural Science Foundation of Guangdong Province (015036) is gratefully acknowledged.

References

1. Wei Y-C, Hudson SM (1995) J Macromol Sci Rev Macromol Chem Phys C 35: 15
2. Ober CK, Wegner G (1997) Adv Mater 9: 7
3. Okuzaki H, Osada Y (1994) Macromolecules 27: 502
4. Okuzaki H, Eguchi Y, Osada Y (1994) Chem Mater 6: 1651
5. Winnik FM, Regismond STA (1998) In: Kwak J (ed) Polymer–surfactant systems. Dekker, New York, pp 269–315
6. Zana R (1998) In: Kwak J (ed) Polymer–surfactant systems. Dekker, New York, p 405–454
7. Morishima Y (1990) Prog Polym Sci 15: 949
8. Winnik FM (1993) Chem Rev 93: 587
9. Chu DY, Thomas JK (1986) J Am Chem Soc 108: 6270
10. Choi LS, Kim O (1994) Langmuir 10: 57
11. Kogej K, Skerjanc J (1999) Langmuir 15: 4251
12. Almgren M, Hansson P, Mukhtar E, Stam JV (1992) Langmuir 8: 2405
13. Hansson P, Almgren M (1994) Langmuir 10: 2115
14. Saito S (1979) Colloid Polym Sci 257: 266
15. Saito S (1993) J Colloid Interface Sci 158: 77
16. Vasilescu M, Anghel DF, Almgren M, Hansson P, Saito S (1997) Langmuir 13: 6951
17. Zhou S, Burger C, Yeh F, Chu B (1998) Macromolecules 31: 8157
18. Zhou S, Chu B (2000) Adv Mater 12: 545
19. Ren B, Tong Z, Gao F, Liu X, Zeng F (2001) Polymer 42: 7291
20. Wang C, Sun Q, Tong Z, Liu X, Zeng F, Wu S (2001) Colloid Polym Sci 279: 664
21. Wang C, Sun Q, Tong Z, Liu X, Zeng F, Gao F (2001) Chem J Chin Univ 22: 1265 (in Chinese)
22. Morishima Y, Tominaga Y, Kamachi M, Okada T, Hirata Y, Mataga N (1991) J Phys Chem 95: 6027
23. Morishima Y, Tominaga Y, Nomura S, Kamachi M (1992) Macromolecules 25: 861
24. Wang C, Tong Z, Zeng F, Ren B, Liu X (2002) Colloid Polym Sci (in press)
25. Hayakawa K, Satake I, Kwak JCT (1994) Colloid Polym Sci 272: 876
26. Chandar P, Somasundaran P, Turro NJ (1988) Macromolecules 21: 950
27. Morawets H (1988) Science 240: 172
28. Webber SE (1990) Chem Rev 90: 1469
29. Winnik FM (1990) Polymer 31: 2125
30. Kramer MC, Steger JR, Hu Y, McCormick CL (1996) Macromolecules 29: 1992
31. Yamamoto M, Mizusaki M, Yoda K, Morishima Y (1998) Macromolecules 31: 3588
32. Kalyanasundaram K, Tomas JK (1977) J Am Chem Soc 99: 2039
33. Morishima Y, Mizusaki M, Yoshida K, Dubin PL (1999) Colloids Surf A 147: 149

Progr Colloid Polym Sci (2003) 122: 8–15
DOI 10.1007/978-3-540-36114-5

Stella Dragan
Simona Schwarz

Dependence of the aggregation mode of two bidentate azo dyes in polycation/dye multilayers on the dye structure and the polycation conformation

S. Dragan (✉)
"Petru Poni" Institute of Macromolecular Chemistry, Aleea Grigore Ghica Voda 41 A, 6600 Iasi, Romania
e-mail: sdragan@icmpp.tuiasi.ro
Tel: +40-232-217454
Fax: +40-232-211299

S. Schwarz
Institute of Polymer Research, Hohe Strasse 6, 01069 Dresden, Germany

Abstract In this paper we report the growth particularities of some polycation/azo dye multilayers built up by the alternate adsorption of one polycation with quaternary ammonium salt groups in the backbone (95 mol % of *N,N*-dimethyl-2-hydroxypropyleneammonium chloride repeat units) (PCA_5) and two bidentate azo dyes which differ from each other only by the position of sulfonic groups, Crocein scarlet MOO (CSMOO) and Ponceau SS (PSS). The multilayer build up was monitored by UV–vis spectroscopy. The polycation/azo dye binding mode in the multilayer depended on both the dye structure and the polycation conformation. H-aggregation was suggested for CSMOO, PCA_5 being adsorbed from both saltless aqueous solution and 0.2 M low-molecular-weight salt (NaCl and Na_2SO_4) aqueous solution, while for PSS H-aggregation seemed to occur only when PCA_5 was adsorbed from saltless aqueous solution. J-aggregation of PSS on the surface was suggested when PCA_5 was adsorbed from 0.2 M Na_2SO_4 aqueous solution. The aggregation mode of the dye in the multilayer was correlated with the PCA_5/dye interaction mode in the aqueous solution. Electrokinetic measurements were used to verify the dependence of the driving force for the PCA_5/azo dye multilayer growth and the multilayer stability on the dye structure.

Keywords Dye binding mode · Electrokinetic measurements · Metachromasy · Polycation/azo dye multilayers · UV–vis spectroscopy

Introduction

Layer-by-layer self-assembly of different components containing opposite charges, on a solid substrate, proved to be a very versatile tool to achieve thin films ordered at a molecular level, suitable for chemical sensors and devices [1, 2, 3, 4, 5, 6, 7, 8, 9]. The mechanism of the multilayer growth and the stability of the polyelectrolyte and hybrid organic–inorganic nanostructures are still being investigated with growing interest for both fundamental and practical aspects [10, 11].

Molecular thin films including reversible photoreactive systems are promising materials for optical information storage and switching applications in microelectronics and optoelectronics. Azobenzene chromophores are frequently utilized as photoreactive components in the multilayer thin films owing to their property to photoisomerize from the thermodynamical more stable trans configuration (E form) to the cis form (Z form) by UV light irradiation ($\lambda < 370$ nm). The reverse process takes place by thermal treatment and/or by visible light irradiation ($\lambda > 450$ nm) [12, 13, 14, 15, 16]. Azobenzene substituents could be part of a cationic or anionic bipolar amphiphile [16, 17], a polymer with azobenzene in the side chains [18, 19, 20, 21, 22] or in the backbone [23, 24]. Starting with Cooper et al. [25], who demonstrated that

Congo red and copper phthalocyanine tetrasulfonic acid could be assembled with a cationic polypeptide, multicharged azo dyes have also been successfully utilized to build up electrostatic self-assembled thin films [26, 27, 28, 29, 30]. Ariga et al. [26] have demonstrated that the stability of the polycation/dye assembly depended on the polycation structure and the aggregation mode of the dye on the positive surface depended on the dye structure, being independent of the dye aggregation in solution. Recently, Advincula et al. [30] have investigated the molecular assembly process of different multicharged azo dyes with two types of polycations: poly(diallyldimethylammonium chloride) and poly(allylamine hydrochloride). Among the other dyes, Direct red 80 (DR80) has shown significant photoalignment properties [30]. However, details regarding the influence of dye structure and neutral salts on the growth and properties of the polycation/multicharged azo dyes nanostructures are very few [26, 28, 30].

In our previous investigations we followed the growth and stability of some polycation/azo dye thin films, using one polycation containing 95 mol% of *N*,*N*-dimethyl-2-hydroxypropyleneammonium chloride units in the backbone (PCA_5) and two multicharged azo dyes (Direct blue 1 and DR80), with respect to the counterion nature (Cl^-, Br^-, SO_4^{2-}) and the counterion concentration used in the polycation adsorption step [31]. For the present study we selected two bidentate azo dyes which have the same basic structure of 3-hydroxy-4-[4-phenylazo]phenylazonaphthalen and two sulfonic groups distributed in different positions. They were the disodium salts of 3-hydroxy-4-[4-phenylazo]phenylazo-5,7-naphthalendisulfonic acid, Crocein scarlet MOO (CSMOO), and of 3-hydroxy-4-[4-phenylazo]phenylazo-2,7-naphthalendisulfonic acid, Ponceau SS (PSS). Our attention was focused on how the position of sulfonic groups and the polycation conformation could influence the PCA_5/azo dye multilayer construction and the dye aggregation mode. The polycation conformation was controlled by the polycation adsorption from both saltless aqueous solution and 0.2 M low-molecular-weight salt (LMWS) aqueous solution (NaCl or Na_2SO_4). UV–vis spectroscopy and electrokinetic measurements were employed to monitor the polycation/azo dye multilayer build up and stability. Special attention was paid to the dependence of the dye aggregation mode in the multilayer on the dye structure and the polycation conformation compared with the PCA_5/dye interaction in dilute aqueous solutions.

Experimental

Materials

The molecular structures of the azo dyes and of the polycation used in this study are shown in Fig. 1.

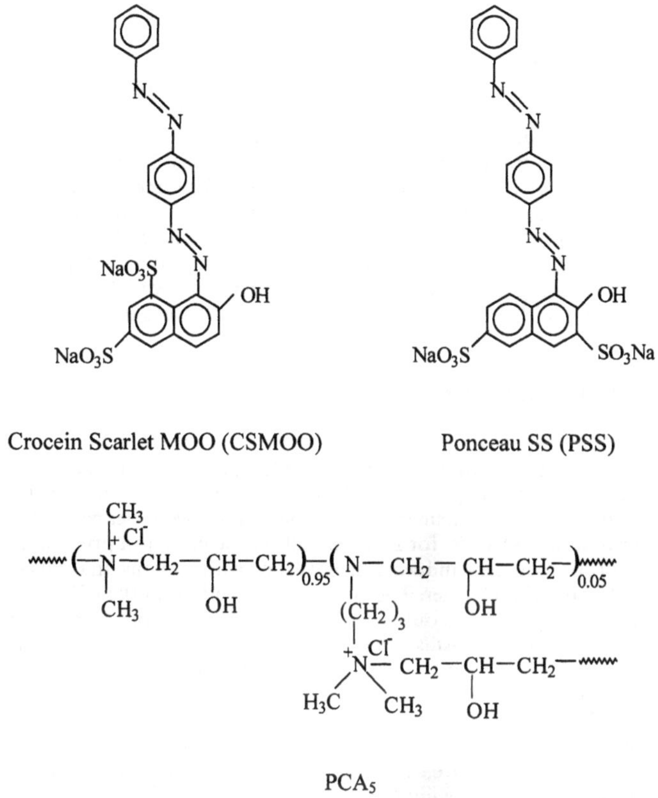

Fig. 1 The structures of the dyes and the polycation used

CSM00 and PSS (both from Aldrich) were used after being recrystallized three times from an aqueous methanol solution (methanol/water 70/30, v/v). The polycation (PCA_5) was synthesized by the condensation polymerization of epichlorohydrin with dimethylamine and *N*,*N*-dimethyl-1,3-diaminopropane according to a method presented in detail elsewhere [32]. Polycation samples were carefully purified by dialysis against distilled water until Cl^- ions were absent from the dialyzate (about 3 days); the absence of Cl^- ions was checked with 0.02 N $AgNO_3$ aqueous solution. The dilute solutions were concentrated using a rotary evaporator and then the polymer was precipitated with acetone and purified again by reprecipitation with acetone from methanol solution. The polycation was vacuum-dried on P_2O_5 at room temperature and then characterized by elemental analysis and intrinsic viscosity in 1 M NaCl aqueous solution, at 25 °C. The ionic chlorine content, Cl_i, was measured by potentiometric titration with 0.02 N $AgNO_3$ and the total chlorine content, Cl_t, was performed by a combustion method (Schöniger technique). Calcd.: N 10.49; Cl_i 25.26. Found: N 11.00; Cl_t 23.58; Cl_i 23.22. The viscometric measurements were carried out with a Ubbelohde viscometer with internal dilution, at 25 °C. $[\eta]_{1MNaCl} = 0.46$ dl/g.

Poly(sodium styrenesulfonate) (NaPSS) with a molecular weight of 70,000 g/mol, from Aldrich, was used as received. NaCl and Na_2SO_4 were purchased from Fluka. They had a purity of 99.5 and 99%, respectively, and were used without further purification.

Solutions

All the solutions were prepared at room temperature 1 day before use. The polymer concentrations were quoted with respect to the monomer repeat unit. A stock solution of PCA_5 with a concentration of 10 unit mM was prepared first. The polycation concentration was determined by titration using a PCD 02 particle charge detector (Mütek, Germany). To obtain the polycation with a concentration of 10 unit mM in LMWS aqueous solution, 1 vol polycation aqueous solution with a concentration of 20 unit mM was mixed with the same volume of the corresponding LMWS aqueous solution having a concentration 2 times higher than the final concentration. The salt concentration was 20 times higher than the polycation concentration and therefore one may assume that whole ion exchange of the initial counterion (Cl^-) with the new counterion (SO_4^{2-}) took place after about 24 h. NaPSS with a concentration of 10 unit mM was used as an anionic component for the background preparation in combination with PCA_5. Dye solutions with a concentration of about 1 mM were used.

Multilayer deposition

Glass and quartz substrates with dimensions of 76 × 26 × 1 mm were used to follow the build up of PCA_5/azo dye multilayers by UV–vis spectroscopy. Only glass substrates were used for the electrokinetic measurements. All the substrates were cleaned in hot chromic acid at 80 °C for 2 h, in an ultrasonic bath, just prior to the adsorption process and were then washed with copious amounts of Millipore Milli-Q water. The alternate adsorptions of the PCA_5 and azo dyes were carried out according to the Decher approach [1], by manually dipping substrates in open glass beakers of 100-ml volume, without stirring, at room temperature for 20 min. The polycation was adsorbed either from saltless aqueous solution or from LMWS aqueous solutions. The dyes were adsorbed only from saltless aqueous solutions. The solution pH for all deposition steps was about 5.5. After each layer had been deposited, the substrates were rinsed with Millipore water three times for 1 min. The substrates were not dried between the adsorption steps.

UV–vis spectroscopy

The UV–vis spectra of the dyes in the presence of the polycation in dilute aqueous solutions were recorded using a SPECORD M42 spectrophotometer. Portions of 10 ml of the polycation aqueous solutions, with different concentrations, were mixed with 10 ml of an aqueous solution of the dye with a concentration of 2×10^{-5} M. The UV–vis spectra of the mixed solutions were recorded after 30 min of stirring, as already described [33, 34]. The UV–vis spectroscopic measurements on the slides with multilayers were performed with a PerkinElmer Lambda 800 device. All measurements were carried out on dry slides.

Electrokinetic measurements

The electrokinetic measurements were performed using an electrokinetic analyzer (Anton Paar, Graz, Austria), by measuring the streaming potential between two electrodes. Two glass substrates were mounted tightly between the two Ag/AgCl electrodes in the measuring cell. A 10^{-3} M aqueous solution of KCl was used as a test solution for all measurements. The zeta potential, ζ, was calculated from the streaming potential determined with respect to the pressure difference and from the specific conductivity of the solution [35], according to the following equation:

$$\zeta = (\eta/\varepsilon_0\varepsilon_r) \times (\Delta U/\Delta p) \times \kappa,$$

where ζ is the zeta potential, ΔU is the streaming potential, Δp is the pressure, κ is the conductivity, η is the dynamic viscosity of the test liquid; ε_r is the relative permittivity of the solution and ε_0 is the dielectric constant.

Fairbrother–Mastin software was used for data evaluation.

Results and discussion

Polycation/azo dye interaction in dilute aqueous solutions

In dilute solutions, the azo dyes used in this study (Fig. 1) showed metachromatic behavior in the presence of PCA_5. That is, some UV–vis spectral changes accompanied the binding of the dye to the polycation chain as a consequence of the π–π interactions between the adjacent dye molecules supported by hydrophobic effects. As the anionic dye molecules are attracted to the oppositely charged sites on the polycation chain they come into close proximity with each other. As a result, spectral shifts could be observed to longer (bathochromic effect) or to shorter (hypsochromic effect) wavelengths depending on the mutual orientation of the dipole moments [36, 37]. Thus, the characteristic wavelengths of the nonaggregated dye molecules used in this study were found at 510 nm for CSMOO and at 514 nm for PSS, but some spectral shifts appeared in the presence of the polycation PCA_5. A blueshift of the characteristic maximum from 510 nm to about 500 nm and a small shoulder at about 550 nm were seen in the CSMOO spectrum in the presence of PCA_5 (Fig. 2a). A redshift from 514 nm to about 530 nm and a

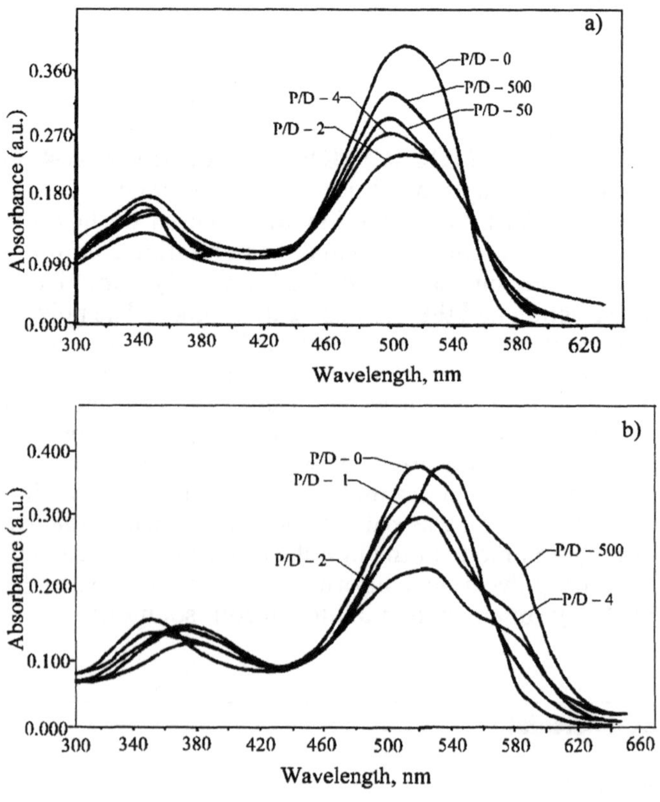

Fig. 2 Spectral changes of the azo dye in aqueous solution in the presence of PCA_5 at different *P*/*D* molar ratios: **a** CSMOO; **b** PSS

distinct shoulder at about 580 nm were found for PSS in the presence of PCA_5 (Fig. 2b). The small blueshift of the characteristic maximum in the case of CSMOO suggests a tendency for better organization of the bound dye molecules. On the other hand, the definite redshift of the maximum in the case of PSS shows that J-aggregation [37] of the dye molecules occurs by interaction with PCA_5 in aqueous solution. No significant changes took place with the maximum at 350 nm, which is present in the UV–vis spectra of both CSMOO and PSS in the presence of PCA_5. As a general trend, the intensity of the characteristic maxima decreased with the increase of the polycation/dye molar ratio, P/D, the minimum being found at a P/D of about 2, for both dyes. By increasing the polymer concentration, the intensity of the characteristic maximum increased again because the distance between the bound dye molecules increased and, therefore, their interaction possibilities diminished, but the metachromatic effect was still observed at $P/D = 500$.

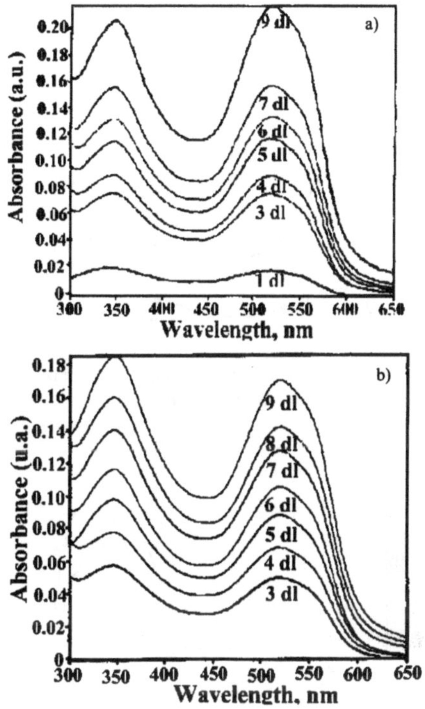

Fig. 3a,b UV–vis spectra of the $(PCA_5/CSMOO)_n$ multilayers at various layer pair number (*dl*) adsorbed on glass substrates which had a $(PCA_5/NaPSS)_2$ precursor; PCA_5 was adsorbed from 0.2 M Na_2SO_4 aqueous solution. **a** CSMOO – the last adsorbed layer. **b** PCA_5 – the last adsorbed layer

Polycation/azo dye self-assembled multilayers

The multilayer build up by the alternate deposition of PCA_5 and both azo dyes was followed by UV–vis spectroscopy. A $(PCA_5/NaPSS)_2$ precursor film had previously been prepared. It is known that the dyes aggregate as a consequence of their flat geometries. The aggregation level and mode depends on the dye structure and the environmental characteristics. Maximum contact between the dye molecules results when the dye aggregation is mainly side by side, a hypsochromic spectral shift being characteristic (H-aggregation) [37]. The spectral shifts to longer wavelengths relative to the monomeric dye absorption are characteristic of the so-called J-aggregates [37]. The dependence of the UV–vis spectra of the $(PCA_5/CSMOO)_n$ multilayers on the layer pair number for the polycation deposition from 0.2 M Na_2SO_4 aqueous solution is shown in Fig. 3a. The UV–vis spectra of the $(PCA_5/CSMOO)_n$ multilayers, with PCA_5 adsorbed from saltless aqueous solution, were exactly the same as those presented in Fig. 3a, only the intensity values were much lower (results not shown). As one can see, the UV–vis spectra of the $(PCA_5/CSMOO)_n$ multilayers are different compared with those found for CSMOO in dilute aqueous solutions in presence of PCA_5 (Fig. 2a). A bathochromic shift from 510 nm to about 516 nm and a definite increase of the maximum at 350 nm were observed with the increase of the layer pair number. The UV–vis spectral changes from the aqueous solution to the adsorbed layer indicated a different binding mode of CSMOO adsorbed in the multilayer, irrespective of the polycation conformation (PCA_5 adsorbed from either saltless aqueous solution or 0.2 M Na_2SO_4 aqueous solution). H-aggregation (side by side) of CSMOO molecules in the $(PCA_5/CSMOO)_n$ multilayer is suggested by the monotonous increase of the maximum at 350 nm [37].

On the other hand, every new polycation layer adsorbed as the topmost layer led to a small decrease of the maximum at 516 nm and to a slight increase of the maximum at 350 nm (Fig. 3b). These changes could be caused by: the removal of part of the dye in the polycation adsorption step and the increase in the organization level of the CSMOO molecules in the $(PCA_5/CSMOO)_n$ multilayer when a new polycation layer was adsorbed.

Figure 4 illustrates the changes of the UV–vis spectra of the $(PCA_5/PSS)_n$ multilayers with changing layer pair number for the polycation deposition from either saltless aqueous solution (Fig. 4a) or 0.2 M Na_2SO_4 aqueous solution (Fig. 4b). Unlike CSMOO, the polycation adsorption medium had a clear influence on the PSS adsorption mode in the multilayer. Thus, when the polycation was adsorbed from saltless aqueous solution (Fig. 4a), the UV–vis spectra of the $(PCA_5/PSS)_n$ multilayers were similar to those characteristic for the $(PCA_5/CSMOO)_n$ multilayers. In contrast, the UV–vis spectra of the $(PCA_5/PSS)_n$ multilayers, with PCA_5 adsorbed from 0.2 M Na_2SO_4 aqueous solution (Fig. 4b), were similar to those characteristic for PSS binding to PCA_5 in dilute aqueous solution (Fig. 2b).

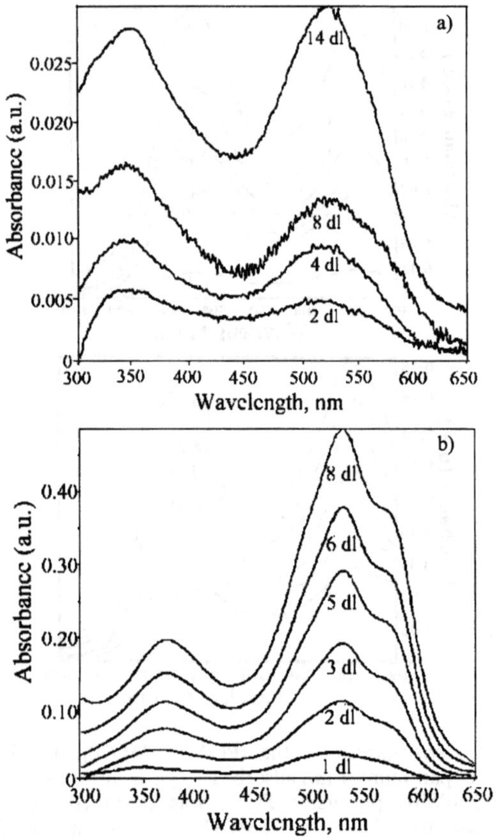

Fig. 4 Changes in the UV–vis spectra of the $(PCA_5/PSS)_n$ multilayers with changing layer pair number adsorbed on glass substrates which had a $(PCA_5/NaPSS)_2$ precursor: **a** PCA_5 was adsorbed from saltless aqueous solution; **b** PCA_5 was adsorbed from 0.2 M Na_2SO_4 aqueous solution

A regular bond of both dyes to the polycation layer was suggested by the position of the characteristic maxima, which remained unchanged irrespective of the layer pair number. Taking into account the differences between the UV–vis spectra of the $(PCA_5/CSMOO)_n$ multilayers (Fig. 3a) and the UV–vis spectra of this dye in the presence of PCA_5 in dilute aqueous solutions (Fig. 2a), one can assume that the CSMOO binding mode in the multilayer is somewhat different compared with PCA_5/CSMOO interaction in dilute aqueous solutions. The CSMOO binding in the $(PCA_5/CSMOO)_n$ multilayers could be a combination between the electrostatic and hydrophobic interactions. We suppose that, because of their proximity, both sulfonic groups of the adsorbed CSMOO molecules are involved from the beginning in the interaction with the polycation layer, but the charge reversal of the multilayer, with every layer adsorbed, requires a further dye amount to be adsorbed. This process could be achieved by the hydrophobic interactions between the azobenzene-containing parts of the dye molecules. The steady increase of the maximum at 350 nm led us to assume that an H-aggregation mode of the dye is possible in this case. This interaction mode could explain the ability of the CSMOO as the last adsorbed layer to bind a new polycation layer mainly by electrostatic interactions.

PSS binding to the PCA_5 in the multilayer appeared to be similar as for CSMOO only when the polycation was adsorbed from saltless aqueous solution (Fig. 4a), but J-aggregation occurred when the polycation was adsorbed from LMWS aqueous solutions (Fig. 4b). To explain this different behavior, we suppose that only one sulfonic group of PSS was bound on the polycation layer matrix and the other remained free and reversed the charge of the surface, which was thus able to adsorb a new polycation layer. A J-aggregation mode was previously assumed for the polycation/azo dye multilayer built up with multicharged azo dyes DB1 and DR80, since only bathochromic shifts were observed for those dyes [31]. J-aggregates were observed by Ariga et al. [26] when Congo red was adsorbed on a positive surface and also by Advincula and coworkers [30] for the polycation/dye multilayers built up with DR80 as multicharged azo dye.

The dependence of the absorbance of the $(PCA_5/\text{azo dye})_n$ multilayers on the layer pair number, with PCA_5 adsorbed from different solutions on the glass slides, is plotted in Fig. 5. A relatively linear increase of the absorbance was found when the PCA_5 was adsorbed from both saltless aqueous solution and LMWS aqueous solutions. The same behavior was observed on the quartz slides as substrates, the only difference being the absorbance values, which were lower than the values found for the adsorption on the glass substrates at the same layer pair number and the same adsorption conditions. This difference was explained by the lower initial charge density of the quartz slides compared with those of glass.

The absorbance variation in Fig. 5 reflects that, for a certain system, every deposition step brings the same amount of material to the surface, which agrees with results obtained by other authors [27, 29, 30]. No differences in the amount of dye adsorbed per layer pair, with respect to the dye structure, were observed when PCA_5 was adsorbed from saltless aqueous solution. However, the amount of dye adsorbed in the multilayer increased differently when PCA_5 was adsorbed from 0.2 M LMWS aqueous solution. The counterions in excess by their screening effect led to the decrease of the electrostatic repulsion between the positive charges of the polycation and to fewer uncompensated free charges. As a consequence, a higher amount of the polycation in a more coiled conformation should be adsorbed to compensate the negative charges on the surface, and a higher amount of the dye is adsorbed on such a layer [28, 30, 31]. The clear influence of the counterion on the amount of dye adsorbed per layer pair is revealed by the results collected in Fig. 5. The absorbances measured for $(PCA_5/\text{dye})_n$ multilayers

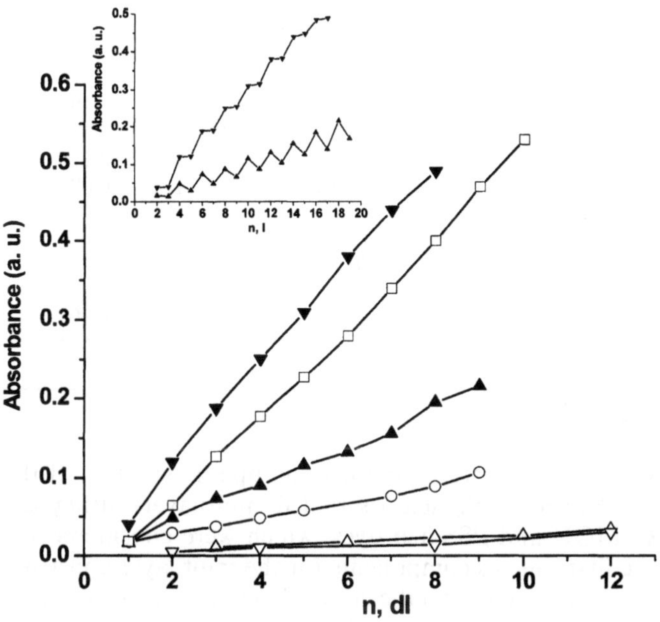

Fig. 5 Absorbance at the characteristic wavelengths of the (PCA_5/azo dye)$_n$ multilayers (516 nm for the PCA_5/CSMOO multilayers and 527 nm for the PCA_5/PSS multilayers) with respect to the layer pair number (dl) adsorbed on glass substrates which had a (PCA_5/NaPSS)$_2$ precursor. The azo dye and the polycation adsorption conditions were as follows: CSMOO, water (*open up triangles*); PSS, water (*open down triangles*); CSMOO, 0.2 M NaCl aqueous solution (*circles*); PSS, 0.2 M NaCl aqueous solution (*squares*); CSMOO, 0.2 M Na_2SO_4 aqueous solution (*closed up triangles*); PSS, 0.2 M Na_2SO_4 aqueous solution (*closed down triangles*). *Inset*: absorbance at 516 nm for the (PCA_5/CSMOO)$_n$ multilayers (*closed up triangles*) and at 527 nm for the (PCA_5/PSS)$_n$ multilayers (*closed down triangles*), with respect to the number of layers adsorbed on the glass substrates which had a (PCA_5/NaPSS)$_2$ precursor, PCA_5 being adsorbed from 0.2 M Na_2SO_4 aqueous solution; every layer was measured

with PCA_5 adsorbed from 0.2 M Na_2SO_4 aqueous solutions were higher than those corresponding to the multilayers built up with PCA_5 adsorbed from 0.2 M NaCl. These results are in good agreement with the influence of the counterion on the PCA_5 conformation [38]. From our previous investigation we know that when the counterions in excess are Cl^- or SO_4^{2-} the specific influence of the nature of the counterion on the PCA_5 conformation is observed only for concentrations lower than 0.5 M LMWS. Therefore, we selected for the present study the 0.2 M LMWS concentration. The higher absorbances found when the multilayers were constructed with PCA_5 adsorbed from 0.2 M Na_2SO_4 aqueous solutions, for both dyes, were explained by the preferential binding of the counterions with a greater valence, a higher amount of the polycation being adsorbed in these conditions.

Competition between the water-soluble polyelectrolyte/dye complex and the dye bound in a polycation/dye multilayer is possible [26, 28] as some of the dye is adsorbed by interactions other than electrostatics. The inset in Fig. 5 shows the variation of the absorbance at 516 nm for the (PCA_5/CSMOO)$_n$ multilayers and at 527 nm for the (PCA_5/PSS)$_n$ multilayers, after every adsorbed layer, when the polycation was adsorbed from 0.2 M Na_2SO_4 aqueous solution. Every polycation adsorption step was accompanied by a decrease of the absorbance in the case of CSMOO as the anionic component, but no decrease was found in the case of PSS. These results confirmed that some of the adsorbed CSMOO molecules (the hydrophobically adsorbed molecules) were removed in the polycation adsorption step, but that no dye molecules left the substrate in the case of PSS (molecules which were only electrostatically adsorbed).

The electrokinetic measurements can give information on the surface charge and on the stability of the polycation/azo dye multilayers with respect to the dye structure. The values of ζ at various pH for the (PCA_5/azo dye)$_5$ multilayers when the polycation was adsorbed from 0.2 M Na_2SO_4 aqueous solution are illustrated in Fig. 6.

In the basic range of Fig. 6, the ζ values are more negative in the case of CSMOO as the last layer adsorbed than in the case of PSS. The almost linear change of ζ with pH in the case of CSMOO suggests a more hydrophobic surface than in the case of PSS and this aspect also supports the different aggregation mode of these dyes. The surface charge was almost independent of the dye structure in the acidic range. The changes of the surface charge by the adsorption of a new polycation layer on the

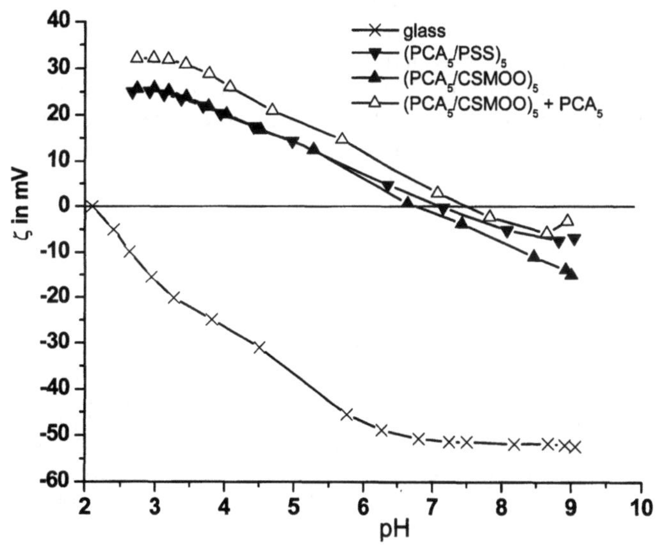

Fig. 6 The effect of pH on the ζ potential for the assembly (PCA_5/NaPSS)$_2$ + (PCA_5/azo dye)$_5$ when PCA_5 was adsorbed from 0.2 M Na_2SO_4 aqueous solution

Table 1 Stability of the (PCA_5/azo dye)$_5$ multilayers under the ζ potential measurement conditions. PCA_5 was adsorbed from 0.2 M Na_2SO_4 aqueous solution

Experiment	Azo dye	*n* (dl)	Absorbance units	
			Before ζ potential measurement	After ζ potential measurement
1	Ponceau SS	5	$A_{527} = 0.289$	$A_{527} = 0.276$
2	Crocein scarlet MOO	5	$A_{514} = 0.116$	$A_{514} = 0.088$
			$A_{350} = 0.114$	$A_{350} \cong 0$
3	Crocein scarlet MOO	5+PCA_5	$A_{516} = 0.088$	$A_{516} = 0.073$
			$A_{350} = 0.098$	$A_{350} = 0.105$

(PCA_5/CSMOO)$_5$ multilayers are also shown in Fig. 6. The ζ values were pushed to higher positive values, at the same pH, and the pH corresponding to the isoelectric point was higher, for the multilayer with PCA_5 as the last adsorbed layer. The stability of the (PCA_5/azo dye)$_5$ assemblies under the ζ measurement conditions, with respect to the dye structure, was revealed by UV–vis spectroscopy. The absorbance values at the characteristic wavelength for some multilayers, built up with PCA_5 adsorbed from 0.2 M Na_2SO_4 aqueous solution, are presented in Table 1. These values were measured after the ζ measurement had been accomplished, and are compared with the absorbance values before the ζ measurement.

Very small differences were found in the case of PSS as the last adsorbed layer. These results demonstrate the high stability of this assembly under the ζ measurement conditions. Different behavior was seen in the case of the (PCA_5/CSMOO)$_5$ assemblies. The maximum at 350 nm diminished almost completely when the dye was the last adsorbed layer. On the other hand, good stability of the multilayer nanoarchitecture was evidenced when PCA_5 was the last adsorbed layer, the absorbance at both characteristic wavelengths of the multilayer being included (Table 1). That is, the order of the hydrophobically aggregated CSMOO molecules was disturbed under the ζ measurement conditions only when the dye was the last adsorbed layer.

Conclusions

By the present study on the build up and stability of the (PCA_5/azo dye)$_n$ self-assembled multilayers using the CSMOO and PSS azo dyes, which were considered for the first time as components for the multilayer construction, we have shown the clear influence of the position of sulfonic groups on both the dye binding mode and the multilayer stability. The changes observed by UV–vis spectroscopy, from the solution spectrum to that of the adsorbed dye layer, indicated molecular rearrangements upon adsorption. The H-aggregation mode supposed for CSMOO, with PCA_5 adsorbed both from saltless aqueous solution and LMWS aqueous solution, has been supported by both UV–vis spectroscopy and the ζ measurement. Unlike CSMOO, the dye binding mode of PSS in the (PCA_5/PSS)$_n$ multilayers depended on the polycation conformation. H-aggregation seemed to occur with PCA_5 adsorbed in a flat conformation and J-aggregates were found with PCA_5 adsorbed from LMWS aqueous solutions. By these results, we have demonstrated that polycation/azo dye multilayers with different molecular arrangements can be constructed, taking into account the position of the sulfonic groups of the dye and the polycation conformation.

Acknowledgement Financial support from SMWK and DAAD is gratefully acknowledged.

References

1. (a) Decher G, Hong J-D (1991) Makromol Chem Macromol Symp 46:321; (b) Decher G, Hong J-D, Schmitt J (1992) Thin Solid Films 210/211:831; (c) Lvov Y, Decher G, Sukhorukov G (1993) Macromolecules 26:5396; (d) Decher G (1997) Science 277:1232
2. (a) Ferreira M, Rubner MF (1995) Macromolecules 28:7104; (b) Yoo D, Shiratori SS, Rubner MF (1998) Macromolecules 31:4308
3. Caruso F, Niikura K, Furlong DN, Okahata Y (1997) Langmuir 13:3427
4. Lowack K, Helm CA (1998) Macromolecules 31:823
5. Wang L, Fu Y, Wang Z, Fan Y, Zhang X (1999) Langmuir 15:1360
6. Dubas ST, Schlenoff JB (1999) Macromolecules 32:8153
7. Hao E, Lian T (2000) Chem Mater 12:3392
8. Schüler C, Caruso F (2000) Macromol Rapid Commun 21:750
9. Bertrand P, Jonas A, Laschewsky A, Legras R (2000) Macromol Rapid Commun 21:319
10. (a) Arys X, Laschewsky A, Jonas AM (2001) Macromolecules 34:3318; (b) Glinel K, Laschewsky A, Jonas AM (2001) Macromolecules 34:5267
11. (a) Schlenoff JB, Dubas ST (2001) Macromolecules 34:592; (b) Dubas ST, Schlenoff JB (2001) Macromolecules 34:3736
12. Rau H (1990) Azo Compounds In: Dürer H, Laurent HB (eds) Photochromism – molecules and systems. Elsevier, Amsterdam, pp 165–192

13. Knobloch H, Orendi H, Büchel M, Seki T, Ito S, Knoll W (1994) J Appl Phys 76:8212
14. Menzel H, Rüther M, Stumpe J, Fischer T (1998) Supramol Sci 5:49
15. Imai Y, Naka K, Chujo Y (1999) Macromolecules 32:1013
16. (a) Toutianoush A, Tieke B (1998) Macromol Rapid Commun 19:591; (b) Toutianoush A, Saremi F, Tieke B (1999) Mater Sci Eng C 8–9:343
17. Hong J-D, Park E-S, Park A-L (1999) Langmuir 15:6515
18. (a) Wang X, Balasubramanian S, Kumar J, Tripathy SK (1998) Chem Mater 10:1546; (b) Balasubramanian S, Wang X, Wang HC, Yang K, Kumar J, Tripathy SK (1998) Chem Mater 10:1554
19. Lenahan KM, Wang Y-X, Liu Y, Claus RO, Heflin JR, Marciu D, Figura C (1998) Adv Mater 10:853
20. Dante S, Advincula R, Frank CW, Stroeve P (1999) Langmuir 15:193
21. (a) Laschewsky A, Mayer B, Wischerhoff E, Arys X, Jonas A, Kauranen M, Persoons A (1997) Angew Chem Int Ed Engl 36:2788; (b) Fischer P, Laschewsky A (2000) Macromolecules 33:1100
22. (a) Tuo X, Chen Z, Wu L, Wang X, Liu D (2000) Polym Prepr Am Chem Soc Div Polym Chem) 41:11405; (b) He X, Chen Z, Wu L, Wang X, Liu D (2000) Polym Prepr Am Chem Soc Div Polym Chem 41:1358
23. Cochin D, Passmann M, Wilbert G, Zentel R, Wischerhoff E, Laschewsky A (1997) Macromolecules 30:4775
24. Fischer P, Laschewsky A, Wischerhoff E, Arys X, Jonas A, Legras R (1999) Macromol Symp 137:1
25. Cooper TM, Campbell AL, Crane RL (1995) Langmuir 11:2713
26. Ariga K, Lvov Y, Kunitake T (1997) J Am Chem Soc 119:2224
27. Yoo D, Wu A, Lee J, Rubner MF (1997) Synth Met 85:1425
28. Linford MR, Auch M, Möhwald H (1998) J Am Chem Soc 120:178
29. (a) Laschewsky A, Mayer B, Wischerhoff E, Arys X, Jonas A (1996) Ber Bunsenges Phys Chem 100:1033; (b) Koetse M, Laschewsky A, Verbiest T (1999) Mater Sci Eng C 10:107
30. (a) Advincula RC, Fells E, Park M (2001) Chem Mater 13:2870; (b) Wang Y, Wallace E, Walton A, Bathia G, Park M-K, Advincula R (2000) Polym Prepr Am Chem Soc Div Polym Chem 41:1016
31. Dragan S, Schwarz S, Eichhorn K-J, Lunkwitz K (2001) Colloids Surf A 195:243
32. Dragan S, Ghimici L (1991) Angew Makromol Chem 192:199
33. Dragan S, Cristea M, Airinei A (1997) Macromol Rapid Commun 18:541
34. Dragan S, Ghimici L, Cristea M, Airinei A (1999) Acta Polym 50:260
35. (a) Schwarz S, Eichhorn K-J, Wischerhoff E, Laschewsky A (1999) Colloids Surf A 159:491; (b) Fadel H, Schwarz S, Lunkwitz K, Jacobasch H-J (1998) Angew Makromol Chem 263:79
36. Kugel R (1993) Adv Chem Sci 236:507
37. Buss V, Eggers L (2000) In: Lindon JC, Tranter GE, Holmes JL (eds) Encyclopedia of spectroscopy and spectrometry. Academic, New York, pp 388–396
38. Ghimici L, Dragan S, Popescu F (1997) J Polym Sci Part B Polym Phys 35:2571

Progr Colloid Polym Sci (2003) 122: 16–26
DOI 10.1007/978-3-540-36114-5

Dan F. Anghel
Shuji Saito
Alina Iovescu
Adriana Băran
Gabriela Stîngă
Constantin Neamtu

Counterion effect of cationic surfactants on the interaction with poly(acrylic acid)

Abstract In the interaction of poly(acrylic acid) (PAA) and some cationic surfactants, the effects of counterions on the critical aggregation concentration (Cac or T_1) and other critical points were studied by surface tension, conductance, pH and viscosity measurements in unbuffered systems. The surfactant cations were *n*-dodecylammonium, *n*-dodecyltrimethylammonium and *n*-decylammonium (DeA), the latter being mainly employed. The inorganic counterions were chloride, iodide and thiocyanate. These anions are in both far positions in the lyotropic series. The organic counterions were short-chain alkyl carboxylate ions (R^-, i.e., acetate, propionate and butyrate). Whereas the critical micelle concentration (cmc) of DeA-Cl was higher than that of DeA-SCN, T_1 in presence of PAA was the opposite. No pH shift was observed below T_1. It was considered that when Cl^- ion is taken as the reference, the strong water-structure-breaking SCN^- ion may promote hydrophobic self-association of surfactant cations and induce low dissociation or ion-pairing. This counterion effect lowers the cmc and raises T_1. The cmc of DeA-acetate was the highest among the three DeA-Rs, but its T_1 was the lowest though the differences were slight. Therefore, the opposite relation between the cmc and T_1 holds for inorganic and the organic counterions as well. The cmcs of DeA-Cl and DeA-acetate were similar, but T_1 for the latter was much lower. By addition of DeA-Rs, the pH rose even below T_1. The carboxylate ion positions between F^- and Cl^- in the lyotropic series because of its strong electrical field, and the degree of dissociation of the surfactant as salt is related to T_1. In these T_1 features, in addition to the electrostatic effect, the hydrophobic factor and pH change are involved. Thus, SCN^- and R^-s emphasize significant roles of the counterion in the interaction with oppositely charged polymers. The T_1/cmc ratios for the three DeA-Rs were much lower than that for DeA-Cl, which was lower than that for DeA-SCN. For nonionic surfactants, the T_1/cmc was markedly higher than for the cationic surfactants.

Keywords Poly(acrylic acid) · Cationic surfactants · Counterions · Nonionic surfactants · Complexes

D. F. Anghel (✉) · A. Iovescu · A. Băran
G. Stîngă
"I.G. Murgulescu" Institute of Physical Chemistry, Department of Colloids, Splaiul Independentei 202, Bucharest 77208, Romania
e-mail: adan@chimfiz.icf.ro
Tel.: +4021-312-1147
Fax: +4021-312-1147

S. Saito
Nigawa-Takamaru 1-12-15, Takarazuka 665-0063, Japan

C. Neamtu
Institute of Chemical Research, Department of Pesticides, Splaiul Independentei 202, Bucharest 77208, Romania

Introduction

The behavior of surfactants and polymers in their mixed aqueous solution has been widely investigated for a half century with increasing industrial application [1, 2, 3]. The interaction of surfactants and polymers of opposite charge has been studied most extensively, and several reviews on this topic have been published [1, 2, 3, 4, 5, 6]. Although the properties of ionic surfactants, especially of cationics [7], and their interaction with nonionic polymers depends significantly on counterions [8, 9, 10, 11, 12, 13, 14, 15], the counterion effect on the interaction with oppositely charged polymers has been scarcely treated so far [16, 17].

In the interaction with some nonionic polymers, the cationic surfactants are generally weaker than their anionic counterparts, and the reason is still the subject of controversy [1, 2, 3, 18]. However, the surfactant ions are not the only species involved in the different behavior between anionic and cationic surfactants. It has been proved that counterions such as thiocyanate and iodide for alkylammonium surfactants exhibit striking differences in the interaction compared with chloride ion [8, 9, 10]. Compared with chloride ion, which is weakly water structure breaking, those large monovalent anions are considered to be very strong water structure breakers to bulk and local water around hydrophobic solutes as well [19, 20]. Since cosolutes affect each other's behavior through water, it is qualitatively supposed that, arising from such incompatibilities of water structures surrounding surfactants, counterions and nonionic polymers, additional attraction or repulsion occurs among them. In the cases just mentioned, those water-structure-breaking counterions may, in reference to chloride ion, induce such attraction between alkylammonium cations themselves (lowering the critical micelle concentration, cmc) and also cosolute polymers accompanying low dissociation or ion-pairing [11]. This counterion effect is limited when the chain length of alkylammonium ions is short [11].

On the other hand, partially hydrophobic, and therefore, water-structure-making anions, like ethyl sulfate and butyrate, as counterions for dodecylammonium (DoA) do not show by dye solubilization measurements any sign of the previously mentioned special effect in the interaction [11]. Counterions with longer alkyl chains, such as butylsulfate and caproate, cause turbidity with the cationic surfactant, whereas shorter-chain alkyl anions (i.e., ethyl sulfate and butyrate) do not. Thus, the last two counterions are supposed to be also slightly dissociated by surfactant cation but in a different way from the water-structure-breaking ions [11].

For surfactant anion dodecyl sulfate, little difference is observed between Na^+ and K^+ [12] or Li^+ [21] as counterion, but guanidinium ion, a strongly water-structure-breaking cation with reference to Na^+ ion, displays a similar effect as SCN^- counterion for the cationic counterpart. In contrast, strongly water-structure-making hydrophobic cations like tetraalkylammoniums with lengthening alkyl moieties (methyl, ethyl, *n*-propyl, and *n*-butyl) tend to discourage even the strong interaction of the surfactant anion with nonionic polymers [12].

It is considered that these phenomena may be attributed to complex effects of the ionic head group of surfactant and counterion as a salt on the nonionic polymer, specifically in its hydrophobic segments. Thus, by attraction or binding of surfactant salt to the polymer, the surfactant ion and counterion demonstrate together salt properties. This happens locally on the polymer, where they are highly concentrated, i.e., at much lower bulk concentrations than simple inorganic salts [22]. It is also suggested that the generally recognized difference between anionic and cationic surfactants in the interaction with nonionic polymers may be ascribed to such a kind of salt effect [22].

In case of the present interaction of opposite charge, the counterions with the same charge as the polymer are accumulated to and distributed around it only together with surfactant ions, particularly when surfactants form micellar aggregates on the polymer chain. Thus, the counterions are expected to exert their specific influence on the polymer–surfactant complex in its close environment. The effects of anionic counterions Cl^- and I^- or SCN^-, situated in far positions in the lyotropic series, are studied on the interaction of some cationic surfactants with poly(acrylic acid) (PAA). We are interested in these counterion effects mainly on the critical aggregation concentration (cac or T_1), at which the surfactants are abruptly bound micellewise on the polymer below the cmc [3, 18]. In addition, homologous short-chain alkylcarboxylate ions (acetate, propionate and butyrate) are included in the counterions. The carboxylate ion is considered as a stronger electrostrictive anion for water molecule than Cl^- but a little weaker than F^-, which is the strongest because of its smallest size [22]. Thus, the carboxylate ion will be more incompatible with surfactant cations than Cl^- with respect to aqueous environment. Therefore, it is worthy of note in what ways these counterions work on the cmc and T_1.

Finally, a comparison of the interaction of cationic surfactants and poly(oxyethylene)-type nonionic surfactants with PAA is made. These nonionic surfactants are known to interact with PAA by a combination of hydrophobic attraction and hydrogen bonding between oxyethylene and carboxylic groups [23].

Experimental

Materials

PAA with a degree of polymerization of about 2,100 was the same as previously employed [24]. Dodecylammonium chloride (DoA-Cl) was prepared from reagent grade *n*-dodecylamine (Fluka) and reagent grade HCl (Merck), and was recrystallized three times from water and twice from ethanol. DoA-Cl was proved to be pure with no minimum in the surface tension isotherm. The cmc (1.5×10^{-2} M at 25 °C, Table 1) determined by surface tension was in accordance with the literature value [25]. Dodecylammonium thiocyanate (DoA-SCN) was converted from DoA-Cl by repeated precipitations from concentrated aqueous NH_4SCN solution [8, 13, 14, 15]. The surface tension isotherm confirmed its purity. The cmc (7×10^{-3} M at 25 °C, Table 1) nearly agreed with the literature value at 30 °C [8, 11, 13, 14, 15]. Reagent grade *n*-dodecyltrimethylammonium chloride (DoTA-Cl) (Fluka) was used without further purification. Dodecyltrimethylammonium iodide (DoTA-I) was prepared in this laboratory from *n*-dodecylamine first dimethylated with formaldehyde in the presence of formic acid and was then reacted with methyl iodide by heating. Methyl iodide was obtained by heating methanol and iodine in the presence of red phosphorus [26], and was purified by distillation. DoTA-I was recrystallized twice from acetone. Its surface tension isotherm had no minimum. The cmc was 5×10^{-3} M at 25 °C. Decylammonium chloride (DeA-Cl) was prepared from *n*-decylamine (Wako Pure Chemical) and HCl, and was recrystallized in the same manner as DoA-Cl. Decylammonium thiocyanate (DeA-SCN) was prepared from DeA-Cl in the same way as DoA-SCN. The melting points of DeA-Cl and DeA-SCN were 172 (decomposed) and 166–168 °C, respectively. Their surface tension isotherms had no minima. Decylammonium acetate (DeA-Ac), decylammonium propionate (DeA-Pr) and decylammonium butyrate (DeA-Bu) were prepared from *n*-decylamine by neutralization with reagent grade acetic (Reactivul, Romania), propionic and butyric (both Merck) acids, respectively, in *n*-hexane. DeA-Ac and DeA-Pr were recrystallized twice and once, respectively, from *n*-hexane. The surface tension isotherms for DeA-Ac and DeA-Pr presented minima (a deeper one for the former), and the shape of the minimum for DeA-Ac and DeA-Pr did not change on further purification. DeA-Bu crystals were only washed once with *n*-hexane, but the surface tension isotherm had no minimum. These facts allowed us to assume that DeA-Ac and DeA-Pr may be purified well, and another reason than a presumed impurity should be responsible for the respective surface tension minimum. This will be explained in the Results and discussion. The melting points of DeA-Ac, DeA-Pr and DeA-Bu were 58.5, 42 and 12 °C, respectively. Acetone, ethanol and *n*-hexane were of reagent grade (Reactivul, Romania). Water had an electrical conductivity lower than 1.5 μS/cm.

Methods

Surface tension measurements were made as previously described [24]. The pH measurements were made with an Orion pH meter (model 420A) with an error of less than 0.5%. The electrical conductance and viscosity measurements were carried out as before [24, 27, 28]. In the case of partially precipitated systems, the viscosity measurements were done for the supernatant. Throughout the measurements, the polymer concentration was kept constant at 1.02×10^{-2} M in accordance with the previous studies on PAA–nonionic surfactant interaction [24, 29]. Although the interaction of ionic surfactants with PAA is significantly affected by pH [30, 31], no pH control was performed because addition of buffer salts disturbs the effect of various counterions of cationic surfactants. All the measurements were carried out at 25 °C.

Table 1 Critical micelle concentration (*cmc*) of dodecylammonium chloride (*DoA-Cl*), dodecylammonium thiocyanate (*DoA-SCN*), dodecyltrimethylammonium chloride (*DoTA-Cl*), dodecyltrimethylammonium iodide (*DoTA-I*) and octa(oxyethylene) mono(*n*-dodecyl) ether (*Do-E*$_8$) and their critical concentrations, T_n, in interaction with 1.02×10^{-2} M poly(acrylic acid) (*PAA*), at 25 °C

Surfactant	Method								
	Surface tension						Electrical conductance		
	cmc (M)	T_1 (M)	T_2' (M)	T_2 (M)	$T_2' - T_1$ (M)	T_2 − cmc (M)	cmc (M)	T_1 (M)	T_2 (M)
DoA-Cl	1.5×10^{-2}	7×10^{-5}	1×10^{-2}	$(2-3) \times 10^{-2}$	9×10^{-3}	$(5-15) \times 10^{-3}$	1.5×10^{-2}	1×10^{-4}	3×10^{-2}
DoA-SCN	7×10^{-3}	7×10^{-5}	9×10^{-4}	1.5×10^{-2}	2×10^{-3}	8×10^{-3}	7×10^{-3}	1×10^{-4}	$(1.2-1.5) \times 10^{-2}$
DoTA-Cl	2×10^{-2} [a]	2×10^{-4}	4×10^{-3}	3×10^{-2}	3.8×10^{-3}	1×10^{-2}	–	–	–
DoTA-I	5×10^{-3}	2×10^{-4}	2.1×10^{-3}	$(1-1.5) \times 10^{-2}$	1.9×10^{-3}	$(5-10) \times 10^{-3}$	–	–	–
Do-E$_8$ [b]	8×10^{-5}	6.3×10^{-5}	8.4×10^{-4}	2.8×10^{-3}	7.8×10^{-4}	2.7×10^{-3}	–	–	–

[a] From Ref. [7]
[b] From Ref. [29]

Results and discussion

DoA-Cl and DoA-SCN

By addition of small amounts of cationic surfactants, the PAA solution remained transparent. With increasing additive concentration, clouding or precipitation occurred, and at high surfactant concentration the systems tended to be clear again (resolubilization).

The DoA-Cl and DoA-SCN systems without and with PAA were studied by means of surface tension. The results are presented as a function of the logarithm of surfactant concentration in Fig. 1. For the polymer-free system, the descending part of the surface tension isotherm intercepted the horizontal part, and this point is the cmc. In the presence of PAA, the curve abruptly broke at the cac or T_1, and the systems were still transparent. Beyond T_1 a plateau region appeared until T_2'. In this region clouding occurred, or a precipitate in equilibrium with a clear phase appeared. Above T_2', the amount of precipitate gradually decreased and resolubilization took place. In the PAA + DoA-Cl system, with the increase in surfactant concentration, the precipitate dispersed gradually and diminished around 5×10^{-2} M, as observed previously [32]. The Cl^- system looked clearer than the SCN^- system. Finally the curve joined with the surface tension isotherm without PAA at a surfactant concentration denoted as T_2. In both the Cl^- and the SCN^- systems, the isotherms with PAA suggest cooperative interaction which starts at T_1, a surfactant concentration far below the respective cmc. Such a lowering of the cac with PAA is usually found in the interaction between polymers and oppositely charged surfactants [5]. In the present cases, it means that the cationic surfactants are preferably attracted to PAA initiated by electrostatic force and then grow micellewise on it rather than forming free micelles by themselves.

The cmc, T_1, T_2' and T_2 are summarized in Table 1. One can see that whereas the cmc, T_2' and T_2 are greater for the Cl^- system than for the SCN^- system, T_1 is unchanged by the counterions.

Electrical conductance measurements were also carried out on these systems. In the absence of PAA, the results plotted against surfactant concentration indicated linear relations that crossed sharply at a point which is taken as the cmc (data not shown). The micellar degree of dissociation, α, was calculated as the ratio of the slopes of the straight portion of the conductance above and below the cmc [33]. The α values obtained for DoA-Cl and DoA-SCN micelles were 0.24 and 0.15, respectively. This α value for DoA-Cl is within the range of the two literature values of 0.13 and 0.44 [34]. It results that the DoA-SCN micelles are stabilized by a larger number of counterions than the DoA-Cl micelles.

The electrical conductance data obtained with PAA are shown in Fig. 2. For each mixed system, the curve revealed two breakpoints at low and high concentrations, which respectively coincided within experimental error with T_1 and T_2 derived from surface tension measurements. The values of these critical points are shown in Table 1. Since the surfactant binding modifies the dissociation of PAA and the conductance change is more complicated in the presence of polymer, the same procedure as just described is not valid to determine the degree of dissociation of bound micelles.

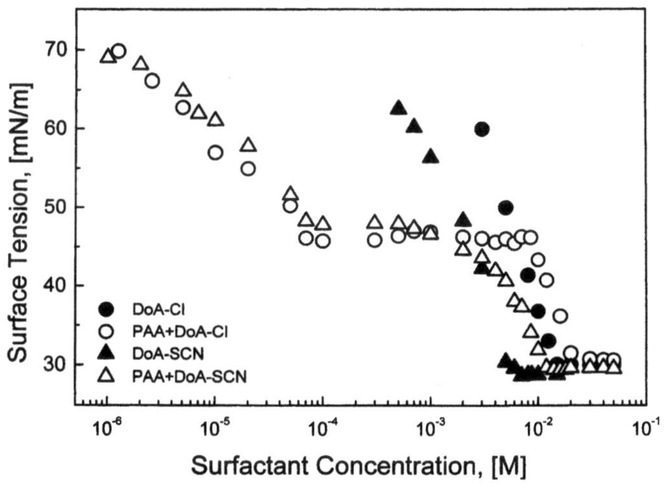

Fig. 1 Surface tension isotherms of dodecylammonium chloride (*DoA-Cl*) and dodecylammonium thiocyanate (*DoA-SCN*) without and with 1.02×10^{-2} M poly(acrylic acid) (*PAA*)

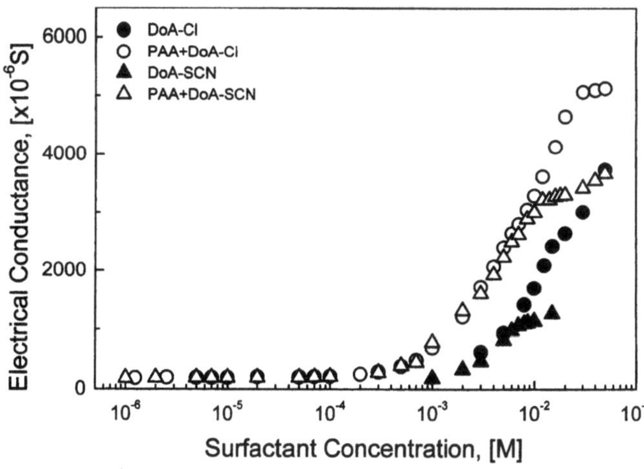

Fig. 2 Electrical conductance of DoA-Cl and DoA-SCN without and with 1.02×10^{-2} M PAA against surfactant concentration

DoTA-Cl and DoTA-I

The PAA + DoTA-Cl system showed cloudiness but no precipitate in the dilute region, and was transparent over 2×10^{-2} M DoTA-Cl, as observed previously [32]. The PAA + DoTA-I system remained turbid without precipitation even at high concentrations. A similar tendency was observed by Hansson and Almgren [35] regarding DoTA-Cl and DoTA-Br in the interaction with Na-PAA.

The surface tension isotherms for the DoTA-Cl and DoTA-I systems in the absence and presence of PAA are shown in Fig. 3, and the cmc, T_1, T_2' and T_2 are also included in Table 1. Although the cmc, T_2' and T_2 for the Cl^- system are higher than those of the I^- system, the T_1 values for both systems are the same. Like SCN^-, I^- is also classified as a strongly water-structure-breaking anion [19, 20]. In this respect, I^- works similarly as SCN^- compared to Cl^- as a counterion for the DoA surfactants.

DeA-Cl and DeA-SCN

No evidence of counterion effect on T_1 was found in the DoA and DoTA surfactants and this may be due to their very low T_1 values. In such a dilute range, the counterions seem to behave in a similar way to each other without exerting any specific function on surfactant cations in the presence of PAA. To find out if the nature of the counterions has any influence on T_1, the effects of DeA-Cl and DeA-SCN, which have higher cmcs than the respective DoA surfactants and thus are expected to have higher T_1s, were investigated for their interaction with PAA.

The surface tension isotherms are illustrated in Fig. 4. Their cmcs and critical points of interaction are summarized in Table 2. As expected, the counterion effect appeared unmistakably in these systems: T_1 for the Cl^- system was lower than for the SCN^- counterpart, whereas its cmc, T_2' and T_2 were higher than those for the SCN^- system. The appearance of the systems was like in PAA + DoA surfactant systems but only shifted to a higher-concentration region.

Measurements of the pH of PAA + DeA-Cl and PAA + DeA-SCN systems confirmed these findings. The polymer pH changes induced by the surfactants are illustrated in Fig. 5. In these curves, the beginning of the steep pH lowering is considered to be T_1. Below T_1 the pH, i.e., dissociation of PAA, is almost unchanged. The point where the pH remains almost constant at a low value is taken as T_2. No clear T_2' was observed. The T_1 and T_2 values are included in Table 2. Both points agree with the surface tension data within experimental error. As the pH at 3×10^{-2} M DeA-Cl is 4.74, the pH decrease at high surfactant concentrations in mixed systems is due to the production of HCl [36] or HSCN as a result of electrostatic binding to PAA. Both HCl and HSCN are strong acids (see also later). This result shows that from T_1 to T_2 the cationic surfactants are bound to PAA both electrostatically and hydrophobically in a micellar form. The small differences in pH between the downward curves for both surfactants may be due to a shift of T_1.

The relative viscosity of both systems shown in Fig. 6 began to decrease at a concentration much lower than T_1. This is a sign of polymer coiling due to diminution of electrical repulsion on the PAA chain, arising from the free surfactant salt in the solution just like salt addition. The hydrophobic attraction by monomeric

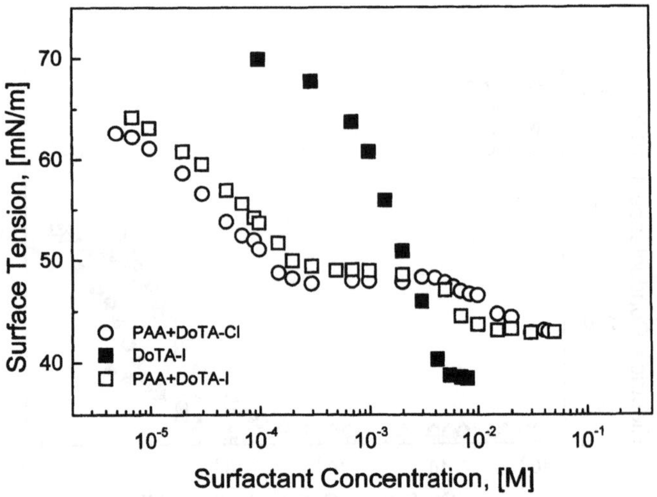

Fig. 3 Surface tension isotherms of dodecyltrimethylammonium chloride (*DoTA-Cl*) and dodecyltrimethylammonium iodide (*DoTA-I*) with 1.02×10^{-2} M PAA and of DoTA-I

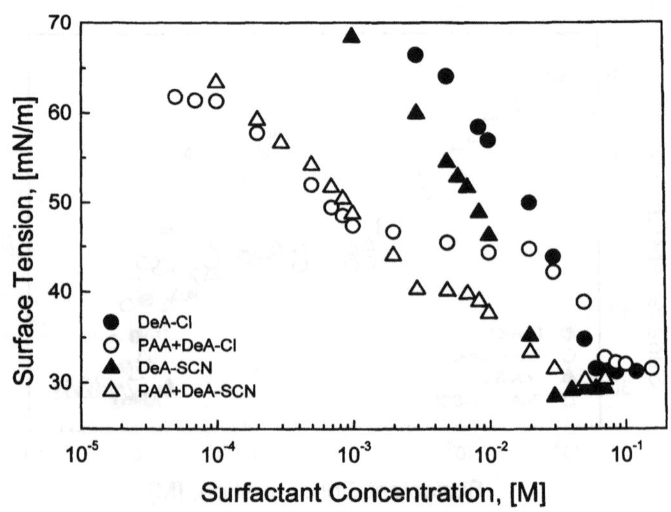

Fig. 4 Surface tension isotherms of decylammonium chloride (*DeA-Cl*) and decylammonium thiocyanate (*DeA-SCN*) without and with 1.02×10^{-2} M PAA

Table 2 cmc of decylammonium chloride (*DeA-Cl*), decylammonium thiocyanate (*DeA-SCN*), decylammonium acetate (*DeA-Ac*), decylammonium propionate (*DeA-Pr*), decylammonium butyrate (*DeA-Bu*), and octa(oxyethylene) mono(*n*-decyl) ether (*De-E*$_8$) and their critical concentrations, T_n, in interaction with 1.02×10^{-2} M PAA, at 25 °C

Surfactant	Method								
	Surface tension						pH		Relative viscosity
	cmc (M)	T_1 (M)	T_2' (M)	T_2 (M)	$T_2' - T_1$ (M)	T_2 − cmc (M)	T_1 (M)	T_2 (M)	T_2 (M)
DeA-Cl	6×10^{-2}	1×10^{-3}	2×10^{-2}	7×10^{-2}	1.9×10^{-2}	1×10^{-2}	1×10^{-3}	7×10^{-2}	7×10^{-2}
DeA-SCN	3×10^{-2}	3×10^{-3}	7×10^{-3}	$(4-5) \times 10^{-2}$	4×10^{-3}	$(1-2) \times 10^{-2}$	$(1-2) \times 10^{-3}$	5×10^{-2}	5×10^{-2}
DeA-Ac	5×10^{-2}	4×10^{-4}	3×10^{-2}	8×10^{-2}	3×10^{-2}	3×10^{-2}	–	–	–
DeA-Pr	4×10^{-2}	4×10^{-4}	2×10^{-2}	7×10^{-2}	2×10^{-2}	3×10^{-2}	–	–	–
DeA-Bu	4×10^{-2}	7×10^{-4}	1×10^{-2}	6×10^{-2}	9.3×10^{-3}	2×10^{-2}	–	–	–
De-E$_8$ [a]	1×10^{-3}	7×10^{-4}	1.5×10^{-3}	4×10^{-3}	8×10^{-4}	3×10^{-3}	–	–	3×10^{-4}

[a] From Ref. [24]

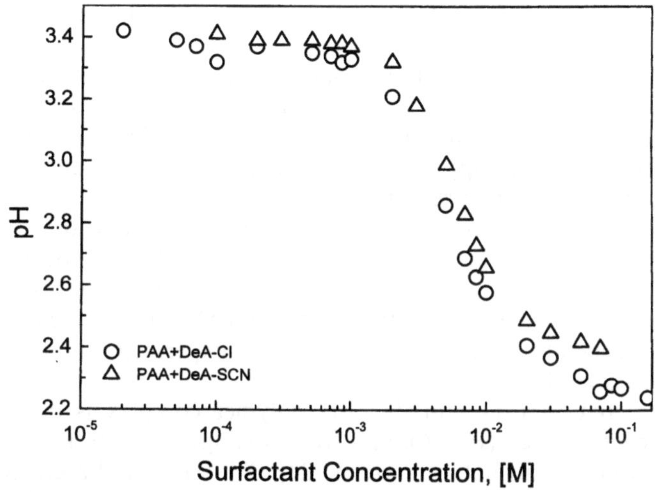

Fig. 5 The effect of surfactant concentration on the pH of DeA-Cl and DeA-SCN with 1.02×10^{-2} M PAA

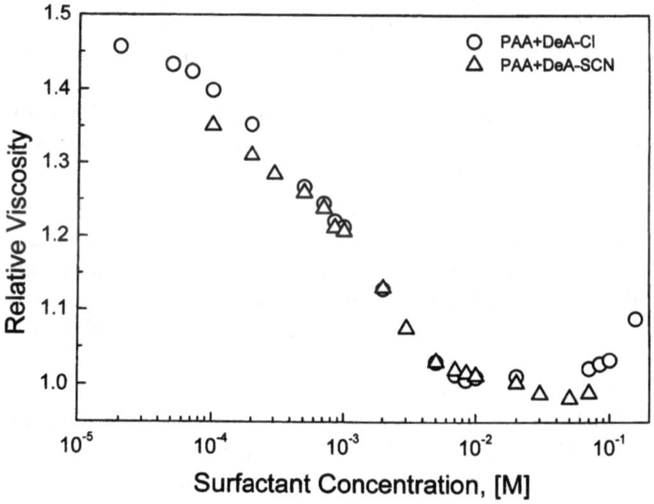

Fig. 6 Relative viscosity of DeA-Cl and DeA-SCN with 1.02×10^{-2} M PAA against surfactant concentration

surfactant binding is considered very small from the little pH change in Fig. 5. Thus T_1 at which the micellar aggregation begins on the polymer was not detected distinctively. The beginning of the small relative viscosity increase in the Cl^- system in the high-concentration region is identified with T_2. Above T_2, polymer uncoiling by further micellewise surfactant binding, without Coulombic attraction, takes place. Below T_2', no significant difference in viscosity behavior was observed between the Cl^- and SCN^- systems.

Binana-Limbele and Zana [16] found that in the interaction of a hydrophobically modified anionic polyelectrolyte with decyltrimethylammonium chloride, bromide and iodide, the T_1 value rises as the counterion is changed from Cl^- to Br^- to I^-, whereas the cmc decreases in this order. They considered that the halide ions and polyanion units compete for complexation with the cationic surfactants and the tendency for complexation increases in the order just described. Their finding regarding the opposite relation between the cmc and T_1 agrees with the present cases.

The T_2 − cmc value is the saturated amount of surfactant bound to PAA under the influence of its negative charge. This quantity is much bigger than the effective charge of PAA (see later) and is nearly independent of counterion within experimental error (Tables 1, 2).

Between T_1 and T_2', the polymer complex is compactly coiled (Fig. 6), and from analogy with the PAA–nonionic surfactant systems [24], $T_2' - T_1$ may be taken as the saturated amount of surfactant as isolated micellar aggregates on PAA. This quantity for counterion Cl^- is higher than that for SCN^- (Tables 1, 2). This trend of $T_2' - T_1$ is opposite to that of T_1 but the same as that of the cmc. As T_2's in these cases are much higher than the amount of the dissociated carboxylic units of PAA (4×10^{-4} M), it is explained that the binding of surfactants to PAA between T_1 and T_2' is triggered by electrostatic force but increases mainly by micellization-like hydrophobic aggregation. Accordingly, the trend of $T_2' - T_1$ is explained.

$T_2' - T_1$ for the DoA-Cl and DeA-Cl systems are larger than those of the DoA-SCN and DeA-SCN systems (Tables 1, 2). Although no clear differences were observed in pH (Fig. 5) and in viscosity (Fig. 6) between the Cl^- and SCN^- systems, from the pH results (Fig. 5) it is considered that, as from the α values of the free DoA-Cl and DoA-SCN micelles, the micellewise-bound DoA-Cl or DeA-Cl on PAA may also be more dissociated than DoA-SCN and DeA-SCN. Therefore, between T_1 and T_2' the PAA chain with DeA-Cl may be less contracted and offer more space for the surfactant binding. Another possibility is that the number of surfactant molecules in one micelle on PAA may be larger in the Cl^- system than in the corresponding SCN^- system.

The reason for the counterion effects on T_1 may be the different degrees of dissociation of these surfactant micelles, which depend on the specific function of counterion on the hydrophobic cosolutes (see also the following subsection). As elaborated in the Introduction, when Cl^- ion is taken as a reference, I^- and SCN^- may encourage hydrophobic self-attraction of cationics (lowering the cmc) and induce decreasing dissociation (ion-pairing) [8, 11, 22]. This decrease in the effective surfactant concentration may suppress the surfactant binding to PAA, and as a result, T_1 for the system with counterion I^- or SCN^- becomes higher than that with counterion Cl^-. By contrast, Cl^- is more hydrated because of its smaller ionic size than I^- and SCN^-, and tends to discourage micellization of cationics (raising the cmc) and in turn to lower T_1.

In order to evaluate the strength of the polymer-surfactant interaction one should consider to a first approximation both the molar free energy of micellization,

$$\Delta G^0_{\mathrm{m}} = RT \ln \mathrm{cmc} \ ,$$

and the molar free energy of surfactant binding to the polymer,

$$\Delta G^0_{\mathrm{b}} = RT \ln T_1 \ .$$

From these two relations, the molar free energy for the reaction free micelle↔polymer-bound micelle, a convenient measure of the strength of the interaction [37] is derived as

$$\Delta G^0_{\mathrm{PS}} = RT \ln(T_1/\mathrm{cmc}) \ .$$

The values T_1/cmc and $\Delta G^0{}_{\mathrm{PS}}$ for all the systems studied are presented in Table 3. Although the T_1s for DoA-Cl and DoA-SCN, and for DoTA-Cl and DoTA-I, are the same irrespective of counterion, T_1/cmc and thus $\Delta G^0{}_{\mathrm{PS}}$ turn out to be different. Like in the DeA surfactants, in the Cl^- systems, $\Delta G^0{}_{\mathrm{PS}}$ is smaller and, therefore, the binding of the surfactants to PAA is stronger than in the I^- or SCN^- systems.

DeA-Ac, DeA-Pr and DeA-Bu

The effect of short-chain alkylcarboxylate ions (R^-s), i.e., acetate (Ac^-), propionate (Pr^-) and butyrate (Bu^-), as counterions on the interaction of DeA with PAA was investigated by surface tension. The results are displayed in Figs. 7, 8 and 9. The existence of a minimum in the surface tension isotherm makes the determination of the cmc vague and controversial [38].

Table 3 T_1/cmc and free energy, ΔG^0_{PS}, in the interaction with 1.02×10^{-2} M PAA of the surfactants investigated

Surfactant	T_1/cmc	ΔG^0_{PS} (kcal/mol)
DoA-Cl	4.7×10^{-3}	−3.17
DoA-SCN	1×10^{-2}	−2.73
DoTA-Cl	1×10^{-2}	−2.73
DoTA-I	4×10^{-2}	−1.91
Do-E_8	7.9×10^{-1}	−0.14
DeA-Cl	1.7×10^{-2}	−2.41
DeA-SCN	1×10^{-1}	−1.36
DeA-Ac	8×10^{-3}	−2.86
DeA-Pr	1×10^{-2}	−2.73
DeA-Bu	1.75×10^{-2}	−2.40
De-E_8	7×10^{-1}	−0.21

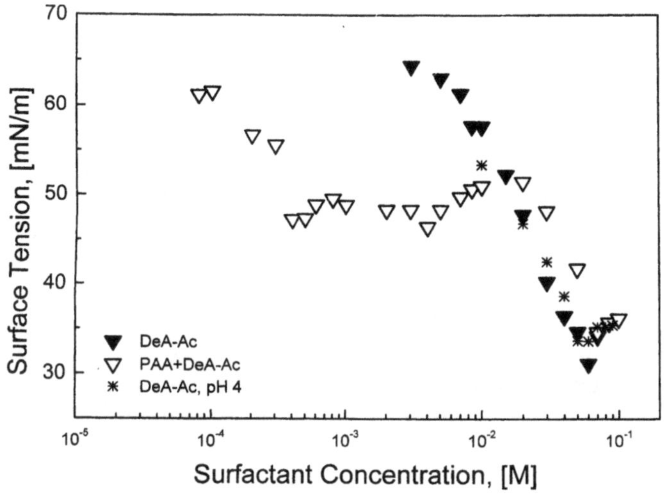

Fig. 7 Surface tension isotherms of decylammonium acetate (*DeA-Ac*) without and with 1.02×10^{-2} M PAA, and of DeA-Ac at pH 4

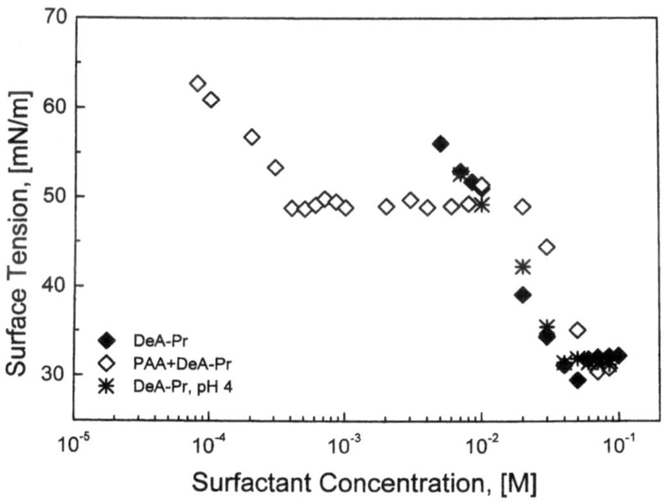

Fig. 8 Surface tension isotherms of decylammonium propionate (*DeA-Pr*) without and with 1.02×10^{-2} M PAA, and of DeA-Pr at pH 4

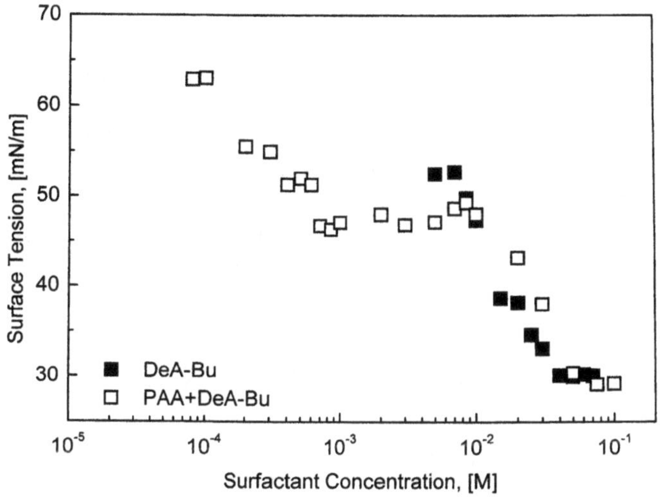

Fig. 9 Surface tension isotherms of decylammonium butyrate (*DeA-Bu*) without and with 1.02×10^{-2} M PAA

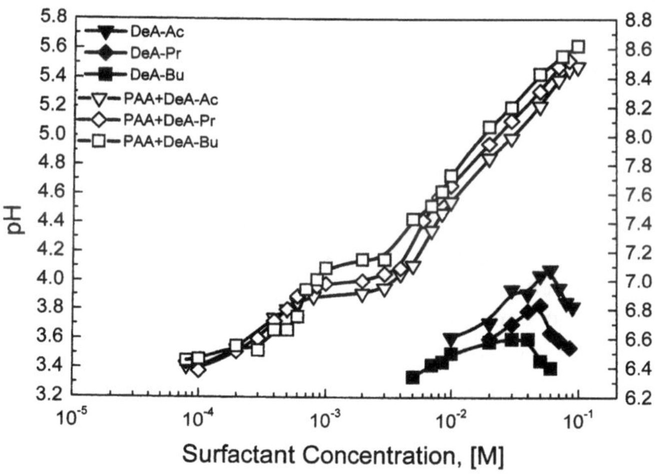

Fig. 10 pH of DeA-Ac, DeA-Pr and DeA-Bu without and with 1.02×10^{-2} M PAA against surfactant concentration

The pH of DeA-R solutions (Fig. 10) showed a peak at the concentration corresponding to the minimum in the respective surface tension isotherm (Figs. 7, 8), though DeA-Bu had no minimum (Fig. 9). Finch and Smith [39] found an optimum pH at which the surface tension of DoA-Ac solutions reaches a minimum, and attributed it to surfactant hydrolysis. In fact, at around pH 4 by addition of about 0.1 M acetic acid, the valley in the DeA-Ac isotherm became much shallower (Fig. 7). This minimum point (5×10^{-2} M) is considered more appropriate as the cmc than the minimum point at higher pH, because the pHs in the present PAA + DeA-Ac system are in the lower region (Fig. 10). Also, such low additive concentrations will not affect significantly the aqueous environment around the hydrophobic moieties. The same applies to DeA-Pr: no minimum was seen at around pH 4 by addition of about 0.1 M propionic acid (Fig. 8). These cmc values are shown in Table 2.

The pHs of the PAA + DeA-R systems below 10^{-1} M surfactant were lower than 5.6 and, therefore, no hydrolysis took place. The critical points obtained are summarized in Table 2. Around the respective T_1, the Ac^- system produced an extremely small amount of precipitate, whereas the Pr^- and Bu^- systems were clear. Between T_1 and T_2', the systems were cloudy and gradually became clearer with increasing surfactant concentration with some precipitate coexisting. Beyond T_2', the precipitation decreased and resolubilization began. In turn, the cloudiness increased and above T_2

only the Ac^- system was increasingly transparent. This reflects a more hydrophilic Ac^- ion than the others.

For a convenient criterion of these counterion specialties, the present cases are inferred from the salt effect (medium effect) on a nonionic polymer solution [22], because the PAA + cationics complexes at T_1 are locally almost nonionic and hydrophobic. The effects of various anions as Na or K salts on the nonionic polymer change widely in the following order including the inorganic lyotropic series:

$$IO_3^- < F^- \leq Ac^- \leq Pr^- \leq Bu^- \leq \text{formate}^- < Cl^- < Br^- = \text{no salt} < NO_3^- < ClO_4^- < I^- < SCN^- \; , \quad (1)$$

the last three being typical water-structure breakers. The effects of formate and R ions are in almost the same degree and all are between F^- and Cl^-. Also in the degrees of association of counterions in DoTA surfactant micelles measured by the electromotive force, the position of Ac^- is between F^- and Cl^- and is closer to F^- than to Cl^- as in the order in Eq. (1) [40]. This may probably hold also for DeA surfactants. It is supposed that the electrical field around the carboxylate head is so intense that the ordered water structure surrounding the short-chain alkyl groups attached is strongly perturbed. This will explain the small differences in the cmcs of the DeA-Rs and also similar cmcs of DeA-Cl and DeA-Ac. The cmcs of DoA-F, DoA-Cl, DoA-Ac, DoA-NO_3, DoA-I and DoA-SCN are 13×10^{-3}, 11×10^{-3}, 10×10^{-3}, 7.5×10^{-3}, 7×10^{-3} and 5.7×10^{-3} M, respectively, at 30 °C [11].

At the same time, the effects of *n*-alkyl sulfate ions as Na salts are [22]:

$$Ac^- < \text{no salt} \leq CH_3SO_4^- \leq C_2H_5SO_4^- \ll C_4H_9SO_4^- < SCN^- \ll C_6H_{13}SO_4^- \ll C_8H_{17}SO_4^- \; . \quad (2)$$

And, for *n*-alkylammonium ions as Cl salts, the order is [22]:

$$NH_4^+ = CH_3NH_3^+ \leq C_2H_5NH_3^+ \leq \text{no salt} \leq C_4H_9NH_3^- < C_6H_{13}NH_3^- < NaC_4H_9SO_4 \ll C_8H_{17}NH_3^+ \ll C_{12}H_{25}NH_3^+ \; . \quad (3)$$

The effects of the orders in Eqs. (2) and (3) are transient from local accumulation (medium effect) to firm binding to the nonionic polymer. These orders (Eqs. 1, 2, 3) suggest that the effects of short-chain alkyl groups are similar, and that the influence of the electrical field of the ion upon the water structure in bulk and around the alkyl group attached depends significantly on the nature of the individual ionic head.

Although DeA-Cl and DeA-Ac have similar cmcs, T_1 for the latter is much lower. This may be not unreasonable from the viewpoint of the large difference in the degrees of dissociation of the Cl^- and Ac^- surfactants as already speculated. The pH of the PAA solution increases by addition of DeA-R even below T_1 as determined by surface tension (Fig. 10), in contrast to the downward trend of DeA-Cl and DeA-SCN only above T_1 (Fig. 5), and this may also contribute to the lowering of T_1 [30, 31]. The bulkiness of R^- may make distinguishing of T_1 rather insensitive. In addition, it seems that, like the cases of DoA and DoTA surfactants (Table 1), the T_1 range may be too dilute to appeal conspicuously to the difference of each counterion.

$T_2' - T_1$ for DeA-Ac is larger than for DeA-Pr and DeA-Bu and even for DeA-Cl (Table 2). Steric hindrance by longer R on PAA may be responsible for the smaller $T_2' - T_1$. Also the number of surfactants in a micelle on the polymer may be considered here. However, as deduced from the order in Eq. (1), perhaps micellar DeA-Ac on PAA may be more dissociated than DeA-Pr, DeA-Bu and also DeA-Cl in this order and, therefore, between T_1 and T_2' more space on PAA may be available for further binding in this order (see earlier).

The appearance of a plateau in the curve in Fig. 10 shows two breakpoints at both ends which are not identified with any of the critical points shown in Table 2. This may be accounted for as follows. Aqueous solutions of DeA-Rs are relatively less acidic (or more neutral), as shown in Fig. 10, in contrast to the cases of DeA-Cl and DeA-SCN, which produce strong acid. For example, an aqueous 10^{-3} M HCl solution has a pH of 2.87; thus, the unbound DeA-R below T_1 raises the pH. As for the coulombically bound ones above T_1, a weak acid HR released will not affect the pH of the PAA solution. At the same time micelles grow on PAA, and they will raise the pH like free DeA-R. As a result, a pH plateau emerges until free DeA-R begins to increase from T_2'.

The higher pH for PAA + DeA-Rs suggests that PAA is more dissociated and less compact. Therefore, more surfactants are electrostatically and hydrophobically bound to PAA than in the case of the inorganic counterparts, especially at higher free surfactant concentrations. This supposition may be explained by the comparison of T_2 − cmc in Table 2. The values for DeA-Rs are coincident with each other within experimental error, as are those for DeA-Cl and DeA-SCN, and they are, in spite of large differences in pH in both systems, only a little larger than those for the inorganic counterparts (Table 2). This may reflect the predominance of hydrophobic micellization around T_2 on PAA.

Comparison of cationic and nonionic surfactants in their interaction with PAA

Both systems resemble each other in that the carboxyls of PAA are blocked by the headgroups of surfactants [23] though in different ways. Two nonionic surfactants, octa(oxyethylene) mono(*n*-dodecyl) ether (Do-E_8) and octa(oxyethylene) mono(*n*-decyl) ether (De-E_8) are compared with DoA and DeA surfactants, respectively, in the interaction with PAA. Their surface tension isotherms and viscosity behavior showed similar patterns. The data taken from the literature are included in Tables 1, 2 and 3 [24, 29]. Evidently, from a free to a PAA-bound micelle, the nonionic surfactants transfer much less favorably than cationic surfactants with the same hydrophobic moieties despite multiple hydrogen bonds.

By assuming reasonably that not all of the oxyethylenes of the surfactants are hydrogen-bonded with PAA below T_2' [29], one Do-E_8 and De-E_8 each occupies less than eight carboxyls of PAA (e.g., five). The $T_2' - T_1$ for them are 7.8×10^{-4} and 8×10^{-4} M, respectively, and both become approximately 4×10^{-3} M. It is interesting that these values happen to be close to those for DoA-SCN and DeA-SCN and DeA-Bu, which are supposed to be in a less dissociated micellar form on PAA between T_1 and T_2'.

Conclusions

Whereas the cmc of DeA-Cl was higher than that of DeA-SCN, T_1 in the presence of PAA was the opposite. The cmc of DeA-Ac was the highest among the three DeA-Rs, but its T_1 was the lowest, though the differences were slight. Thus, the opposite relation between the cmc and T_1 for inorganic counterions also holds for R^-s. In reference to Cl^-, SCN^- and R^-s manifest significant roles of the counterion in the interaction with oppositely charged polymer. The cmcs of DeA-Cl and DeA-Ac were similar, but T_1 for the latter was much lower. The behavior of the T_1s for counterions Cl^-, SCN^- and the R^-s was discussed from the viewpoint of their influences on the water structure around these cosolutes and also from the pH shift. The T_1/cmc values for the three DeA-Rs were much lower than that for DeA-Cl, which was lower than that for DeA-SCN. The T_1/cmc and ΔG^0_{PS} values for nonionic surfactants were markedly higher than those for the cationic surfactants despite multiple hydrogen bonds with the polymeric carboxyls.

References

1. Robb ID (1981) In: Lucassen-Reynders E (ed) Anionic surfactants: physical chemistry of surfactant action. Dekker, New York, pp 109–142
2. Goddard ED (1993) In: Goddard ED, Ananthapadmanabhan KP (eds) Interactions of surfactants with polymers and proteins. CRC, Boca Raton, pp 171–201
3. Rodenhiser AP, Kwak JCT (1998) In: Kwak JCT (ed) Polymer–surfactant systems, vol 77. Dekker, New York, pp 1–20
4. Hayakawa K, Kwak JCT (1991) In: Rubingh DN, Holland PM (eds) Cationic surfactants, physical chemistry. Dekker, New York, pp 189–248
5. Wei YC, Hudson SM (1995) J Macromol Sci Rev Macromol Chem Phys C 35:15
6. Piculell L, Lindman B, Karlström G (1998) In: Kwak JCT (ed) Polymer–surfactant systems, vol 77. Dekker, New York, pp 65–141
7. Ananthapadmanabhan KP (1993) In: Goddard ED, Ananthapadmanabhan KP (eds) Interactions of surfactants with polymers and proteins. CRC, Boca Raton, pp 5–58
8. Saito S, Yukawa M (1969) J Colloid Interface Sci 30:211
9. Saito S, Yukawa M (1969) Kolloid Z Z Polym 234:1015
10. Saito S (1970) J Polym Sci A 1 8:263
11. Saito S, Kitamura K (1971) J Colloid Interface Sci 35:346
12. Saito S, Taniguchi T, Kitamura K (1971) J Colloid Interface Sci 37:154
13. Harada S, Komatsu T, Nakagawa T (1974) Nippon Kagaku Kaishi 662
14. Harada S, Komatsu T, Nakagawa T (1974) Rep Prog Polym Phys Jpn 17:101
15. Harada S, Komatsu T, Nakagawa T (1976) Rep Prog Polym Phys Jpn 19:17
16. Binana-Limbele W, Zana R (1987) Macromolecules 20:1331
17. Zana R (1998) In: Kwak JCT (ed) Polymer–surfactant systems. Dekker, New York, pp 409–454
18. Goddard ED (1993) In: Goddard ED, Ananthapadmanabhan KP (eds) Interactions of surfactants with polymers and proteins. CRC, Boca Raton, pp 123–169
19. Kavanau JL (1964) Water and solute–water interactions. Holden-Day, San Francisco
20. Bockris JOM, Reddy AKN (1971) Modern electrochemistry, vol 1. Plenum, New York
21. Ridell A, Evertsson H, Nilsson S (2002) J Colloid Interface Sci 247:381
22. Saito S (1969) J Polym Sci A 1 7:1789
23. Saito S, Anghel DF (1998) In: Kwak JCT (ed) Polymer–surfactant systems, vol 77. Dekker, New York, pp 357–408
24. Anghel DF, Saito S, Baran A, Iovescu A (1998) Langmuir 14:5342
25. Mukerjee P, Mysels KJ (1971) Critical micelle concentration of aqueous surfactant systems, NSRDS-NBS 36. National Bureau of Standards, Washington, DC
26. Vogel AI (1956) A textbook of practical organic chemistry including qualitative organic analysis, 3rd edn. Longmans, London
27. Raicu V, Baran A, Iovescu A, Anghel DF, Saito S (1997) Colloid Polym Sci 275:372
28. Raicu V, Baran A, Anghel DF, Saito S, Iovescu A, Radoi C (1998) Prog Colloid Polym Sci 109:136
29. Anghel DF, Saito S, Iovescu A, Baran A (1994) Colloids Surf A 90:89
30. Katsuura H, Kawamura H, Manabe M, Maeda H (2001) Colloid Polym Sci 279:858
31. Katsuura H, Kawamura H, Manabe M, Kawasaki H, Maeda H (2002) Colloid Polym Sci 280:30
32. Saito S (1955) Kolloid Z 143:66
33. Rodenas E, Sierra ML (1996) Langmuir 12:1600

34. Fendler JH, Fendler EJ (1975) Catalysis in micellar and macromolecular systems. Academic, New York
35. Hansson P, Almgren M (1994) Langmuir 10:2115
36. Fundin J, Hansson P, Brown W, Lidegran I (1997) Macromolecules 30:1118
37. Lindman B, Thalberg K (1993) In: Goddard ED, Ananthapadmanabhan KP (eds) Interactions of surfactants with polymers and proteins. CRC, Boca Raton, pp 203–276
38. Lin SY, Lin YY, Chen EM, Hsu CT, Kwan CC (1999) Langmuir 15:4370
39. Finch JA, Smith GW (1973) J Colloid Interface Sci 45:81
40. Gaillon L, Lelièvre J, Gaboriaud R (1999) J Colloid Interface Sci 213:287–297

Progr Colloid Polym Sci (2003) 122: 27–36
DOI 10.1007/978-3-540-36114-5

Joachim Koetz
C. Günther
S. Kosmella
E. Kleinpeter
G. Wolf

Polyelectrolyte-induced structural changes in the isotropic phase of the sulfobetaine/pentanol/toluene/water system

Abstract The paper describes the behavior of the sulfobetaine/pentanol/toluene/water system. An isotropic inverse micellar region (L_2 phase) can be observed in the oil corner. The area of the L_2 phase is decreased by adding an anionic polyelectrolyte, i.e. sodium polyacrylate. The incorporation of the cationic polyelectrolyte poly(diallyldimethylammonium chloride) (PDADMAC) induces the extension of the isotropic phase in the direction of the water corner under the formation of an isotropic phase channel. The aim of the work presented here was to study this phase channel at constant surfactant concentration (i.e. 20%) by means of electrical conductivity, ^{1}H NMR self-diffusion experiments, rheology, polarizing microscopy, small-angle X-ray spectroscopy, and micro-differential scanning calorimetry (DSC). Macroscopically, no indication of a phase transition is observed when going from the oil-rich side to the water-rich side. However, above a critical content of the aqueous PDADMAC solution, the diffusion coefficients are changed significantly, freezable water becomes detectable in DSC measurements, and lamellar liquid-crystalline textures become visible after shearing. This behavior can be explained by a shear-induced transition from an L_2 phase to a lamellar liquid-crystalline phase, and is combined by only small viscometric and conductometric effects. The transition from the L_2 to the L_α phase can be explained by a significant change of the curvature of the surfactant film due to Coulombic interactions with the polycation.

Keywords Microemulsion · Phase diagram · Polyelectrolytes · Small-angle X-ray scattering · ^{1}H NMR · Differential scanning calorimetry

J. Koetz (✉) · C. Günther · S. Kosmella
E. Kleinpeter · G. Wolf
Institute of Chemistry,
Colloid Chemistry,
University of Potsdam,
Karl-Liebknecht-Strasse 24/25,
14476 Golm, Germany
e-mail: koetz@rz.uni-potsdam.de
Tel.: +49-331-9775220
Fax: +49-331-9771127

Introduction

According to Shinoda's concept [1] organized solutions can be expected in a solute–solvent system when the solute is amphiphilic. Owing to their amphiphilic constitution, the surfactants show in solution a multiplicity of self-assembled structures which can organize and produce a large variety of phases. Especially ternary mixtures composed of surfactant, water, and oil have attracted much interest since they form thermodynamically stable phases involving self-organized assemblies. If the system contains an additional fourth component, like a cosurfactant, microemulsions can be formed over a wide range of compositions. The size and shape of micelles as well as interactions between microemulsion droplets depends on additional parameters, for example, the temperature.

The different phases in a microemulsion system are known as normal micellar (L_1 phase), reverse micellar (L_2 phase), or bicontinuous phases. In addition to bicontinuous microemulsions, one finds dilute lamellar phases and liquid isotropic phases of randomly connected bilayers called sponge phases. Usually, the regions of the L_1 and L_2 phases are separated in the phase diagram. Only in some special cases has an isotropic connection between the normal micellar and the reverse micellar phase been reported, for example, in cetyltrimethylammonium bromide (CTAB)/methanol/water systems [2], sodium dodeyl sulfate (SDS)/pentanol/water systems [3, 4], as well as in tetradecylmethylamine oxide/benzyl alcohol/water systems [5].

As indicated earlier, the isotropic phase channel in the SDS/pentanol/water system can be extended by adding a noncharged polymer, i.e. poly(vinylpyrrolidone) (PVP) [6]. In presence of oil (e.g. xylene or toluene) the two micellar phases become disconnected again.

However, the oil- and water-soluble PVP induces the formation of a phase channel in the oil-containing system [7]. Nevertheless, in the absence of surfactant–polymer interactions, for example, in the adequate CTAB-based system, no phase channel is formed. It should be reasonable to consider that an extended L_2 phase can only be expected if interactions occur between the surfactant head groups and the functional groups of the polymer.

The study presented here describes the use of a zwitterionic surfactant with a sulfobetaine head group in the pentanol/toluene/water system. To understand the role of Coulombic interactions in such multicomponent systems an anionic as well as a cationic polyelectrolyte were employed. In the latter case, electrostatic attractive interactions between the N function of the polycation and the sulfonate group of the sulfobetaine can be expected.

Materials and methods

Materials

Pentanol (99+%), toluene (99+%), and sulfobetaine (≫97%) were obtained from Fluka, and were used without further purification. Water was purified by the water purification system MODULAB Pure One (Continental). The poly(diallyldimethylammonium chloride) cationic polyelectrolytes PDADMAC1 and PDADMAC4 were synthesized by radical polymerization. The sample with low molecular weight (PDADMAC1) was polymerized in the presence of ethylenediaminetetraacetic acid at 80 °C by using ammonium persulfate as initiator. The PDADMAC4 was polymerized at 38 °C with 2,2-azobis(2-amidinopropane)dihydrochloride initiator. The PDADMAC2 and PDADMAC3 samples were purchased from Aldrich. All the samples were purified by ultrafiltration until the eluted water had a conductivity below 70 μS/cm. The molecular weights of the components were determined by gel permeation chromatography in 0.5 N $NaNO_3$ solution. The results are given in Table 1.

Methods

The phase diagrams were determined optically by titration of the oil/surfactant/alcohol mixture with water or the corresponding aqueous polymer solution. The conductivity measurements were carried out with a microprocessor conductivity meter (LF2000, WTW). The cationic polymer of low molar mass (PDADMAC1) was exclusively used in these experiments. The rheological properties of the mixtures were investigated with a low-stress rheometer (LS 100, Physica). Mixtures of lower viscosity (samples with PDADMAC1 and PDADMAC2) were investigated by using a Couette geometry. The other two samples were investigated with a cone spindle geometry (2° cone angle, 25-mm cone radius). In addition to the shear experiments, oscillation experiments were performed. Small angle X-ray diffraction patterns were obtained from a "Kratky Kompakt Kleinwinkel System" (A. Paar). The birefringence of the samples was checked by a Leica DMLB microscope with crossed polarizers. The differential scanning calorimetry (DSC) measurements were carried out with a Micro-DSC III (Setaram) in a temperature region between −20 and 80 °C. The heating and cooling rate was fixed at 0.175 °C/min. The thermal cycle started at 25 °C. After cooling to −20 °C, the samples were kept frozen for at least 4 h (isotherm ramp), before the heating started. The cycles were repeated several times to ensure the reproducibility of the DSC curves.

The ^{1}H NMR self-diffusion measurements were carried out using a Bruker AMX-300 spectrometer operating at a proton resonance frequency of 300.13 MHz using the LED pulse sequence [8] with bipolar pulse pairs (bpp-LED) [9] at 25 °C. Rectangular gradient pulses of 1.5-ms duration ($\delta/2$) were used. The spacing between pulses of opposite polarity, τ, was 0.5 ms. The diffusion delay, Δ, and the eddy current delay, T_e, were 100 and 5 ms, respectively. Forty 1D slices with the gradient strength, g, ranging from 0.1 to 45 G /cm were recorded in a 2D data matrix. The spectral width was 2,703 Hz.

After zero-filling and the exponential window function, Fourier transformation and phase correction were subsequently applied in F_2 to yield a set of 1D spectra at variable gradient strength. The corresponding signal intensities, $I(g)$, obtained with the bpp-LED pulse sequence are given by Eq. (1):

$$I(g) = \sum_i I_0^i \exp\left[-\gamma^2 g^2 \delta^2 \left(\Delta - \frac{\delta}{3} - \frac{\tau}{2}\right) D^i\right] , \quad (1)$$

where I_0 represents the signal intensity in the absence of gradients, γ is the gyromagnetic ratio of the observed nucleus, and D is the self-diffusion coefficient. The summation runs over all diffusing species i contributing to the signal. D was then determined by fitting Eq. (1) to the experimental signal attenuation, $I(g)$. In most cases a single exponential fit was used; however, for spectral regions where several signals overlapped a biexponential fit was considered. Finally, the diffusion coefficients obtained for each signal of a diffusing species were averaged. The error in D was less than 5%.

Since rapid exchange of protons occurs between water and the terminal OH groups of the alcohol, the apparent water diffusion coefficient $D_{W,obs}$ is given by Eq. (2) [10] if one assumes that proton exchange itself does not contribute significantly to the diffusion process.

$$D_{W,obs} = \frac{2x_W}{2x_W + x_A} D_W + \frac{x_A}{2x_W + x_A} D_A , \quad (2)$$

Table 1 Molecular characterization of the poly(diallyldimethylammonium chloride) (*PDADMAC*) polyelectrolytes by gel permeation chromatography

	M_w (g/mol)	M_w/M_n
PDADMAC1	13,700	–
PDADMAC2	151,000	2.09
PDADMAC3	300,000	2.37
PDADMAC4	518,000	–
Sodium polyacrylate	5,100[a]	–

where x_W and x_A are the mole fractions of water and alcohol, respectively. D_A is the self-diffusion coefficient of alcohol, which can easily be determined from the alkyl chain signals, and D_W denotes the (corrected) self-diffusion coefficient of water.

Results and discussion

Phase behavior

The best way to illustrate the phase behavior of an amphiphilic material in water is to show a phase diagram. The phase diagrams of a number of systems stabilized by conventional surfactants have already been reported on a large scale. Our interests are first of all focused on the phase behavior of the pseudoternary sulfobetaine/(pentanol/toluene)/water system. During the tests, the pentanol/toluene ratio was fixed at 1, whereas particular consideration was given to the isotropic phase. In a second step, polyelectrolytes were added to the system. Figure 1 shows that adding sodium polyacrylate decreases the L_2 phase. At a polymer concentration of 10% by weight, only a very small area of an isotropic phase exists. When the anionic polyelectrolyte is substituted by the cationic PDADMAC, quite different behavior is observed. The effect of PDADMAC is more remarkable, it caused the optically clear phase to extend in the direction of the water corner. It should be mentioned that at a PDADMAC concentration of 20–30 wt% the optically clear phase becomes significantly larger (Fig. 2). The most interesting point is that the optically clear phase is extended to the water corner by increasing the PDADMAC concentration, independent of the molar mass of the PDADMAC used. This opens a way to characterize these systems along a phase channel from the oil to the water corner at constant sulfobetaine concentration (i.e. 20% by weight). It has to be mentioned that along this line (given as a solid line in Fig. 3), macroscopically, no indication of a phase transition is observed. In order to study changes in the behavior along this line ^{1}H NMR self-diffusion, rheological, electrical conductivity, small-angle X-ray scattering (SAXS), and DSC measurements were

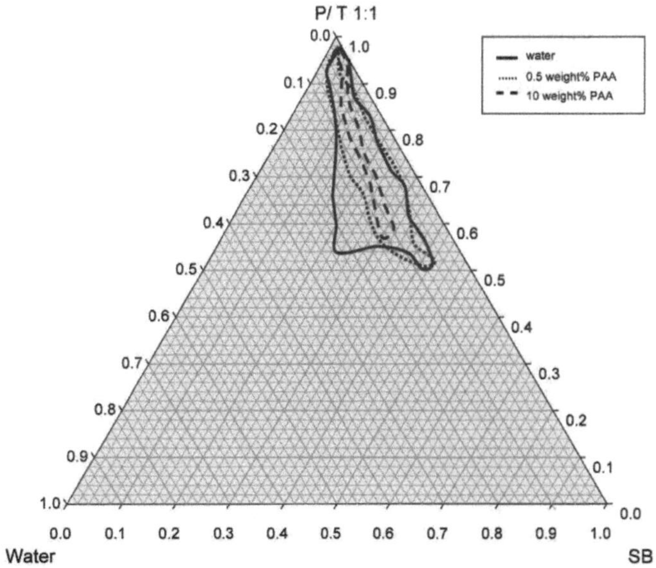

Fig. 1 Phase behavior of the sulfobetaine (*SB*)/(pentanol, *P*/toluene, T)/water system in presence of an anionic polyelectrolyte

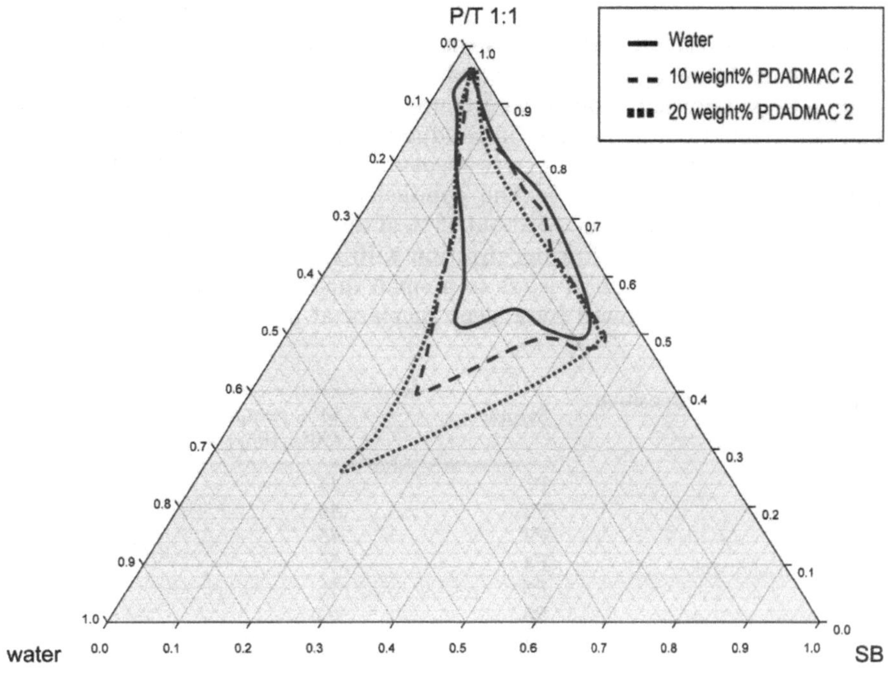

Fig. 2 Phase behavior of the SB/(P/T)/water system in presence of a cationic polyelectrolyte

Fig. 3 Phase behavior of the SB/(P/T)/water system in presence of 30% per weight of polycation in aqueous solution

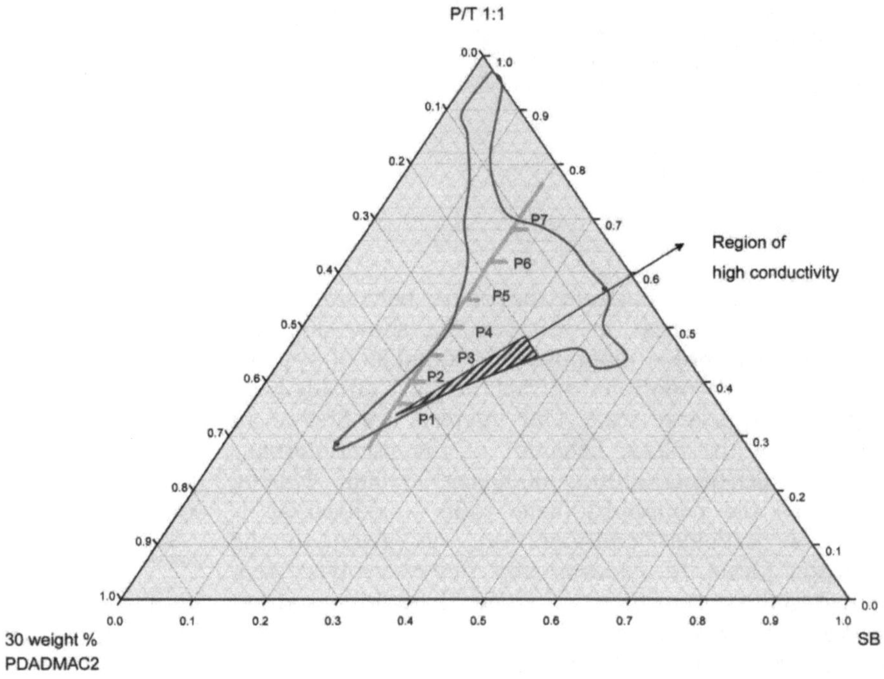

performed. Most of the experiments were carried out on samples P1–P7, which are marked in Fig. 3 (the compositions are listed in Table 2). Additionally, the samples were examined using an optical microscope with crossed polarizers. To determine the percolation boundary, the electrical conductivity at different sulfobetaine concentrations was measured.

Electrical conductivity

The L_2 phase is an "ideal" system for conductivity measurements owing to the low conductivity of the oil. Interactions between droplets and changes in structure can be detected by the increase in conductivity.

The electrical conductivity was measured along the solid line marked in Fig. 3. The conductivity of the system remains constant up to about 35% of the aqueous PDADMAC1 solution. Further titration with the aqueous PDADMAC1 solution leads to a small increase of the conductivity, followed by a steep increase at about 47% (Fig. 4). Such a steep jump in the conductivity is well known as a percolation transition. This transition usually occurs in nonionic surfactant-based systems, because of an increase in temperature or on addition of water by keeping constant the surfactant concentration [11]. A generally accepted idea is that the percolation threshold corresponds to the formation of the first clusters of droplets coming close together.

A dramatic change in the electrical conductivity by several orders of magnitude is observed at high concentrations of surfactant and water near the two-phase region. Additionally performed experiments (Fig. 5), starting at pentanol/toluene surfactant compositions of 62:38, 60:40, and 56:44 show adequate conductivity jumps, when the aqueous PDADMAC solution has reached a content of 21%, in contrast to systems of higher pentanol/toluene concentration. Together with the conductometric results already mentioned, a region of significantly higher conductivity can be marked in the isotropic range of the phase diagram (Fig. 3).

Table 2 Sample composition

Sample	M% pentanol/toluene (50%/50%)	*M*% surfactant	*M*% PDADMAC solution (30% by weight)
P1	34	20	46
P2	40	20	40
P3	45	20	35
P4	50	20	30
P5	55	20	25
P6	62	20	18
P7	68	20	12

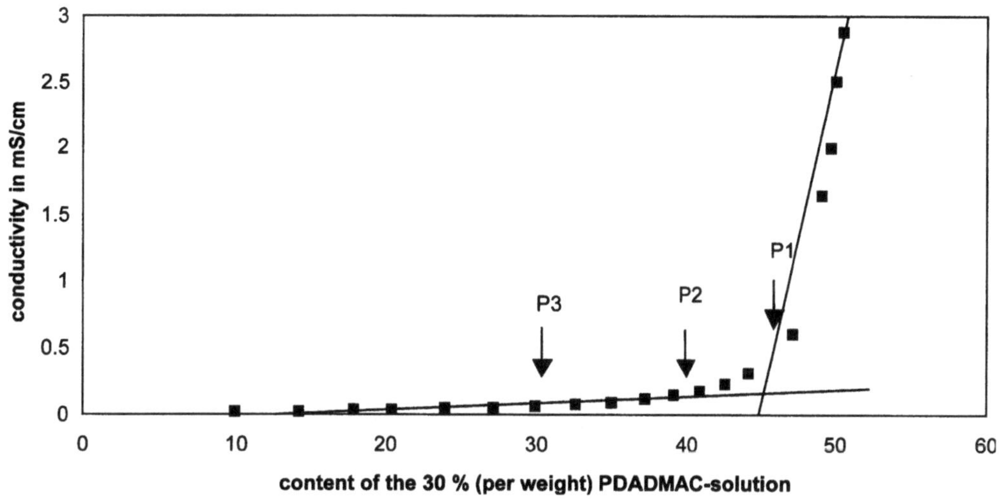

Fig. 4 Conductivity measurements along the solid line in Fig. 3

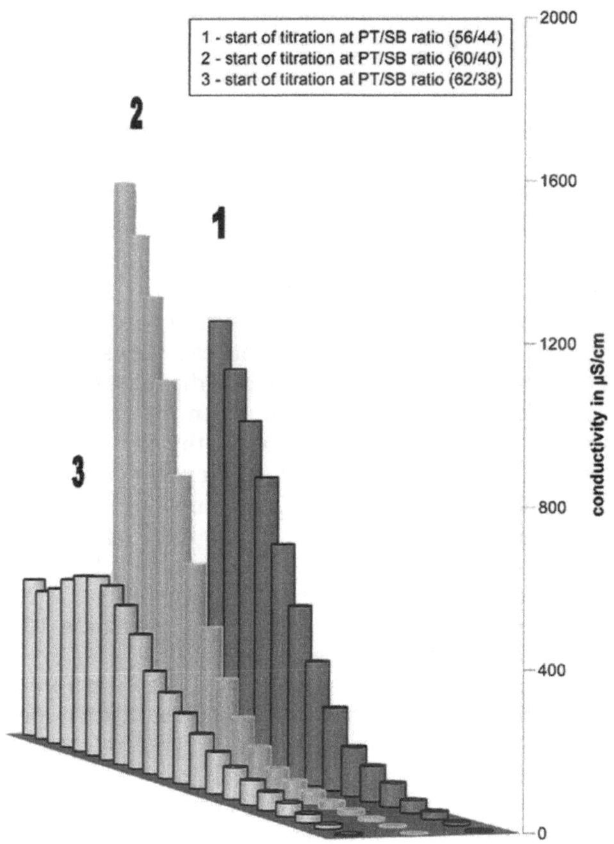

Fig. 5 Conductivity measurements starting at different PT/SB ratios

Rheological experiments

Rheology gives information about the system concerning its response to a mechanical perturbation in terms of elastic deformations and viscous flow. That means rheology proves to be a useful method to get information about the flow behavior and viscoelastic properties of colloidal systems. A characteristic feature of microemulsions is their low viscosity and Newtonian flow behavior. Generally speaking, rheological measurements are of special interest as a sensitive indicator for association phenomena in polymer–surfactant-based systems.

Our systems prepared with PDADMAC2 and PDADMAC3 are characterized by Newtonian flow behavior. The shear viscosities of samples P1–P7 differ between 0.95 and 3.7 PaS. The flow curves at 30% polymer concentration are shown in Fig. 6. A significant viscosity increase is observed only by changing from sample P2 to P1. The dynamic viscosities determined by oscillation measurements are of the same order of magnitude. The storage modulus, G', and the loss modulus, G'', increase with increasing frequency and show behavior usually observed in dilute polymer solutions. This means that at low frequencies the shear is applied slowly and the system has all the time to relax, hereby the energy dissipation takes place and the system does not behave like an elastic substance, namely $G'' > G'$. However, at significantly higher frequencies the storage and loss moduli are of the same order. In all the samples prepared with the low-molecular-weight PDADMAC1 a point of intersection is detected, at frequencies between 10 and 100 Hz. In the presence of PDADMAC3 only samples P1–P3 exhibit such behavior.

However, things are changed by using PDADMAC4, a polycation of significantly higher molar mass. G'' gets larger than G' and no point of intersection is observed. Samples P5–P7 show deviations from Newtonian flow behavior and the viscosity of sample P6 is increased drastically (1 order of magnitude higher). Taking into account that the radius of gyration of the polymer increases with increasing molar mass [12], the results can be explained as follows. When the radius of gyration becomes larger than the droplet radius, the polymer cannot be incorporated into the small microemulsion droplets any longer. Consequently, polymer-induced

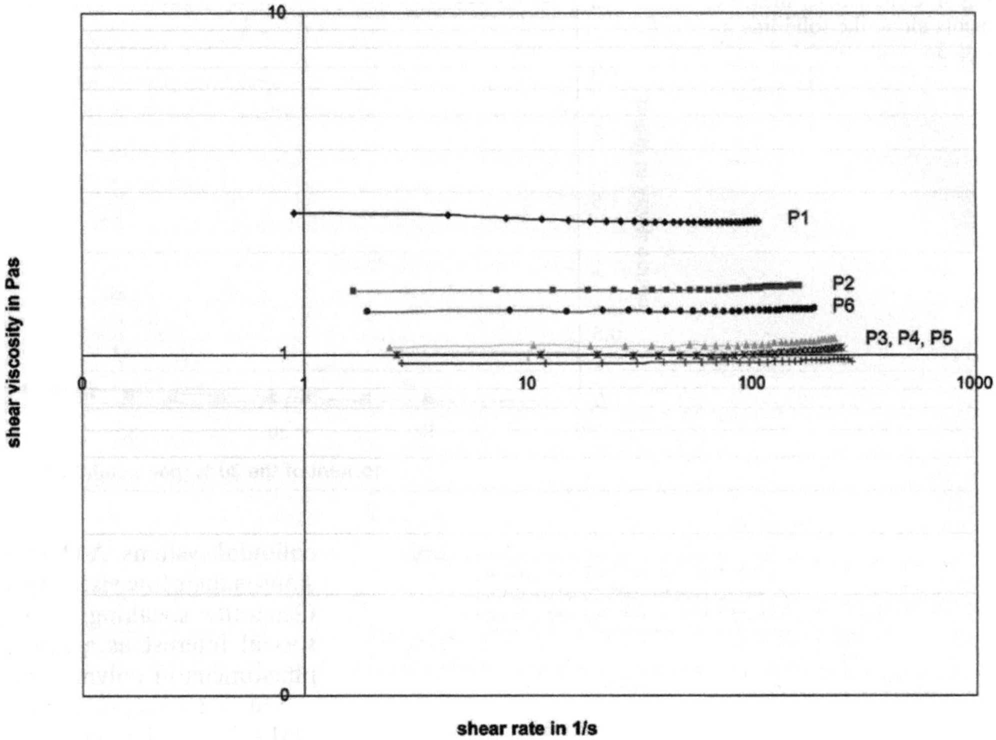

Fig. 6 Shear viscosities for the different samples

cluster formation is observed, indicated by a significant viscosity increase in sample P6.

In the water-rich region of the phase channel (samples P1–P4), PDADMAC4 can be incorporated without problems. Consequently, the shear viscosities observed are quite similar to those of the samples prepared with PDADMAC2 or PDADMAC3.

Differential scanning calorimetry

DSC is widely used for studying binary and multicomponent systems containing surfactants. It is a useful technique to obtain information about the phase diagrams. Transitions between mesophases, mesophase–isotropic liquid transitions, or between crystals and mesophases can be detected. However, additional information (e.g. optical observations with a polarizing microscope, and/or SAXS measurements) is needed to distinguish between the types of phase transition. One principal component in surfactant-based systems is water. The behavior of water is sensitive to the presence of adjacent interfaces. Several types of water have been detected in surfactant-based microstructures such as microemulsions and liquid-crystalline phases. Schulz [13] classified water into free water, interstitial water, surface or interfacial water (physically and/or chemically adsorbed), and internal water (chemically bound water). On the basis of the melting point, Senatra et al. [14] differentiated between free (melts at 0 °C), interphasal (melts at −10 °C), and bound water (melts below −10 °C).

However, in inverse microemulsion droplets, the melting (freezing) point of water can be shifted to subzero temperatures, or completely disappears ("non-freezable water"). Similar effects can be induced by adding polymers, for example, poly(ethylene glycol), to water [15]. The lowering of the melting point is related to the size and shape of the water domains [16].

It has to be mentioned that in the DSC experiments described in the following the subzero region is limited to −20 °C. However, in our micro-DSC experiments a very slow cooling and/or heating rate and long-time (4 h) isotherm ramps at −20 °C were used.

In the heating curve of sample $P6_{water}$ (in the absence of the cationic polyelectrolyte!), an endothermic peak occurs at −16.4 °C (peak top) (Fig. 7). The endothermic peak in the heating curve at −0.5 °C (peak top) can be related to "free" or so called "bulk" water. However, the melting enthalpy (177.8 J/g water) is significantly lower than in pure water (320 J/g water), and it can be concluded that only a part of the water is freezable under the conditions just mentioned.

According to Garti et al. [16] the amount of free water can be determined by the following equation:

$$W_f = \Delta H_{f(exp)} / \Delta H_f^\circ \times 100, \quad (3)$$

where W_f is the weight percentage of free water, $\Delta H_{f(exp)}$ is the measured enthalpy change for the 0 °C peak, and ΔH_f° is the heat of fusion of pure water measured under

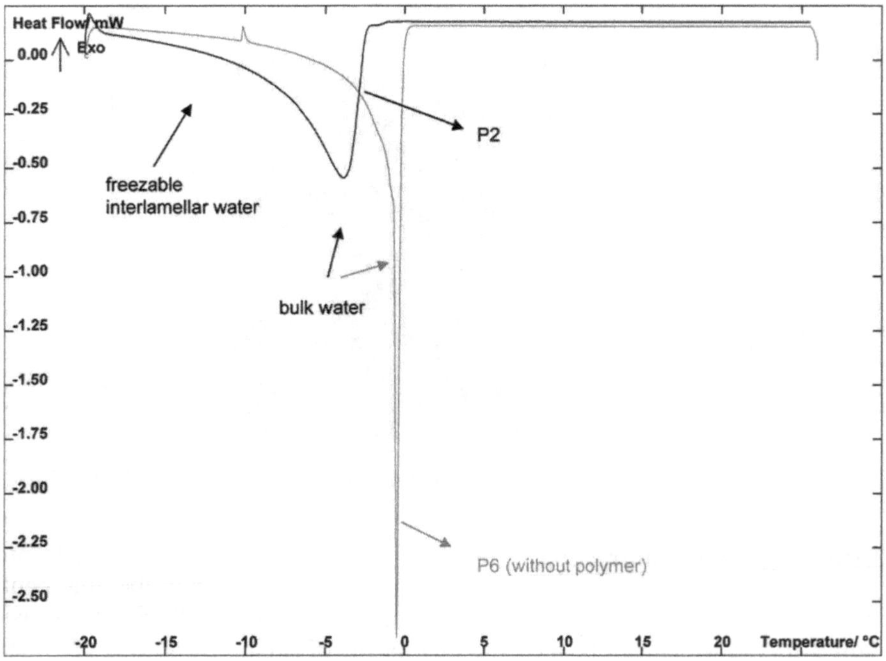

Fig. 7 Differential scanning calorimetry heating curves of samples P2 and P6 (without polymer)

the same experimental conditions. The amount of free water in sample $P6_{(water)}$, calculated using Eq. (1), is 55%.

In sample $P7_{(water)}$ no bulk water can be detected under similar conditions. In samples P6 and P7, prepared with 30% PDADMAC solutions, and also in samples P5 and P4, no water can be detected. Nevertheless, sample P1 again shows an exothermic peak at −18.11 °C (peak top) corresponding with an endothermic peak in the heating curve at −3.3 °C (peak top). That means that in the water-rich section of the phase channel a water peak can be detected again. The main peak at −3.3 °C can be related to "bulk"-like water of the polyelectrolyte solution, and the shoulder between −10 and −20 °C to interphasal and/or bound water (according to Senatra et al. [14]). In samples P2 (Fig. 7) and P3 endothermic peaks were also detected. The peak position is shifted to lower temperatures, and the melting enthalpies are decreased (Table 3). Generally speaking, only in the water-rich part of the phase channel (samples P3–P1) do "bulk"-like water and interphasal and/or internal water appear under the experimental conditions used.

Polarizing microscopy

The most widely used technique of liquid-crystal phase identification is optical polarizing microscopy. Primarily, this method enables the identification of the type of liquid crystal and other mesophases from the optical texture generated. However, the identification of liquid-crystal phases through optical polarizing microscopy is often difficult and requires a lot of experience.

On characterizing the optically clear phase along the line at constant surfactant concentration, a significant change in the optical behavior is observed. Samples P4–P7 are isotropic liquids, but samples P1–P3 show characteristic anisotropic textures after shearing. The polarized micrograph of sample P1 is shown in Fig. 8. The pattern is typical for a lamellar liquid-crystalline system [17]. Nevertheless, the formation of the characteristic Maltese crosses, and "oily strikes" is observed after shearing. That means, the polarizing microscopy indicates a shear-induced transition from the isotropic L_2 phase to an anisotropic phase along the phase channel for samples P3, P2, and P1.

Table 3 Calorimetric data

Sample	Cooling		Heating		
	Peak top (°C)	ΔH_m (J/g)	Peak top (°C)	H_m (J/g)	W_f (wt%)
P1	−18.11	−44.0	−3.3	42.9 ± 2.9	13.2
P2			−3.5	30.1 ± 1.1	9.4
P3			−4.0	25.2 ± 0.3	7.9
P6 (water)	−16.4	−177.8 ± 2.1	−0.5	178.6 ± 2.7	55.8

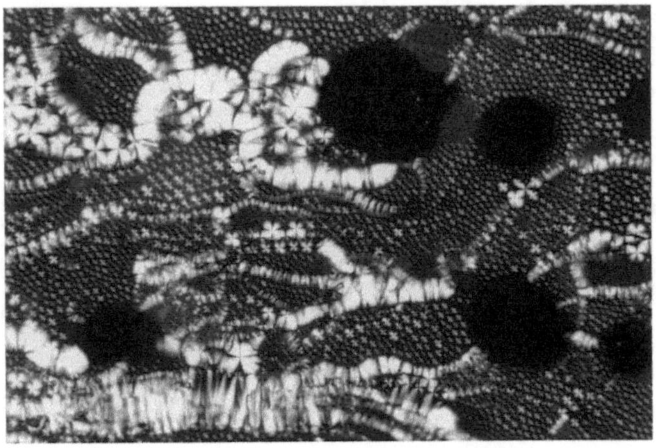

Fig. 8 Polarized micrograph of sample P1 after shearing

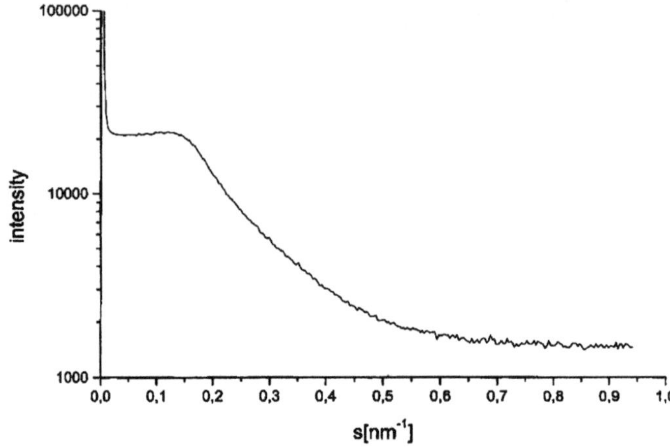

Fig. 9 X-ray diffractogram of sample P1

Small-angle X-ray diffraction

The interlayer spacing is one characteristic feature of a liquid crystal which can be measured by means of small-angle scattering techniques, i.e. SAXS and small-angle neutron scattering experiments. The distinguishing feature between different types of liquid crystals is the interlayer spacing ratio of the first-order, second-order, and third-order peaks. The peak ratios are 1:1/2:1/3 for the lamellar phase and $1:1/\sqrt{3}:1/2$ for the hexagonal phase. In a discontinuous cubic phase the peak ratios are $1:\sqrt{3/8}$ and in a bicontinuous cubic phase $1:\sqrt{6/8}:\sqrt{6/14}$. In general, higher-order peaks are necessary to make distinguish which type of liquid crystal exist.

The X-ray diffractogram of sample P6 displays only one very broad peak, and there is no indication of a corresponding peak of higher order. From the flat maximum of the broad peak, an interlayer spacing of about 5 nm can be calculated in a classical way. However, from the very broad peak and the disappearance of a second-order peak it can be concluded that the order in the system is low. Figure 9 shows that in the diffractogram of sample P1 there is only one very broad peak. The peak position on the *x*-axis is shifted to lower values of the scattering vector. Consequently, the corresponding interlayer spacing of sample P1 is shifted to about 8 nm. However, for sample P1 SAXS measurements indicate a more swollen, but not well-ordered, structure.

^{1}H NMR self-diffusion experiments

The NMR self-diffusion approach is capable of providing accurate, component-resolved diffusion coefficients and has been successfully used to investigate the microstructure and phase transitions in microemulsion systems [18, 19].

In order to monitor the phase structure in the system sulfobetaine/pentanol/toluene/D_2O (+PDADMAC) ^{1}H NMR self-diffusion measurements along the solid line in Fig. 3 were carried out. The resulting self-diffusion coefficients are plotted in Fig. 10 versus the weight fraction of the aqueous phase, $\alpha = w_{aq}/(w_{aq} + w_{P/T})$, at constant surfactant concentration (20% by weight). The self-diffusion coefficient of water was corrected for rapid exchange using Eq. (2). To aid in the structural analysis, the reduced self-diffusion coefficients, D/D_0, of toluene, pentanol, and water were calculated (Fig. 11), where D_0

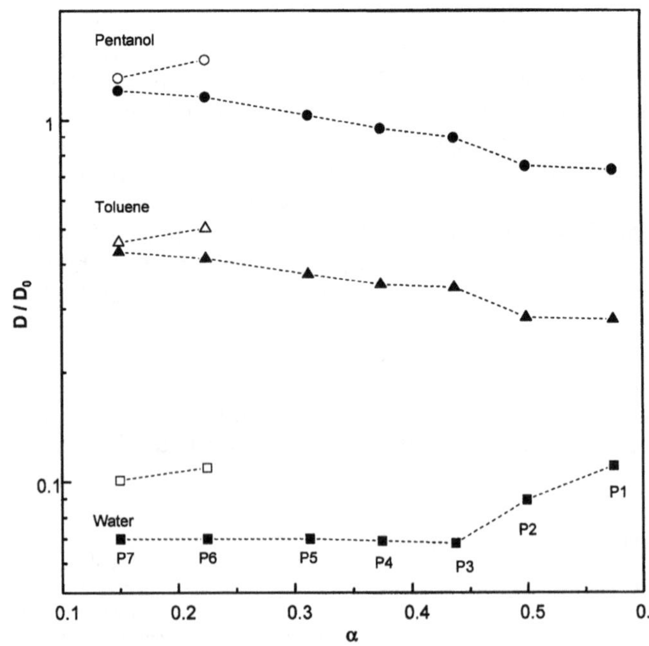

Fig. 10 Self-diffusion coefficients versus the weight fraction of the aqueous phase, $\alpha = w_{aq}/(w_{aq} + w_{P/T})$). *Filled symbols* represent the samples with polymer, *open symbols* represent the samples without polymer

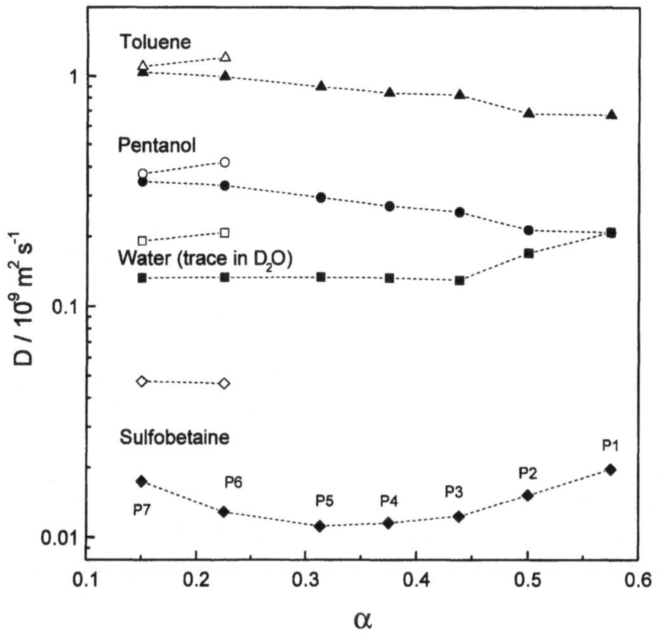

Fig. 11 Reduced self-diffusion coefficients versus α. *Filled symbols* represent the samples with polymer, *open symbols* represent the samples without polymer

represents the self-diffusion coefficient of the neat liquid (Table 4).

The self-diffusion coefficients of the solvents toluene and water in the polymer-free sample $P7_{water}$ are 1.10×10^{-9} and 1.91×10^{-10} m^2/s, respectively. The water diffusion is restricted to a large extent as indicated by its low reduced self-diffusion coefficient (0.10), suggesting that it is confined to the inside of inverse surfactant aggregates; however, it diffuses faster than the surfactant sulfobetaine ($D = 4.7 \times 10^{-11}$ m^2/s). Therefore, it must be assumed that water diffusion is being accelerated by either diffusing as hydration water of pentanol in the continuous phase and/or droplet interactions. The diffusion in the continuous medium toluene is restricted ($D/D_0 = 0.46$) owing to hindrance effects of the dispersed phase, indicating a structural change from spherical to oblatelike structures.

As the water content in the polymer-free system is increased, the surfactant self-diffusion coefficient slightly decreases owing to a small growth of the surfactant aggregate. At the same time, the water, toluene, and

Table 4 Self-diffusion coefficients of neat liquid components at 25 °C

Component	D_0 (10^{-9} m^2/s)	Reference
n-Pentanol	0.286 ± 0.003	[20]
Toluene	2.40 ± 0.02	[21]
Water (trace in D_2O)	1.902	[22]

pentanol self-diffusion coefficients are increased by a small amount too. This may be attributed to a minor change in the size and geometry of the surfactant associates.

In the polymer-modified sample P7 the self-diffusion coefficients of sulfobetaine and water are decreased significantly, whereas toluene and pentanol diffusion is hardly influenced. As expected the polycation PDADMAC is entirely solubilized in the water pool. In turn, the water diffusion decreases upon the addition of PDADMAC. The large decrease of the surfactant self-diffusion coefficient is therefore mainly due to an increase in associate size due to cluster formation in order to entirely solubilize the polycation in the water pool.

If one increases the amount of the aqueous polymer phase, the self-diffusion coefficient of sulfobetaine initially decreases. In order to incorporate additional aqueous phase the surfactant aggregates start to grow and in turn surfactant diffusion drops. However, for sample P3, as the aqueous phase content increases further, an increase of the surfactant diffusion coefficient can be observed. From this point on, the change to an aggregate structure with an extended interface, where surfactant diffusion is less restricted, can be assumed.

The diffusion coefficients of water are approximately the same in samples P3–P7 and the reduced diffusion coefficients are small (0.07), indicating that water is confined to inverse surfactant aggregates. Since the aggregates are growing this suggests that no free water is present within the surfactant aggregates, and is consistent with the DSC results. However, upon the addition of more aqueous polymer phase (P1, P2), water diffusion is increased. Although still restricted to a major extent ($D/D_0 = 0.11$), this, again, indicates a change into an associate structure at P3 where water diffusion is less hindered.

The molecular species in the continuous medium, toluene, and pentanol show rather similar diffusional behavior. As the aqueous phase content is increased the diffusion coefficients decrease. This is mainly due to the hindrance effects of growing surfactant associates. Since the decrease in pentanol diffusion is more pronounced (note the logarithmic scale in Figs. 10, 11) enhanced incorporation of pentanol into the interface layer may also be assumed. However, between P3 and P2 the diffusion coefficients of toluene and pentanol drop to significantly lower values of 0.68 and 0.21×10^{-9} m^2/s, respectively. They remain constant upon the addition of more aqueous phase. This rather abrupt change may be associated with a change in associate structure.

Finally, it has to be mentioned that toluene ($D/D_0 = 0.28$) and water ($D/D_0 = 0.09 - 0.11$) diffusion in P1 and P2 are restricted to a major extent. This suggests that several diffusional barriers are present within the distance monitored by the diffusion experiment. Therefore, a disordered lamellar aggregate structure may be assumed at these compositions.

Conclusions

The investigations reveal that the optically clear region of the phase diagram in the sulfobetaine/pentanol/toluene/water system is decreased by adding an anionic polyelectrolyte and increased after incorporation of the cationic polyelectrolyte PDADMAC. Independent of the molar mass of the PDADMAC the optically clear phase is extended towards the water corner. Investigations along the phase channel from the oil-rich to the water-rich side, at constant surfactant concentration, show characteristic changes of the diffusion coefficients of heavy water by NMR self-diffusion experiments. When the diffusion coefficients of water increase, bulk-like water become detectable by DSC, and anisotropic textures can be visualized by polarizing microscopy after shearing. This implies that at higher water concentrations the phase transition from an isotropic L_2 phase to an anisotropic phase is shear-induced.

The isotropic L_2 phase (P4–P7) in the presence of PDADMAC is characterized by low viscosity, Newtonian flow behavior, low conductivity, and a constant diffusion coefficient for H_2O of 0.15×10^{-9} m^2/s. The broad peak in the SAXS diffraction patterns indicates a low ordered structure with a corresponding interlayer spacing of about 5 nm. Characteristic features of the polymer-modified inverse microemulsions are the disappearance of bulk water and the molar-mass-dependent viscosity increase due to cluster formation of the droplets.

In the extended region of the optically clear phase at higher water content (P1–P3), which only exists in presence of PDADMAC, the diffusion coefficients of H_2O as well as of the surfactant molecules increase, interfacial and bulk water become detectable, and the interlayer spacing (obtained by SAXS) is significantly higher. The rather abrupt changes can be explained by the formation of a disordered lamellar associate structure. Under shear stress, a transition to an anisotropic phase, characterized by the typical optical pattern of a lamellar liquid-crystalline phase, is observed. Generally speaking, PDADMAC tends to favor the formation of a lamellar liquid-crystalline phase (L_α phase). This means that interactions between the sulfobetaine head groups and the functional groups of PDADMAC are of importance not only with regard to the extension of the phase range, but also with regard to the phase transition to an anisotropic phase. When PDADMAC is adsorbed at the sulfobetaine head groups, the surfactant film becomes partly charged, and the curvature is changed. This change of the spontaneous curvature is of major importance with regard to the type of self-assembled associate structure.

At the borderline to phase separation (marked region in Fig. 3) a small region of significantly higher conductivity was observed. In this part of the phase diagram a transition to a bicontinuous phase can be assumed.

Acknowledgement We thank Andreas Thünemann from the Fraunhofer Institute for Applied Polymer Research, Golm, for the SAXS measurements.

References

1. Shinoda K (1991) Langmuir 7:2877
2. Fontell K, Khan A, Lindström B, Maciejewska D, Puang-Ngern S (1991) Colloid Polym Sci 289:727
3. Guerin G, Bellocq, AM (1991) J Phys Chem 92:727
4. Friberg SE, Yang J (1991) In: Rieger MM, Rhein LD (eds) Surfactants in cosmetics. Dekker, New York, p 234
5. Gradzielsky M, Valiente M, Hoffmann H, Engelhaaf S (1998) J Colloid Interface Sci 205:149
6. Koetz J, Beitz T, Tiersch B (1999) J Dispersion Sci Technol 20:139
7. Beitz T, Koetz J, Wolf G, Kleinpeter E, Friberg, SE (2001) J Colloid Interface Sci 240:581
8. Gibbs SJ, Johnson CS Jr (1991) J Magn Reson 93:395
9. Wu D, Chen A, Johnson CS Jr (1995) J Magn Reson A 115:260
10. Goldammer EV, Hertz HG (1970) J Phys Chem 74:3734
11. Meier W (1996) Langmuir 12:1188
12. Dautzenberg H, Görnitz E, Jaeger W (1998) Macromol Chem Phys 199:561
13. Schulz PC (1998) J Therm Anal 51:135
14. Senatra D, Gabrielli G, Caminati G, Zhou Z (1988) IEEE Trans Electr Insul 23:579
15. Antonsen KP, Hoffmann AS (1992) In: Harris JM (ed) Poly(ethylene glycol) chemistry, biotechnical and biomedical applications. Plenum, New York, p 15
16. Garti N, Aserin AES, Tiunova I, Berkovic G (1996) J Colloid Interface Sci 178:60
17. Stegemeyer H (1999) Lyotrope Flüssigkristalle. Steinkopff, Darmstadt, p 45
18. Söderman O, Nydn M (1999) Colloids Surf A 158:273
19. Lindman B, Olsson U, Söderman O (1999) Handbook of microemulsion science and technology. Dekker, New York, p 309
20. Holz M, Heil SR, Sacco A (2000) Phys Chem Chem Phys 2:4740
21. Lindman B, Stilbs P, Moseley ME (1981) J Colloid Interface Sci 83:569
22. Mills R (1973) J Phys Chem 77:685

Progr Colloid Polym Sci (2003) 122:37–46
DOI 10.1007/978-3-540-36114-5

Mandeep S. Bakshi
I. Kaur

Surfactant–polymer aggregates of mixed cationic micelles and anionic polyelectrolytes: a surfactant head group contribution

M.S. Bakshi (✉) · I. Kaur
Department of Chemistry,
Guru Nanak Dev University,
Amritsar 143005, Punjab, India
e-mail: ms_bakshi@yahoo.com
Tel.: +91-183-254685
Fax: +91-183-254685

Abstract Single and mixed-micelle formation by benzyldimethylhexadecylammonium chloride (BHDACl), hexadecyltrimethylammonium bromide (HTAB), and their mixtures in aqueous anionic polyelectyrolytes, viz. carboxymethylcellulose sodium salt, dextran sulfate sodium salt, polystyrene sulfonate sodium salt, and polyacrylic acid sodium salt, were studied with the help of conductivity, κ, viscosity, η, turbidity, τ, and NMR studies. κ showed a single aggregation process for pure and mixed surfactants and the concentration at which it took place was slightly higher than the actual critical micelle concentration in pure water. Hence, it was termed the critical aggregation concentration (cac). Application of the Clint equation revealed that the attractive interactions between BHDACl and HTAB even in the presence of polyelectrolytes were responsible for the large deviation from ideal behavior. Both η_r and τ did not indicate the presence of a cac but instead they showed another aggregation process at much lower concentration than the cac, which was termed the pre-cac. They also demonstrated strong electrostriction effects in the case of BHDACl–polyelectrolyte systems, whereas weak electrostriction effects were observed for HTAB–polyelectrolyte. The former was explained on the basis of weak electrostatic interactions due to the steric hindrances created by an aromatic ring in the BHDACl head group. NMR results showed that the proton resonance of head groups and hydrophobic tails of both surfactants underwent deshielding and shielding effects, respectively. The former was more significant in the case of BHDACl, whereas the latter one was more significant in the case of HTAB. Both factors indicated the formation of compact mixed-micelle-polyelectrolyte aggregates.

Keywords Cationic surfactants · Mixed micelles · Anionic polyelectrolytes · Micelle–polyelectrolyte interactions

Introduction

Surfactant–polymer interactions have gained considerable importance owing to their potential applications both in technological and in scientific fields [1, 2, 3, 4]. A combination of cationic surfactant and anionic polyelectrolyte is an example of a strongly interacting system with extraordinary solution properties [5, 6, 7]. Fundamentally, the driving force of electrostatic interactions depends upon several factors, of which the nature of the cationic surfactant head group is the main contributing factor which screens the electrostatic interactions significantly if it is bulky [8, 9, 10]. For instance, if we compare the electrostatic interactions of tetraalkylammonium head group and pyridinium head group cationic surfactants with the same polymer, the latter would have stronger interactions in comparison to the former [10]. On the other hand, the flexibility of the polymer backbone is another important factor particularly when folding or defolding is introduced upon micelle–polyelectrolyte interactions [8]. The pioneering work of Saito and Yukawa [11, 12, 13] established that the surfactant–polymer ratio is another criterion which governs the overall interactions. At a low ratio when the surfactant is essentially in the monomeric form, the binding is due to the dipolar interactions of the surfactant head group with polar sites on the polymer backbone. At a high ratio, when apart from monomers, micelles are also present, the structure of micelle–polymer complex becomes quite difficult to establish.

The present account deals with strongly interacting surfactant–polymer systems consisting of cationic surfactants, viz. benzyldimethylhexadecylammonium chloride (BHDACl), hexadecyltrimethylammonium bromide (HTAB), and their mixtures with anionic polyelectrolytes such as carboxymethylcellulose sodium salt (CMC), dextran sulfate sodium salt (DX), polystyrene sulfonate sodium salt (PSS), and polyacrylic acid sodium salt (PAA). Though, both BHDACl and HTAB belong to a category of cationic surfactants with identical hydrophobic tails, the presence of unlike head groups is expected to induce a significant difference in the surfactant–polymer interactions. The physiochemical properties of these systems have been explored with the help of conductivity, κ, viscosity, η, turbidity, τ, and NMR studies. κ is a very useful way to evaluate the micellar properties [14, 15, 16] when the micellization of ionic surfactants is involved. η [17, 18] gives major insight into the mode of surfactant–polymer association, particularly when this association is mainly influenced by electrostriction effects. Generally, cationic surfactant–anionic polyelectrolyte interactions lead to the formation of an insoluble salt, which in turn provides the extent of neutralization between the charged sites of opposite polarities. In such a situation, τ helps [19, 20, 21] in exploring the concentration range of both surfactant and polymer where it happens. One can always pinpoint the origin of the driving forces responsible for such kind of association from NMR studies. NMR methods have demonstrated considerable, and sometimes fundamental, contributions to the physiochemical properties of surfactants [22, 23]. The paucity of literature suggests that NMR has rarely been employed to investigate the characteristic properties of the complex formation between oppositely charged polyelectrolytes and ionic surfactants above their critical micelle concentration (cmc). A collective representation of all the results has demonstrated some interesting features of cationic surfactant–anionic polyelectrolyte interactions, which are significantly influenced by head group modifications.

Experimental

Materials

BHDACl, 98% pure from TCI, Japan, and HTAB, 99% pure from Lancaster Synthesis, UK, were used as received. CMC, average molecular weight 100,000, DX, average molecular weight 500,000, PSS, average molecular weight 70,000, and PAA, average molecular weight 30,000, were from TCI, Japan. Triply distilled conductivity water was used in the preparation of all the solutions. All the solutions were prepared by mass with an accuracy of ±0.01 mg. The mole fractions were accurate to ±0.0001 units. The mixed-micelle formation was studied in the presence of 0.001 w/v% of each polymer. An automatic thermostat bath from Shimadzu, Japan, was used for maintaining the temperature within ±0.05 °C. All the measurements were performed at 25 °C after giving sufficient time for stabilization.

Conductometric and viscometric measurements

The conductivity measurements of BHDACl and HTAB mixtures in the presence of polyelectrolytes were carried out over the whole mixing range with the help of a digital conductivity meter (Naina Electronics, Chandigarh, India, model NDC-732) working at a fixed frequency of 1,000 Hz. The error in the conductivity measurements was ±0.5%. Efflux times of the dilute solutions were determined with the help of a Ubbelohde-type suspended-level capillary viscometer sealed in a glass jacket to circulate the thermostated water at 25 °C. The time of flow for water was 175 s. From the ratio of the efflux time of the test solution, t, to that of the reference solution, t_0, the relative viscosity was calculated ($\eta_{rel} = t/t_0$) by ignoring the density corrections for dilute solutions [2, 24]. The measured η_r were accurate to 0.5%, while the precision of these measurements was noted to be ±0.2%.

Turbidity measurements

The turbidity measurements for all the systems were performed using a Nephelo-Turbidity meter, type 131, Systronics, India. This instrument works on the basis of scattering of light by the colloidal particles.

The details of the κ, η_{rel}, and τ, measurements have been reported elsewhere [21].

NMR measurements

^{1}H NMR spectra were run on a Bruker AC200E instrument. All the chemical shifts were measured relative to sodium 4,4-dimethyl-4-silapentane-1-sulfonate, which acted as an internal standard. At certain mole fractions, dark turbid solutions were first filtered and the NMR of the clear solutions was taken. D_2O was used as a solvent instead of water to weaken the water signal for all the solutions and the chemical shifts were measured as a function of added type. Note that the surfactant concentrations were always higher than the critical aggregation concentration (cac) and 0.005w/v% of each polyelectrolyte was used for the NMR studies. The chemical shift differences were only considered in this study.

Results and discussion

Conductivity behavior

The break in the κ plot for BHDACl, HTAB, and their mixtures shows the cac of the respective micelle formation in the presence of polyelectrolytes (Fig. 1a). The values for the pure components and their mixtures were evaluated by linear regression of the κ data both

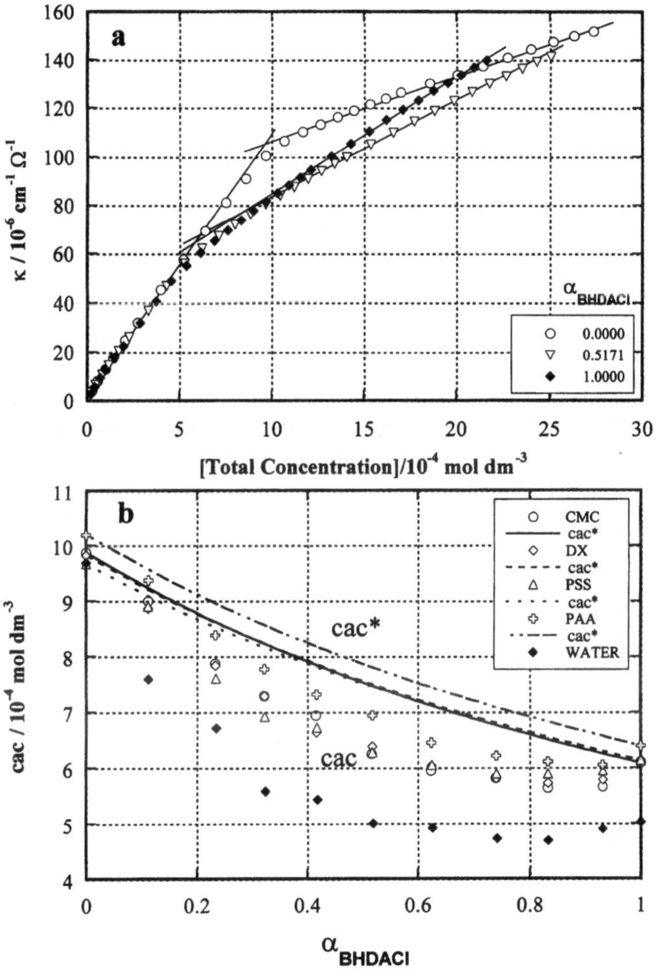

Fig. 1 Conductivity, κ, versus total concentration of benzyldimethylhexadecylammonium chloride (*BHDACl*) and hexadecyltrimethylammonium bromide (*HTAB*) mixtures at some selected mole fractions in the presence of dextran sulfate sodium salt (*DX*). **b** Critical aggregation concentration (*cac*) versus mole fraction of BHDACl for BHDACl + HTAB mixtures in the presence of various polyelectrolytes: carboxymethylcellulose sodium salt (*CMC*), polystyrene sulfonate sodium salt (*PSS*), polyacrylic acid sodium salt (*PAA*) Experimental cac (*symbols*); predicted cac* (*lines*)

in the premicellar and postmicellar regions [25] and the values for the pure components are listed in Table 1 along with their uncertainties. A graphical representation of the cac in aqueous polymer solutions is shown in Fig. 1b. In all cases, the cac values over the whole mole fraction range show a nonlinear variation. The Clint equation in the following form can be used to identify the ideality in the mixed-micelle formation [26]:

$$\frac{1}{\mathrm{cac}^*} = \frac{\alpha_1}{\mathrm{cac}_1} + \frac{(1-\alpha_1)}{\mathrm{cac}_2}, \tag{1}$$

where α_1 is the mole fraction of surfactant 1 (BHDACl) in total mixed solute; cac_1 and cac_2 are the cacs of components 1 and 2, respectively. The cac* values representing the ideal-state values thus calculated are plotted in Fig. 1b. One can see that the cac values in all cases are always lower than the corresponding cac* values, thus suggesting that mixed aggregation in all cases takes place owing to attractive interactions between the components of the BHDACl and HTAB mixtures even in the presence of various polymers. Macdonald et al. [27] also observed similar mixed-micellar behavior of cationic and zwitterionic surfactants in the presence of PSS. Apart from this, the values obtained when polymers are present are higher than those in pure water [28]. This means that micelle formation is taking place at higher surfactant concentration in the event of insoluble salt formation, which reduces the actual amount of surfactant available for the micellization. Similar results have already been reported in the case of micelle formation by anionic and cationic surfactants [21]. An increase in the cac value than actual cmc of alkyltrimethylammonium bromides in the presence of PSS has already been observed by Kogej and Skerjane [29]. It is to be mentioned here that no additional break in the κ plots before the cac was observed, which could suggest the association between the monomeric surfactants and polyelectrolyte macromolecules. Actually this type of association is generally expected for such systems and is mostly visible from other studies, like surface tension [30]. These results were further explored with the help of viscosity and turbidity measurements.

Table 1 Critical aggregation concentration (*cac*) from conductivity, κ, and pre-cac from viscosity, η_r, and turbidity, τ, measurements for benzyldimethylhexadecylammonium chloride (*BHDACl*) and hexadecyltrimethylammonium bromide (*HTAB*) in the presence of various polyelectrolytes

	Carboxymethyl cellulose sodium salt			Dextran sulfate sodium salt			Polystyrene sulfonate sodium salt			Polyacrylic acid sodium salt		
	κ	η_r	τ	κ	η_r	τ	κ	η_r	τ	κ	η_r	τ
HTAB	9.87 ± 0.78	6.49	6.61	9.82 ± 0.18	6.58	6.51	9.67 ± 0.17	7.73	6.21	10.2 ± 0.13	9.54	6.68
BHDACl	6.09 ± 0.27	1.14	4.01	6.14 ± 0.20	2.75	3.81	6.14 ± 0.20	3.37	3.21	6.4 ± 0.23	2.86	3.06

Viscometric and turbidity behavior

Viscometric investigation was applied to further explore the surfactant–polymer interactions. The variation of η_r with respect to α_{BNHACl} is shown in Fig. 2a. As soon as the amount of BHDACl increases in a respective mixture, η_r starts decreasing in the case of pure water and in aqueous DX till $\alpha_{BHDACl} \approx 0.4$, and then becomes almost constant with further increase in α_{BHDACl}. In the presence of other aqueous polyelectrolytes, i.e. CMC, PSS, and PAA, η_r remains more or less constant. A deviation from the ideal behavior (shown by broken lines) especially in the HTAB-rich region of the mixture shows strong negative departures in pure water and in aqueous DX. In contrast, τ (Fig. 2b) significantly increases in this region when CMC is present in comparison to DX. For instance, there is a 4.5% decrease in η_r and a 79% increase in τ for pure HTAB in comparison to 0.3 and 14%, respectively, for pure BHDACl. Thus, a larger τ value can be explained on the basis of stronger HTAB–CMC interactions resulting in the formation of insoluble salts in comparison to lower τ values. The values obtained when PAA and PSS are present seem to lie between these two extremes. Therefore, pure-water-like behavior of aqueous DX especially in the HTAB-rich region of the mixture in Fig. 2a and b suggests weaker HTAB–DX or (HTAB–BHDACl)–DX interactions in comparison to stronger HTAB–CMC or (HTAB–BHDACl)–CMC interactions. The presence of minimum at $\alpha \approx 0.4$ in the η_r plot (Fig. 2a) in fact demonstrates the formation of compact micelles with smaller dimensions owing to maximum synergistic interactions. Interestingly, the disappearance of the minimum when different polyelectrolytes are present, except for DX, explains the predominance of strong surfactant–polyelectrolyte interactions over the mutual synergism between HTAB and BHDACl and which was of course not visible from κ measurements (Fig. 1b). The results of η_r and τ investigations on the micelle formation by BHDACl, HTAB, and their equimolar mixture are shown in Figs. 3 and 4, respectively. The strong electrostriction effects are believed to be responsible for the higher [31, 32, 33] η_r and lower τ of BHDACl in aqueous polymer in comparison to pure water (Figs. 3a, 4a, respectively). Stronger electrostriction effects are generally produced by the electroactive sites available for electrostatic interactions with water dipoles. This may only be possible when opposite polarities of cationic micelles and anionic polyelectrolytes are not completely neutralized and which will produce higher η_r and lower τ values as in the case of BHDACl–polyelectrolyte in comparison to HTAB–polyelectrolyte systems. In contrast, in the case of HTAB, a large decrease in the postmicellar η_r (Fig. 3b) and an increase in τ when CMC and PAA are present (Fig. 4b) may be attributed to a significant decrease in the electrostriction effects upon micelle–polyelectrolyte favorable complexation leading to the insoluble salt formation. Abuin and Scaiano [34] also observed a significant fall in the reduced viscosity of dodecyltrimethylammonium bromide in the presence of PSS. They ascribed it on the basis of the coiling of the polyelectrolyte backbone around surfactant aggregates. Similarly, Bocquenet and Siffert [35] reported a decrease in viscosity in the case of the dodecylbenzene sulfate and polyacrylamide system. Figure 3a and b also shows that a typical monomeric surfactant–polyelectrolyte aggregation process (pre-cac) is represented by the appearance of a minimum in the η_r profiles in each case, which was not visible from the κ plots. It is much lower than the actual cac value of micelle formation in the presence of polyelectrolyte and is the same point where the maximum turbidity is observed (Fig. 4). The values of the pre-cac both from η_r and from τ are compared in Table 1 with those obtained from κ measurements. It can be seen

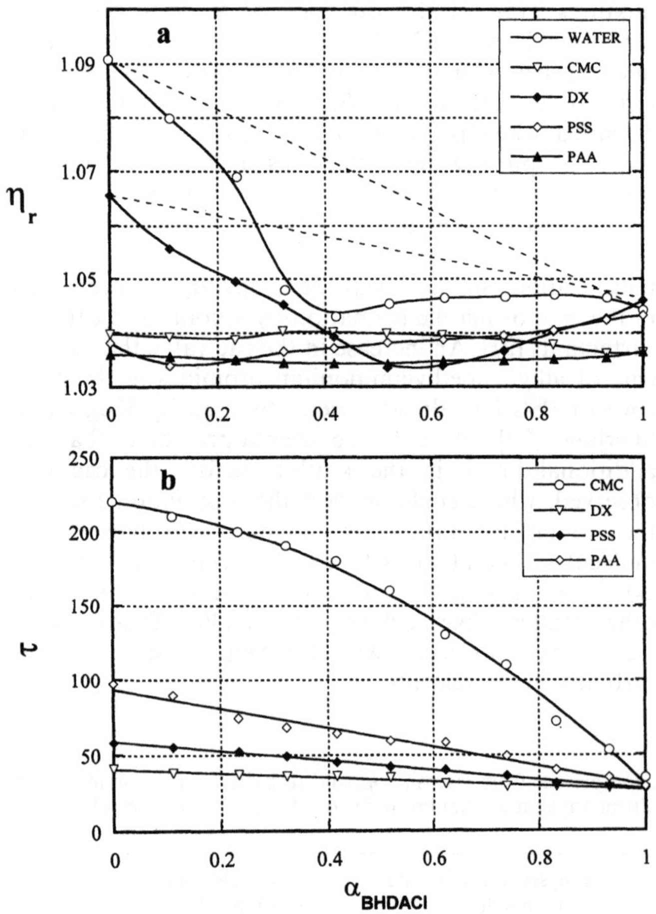

Fig. 2 η_r versus mole fraction of BHDACl for BHDACl + HTAB mixtures in the presence of various polyelectrolytes. The *broken lines* represent ideal mixing. **b** τ versus mole fraction of BHDACl for BHDACl + HTAB mixtures in the presence of various polyelectrolytes

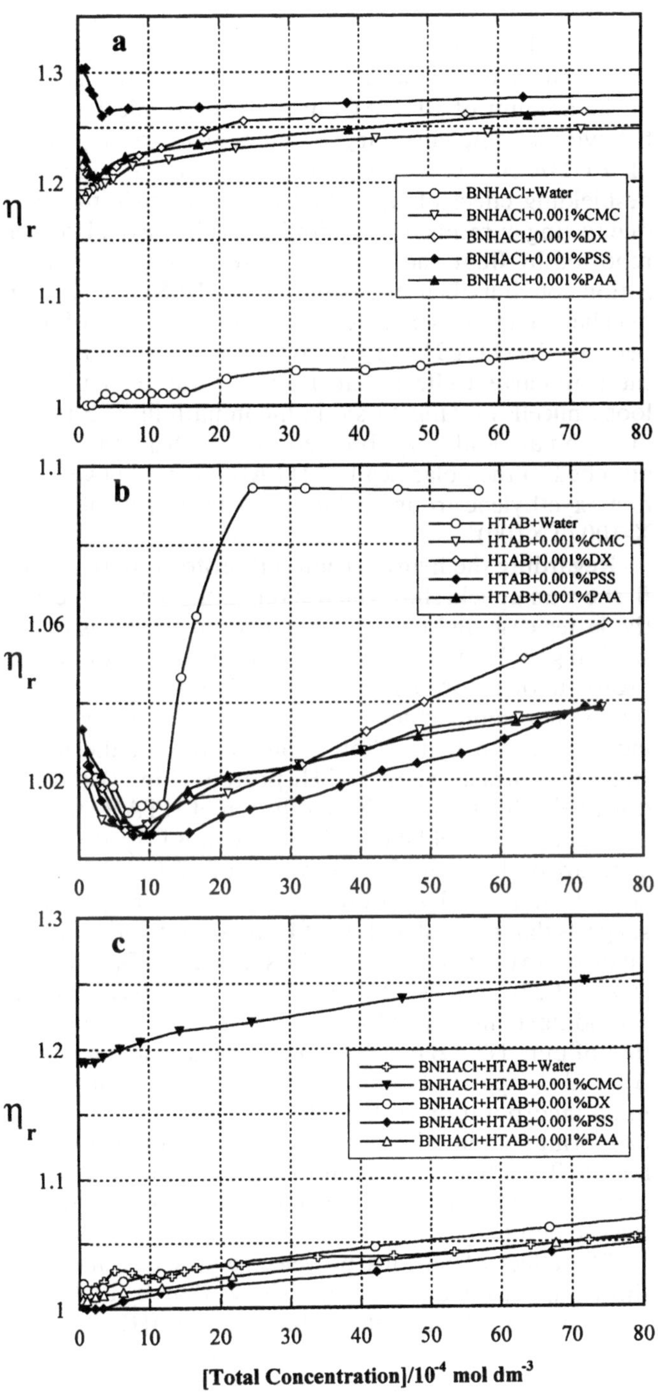

Fig. 3 η_r versus total concentration of BHDACl in pure water and in the presence of CMC, DX, PSS, and PAA. **b** η_r versus total concentration of HTAB in pure water and in the presence of CMC, DX, PSS, and PAA. **(c)** η_r versus total concentration of BHDACl + HTAB at equimolar proportions in pure water and in the presence of CMC, DX, PSS, and PAA

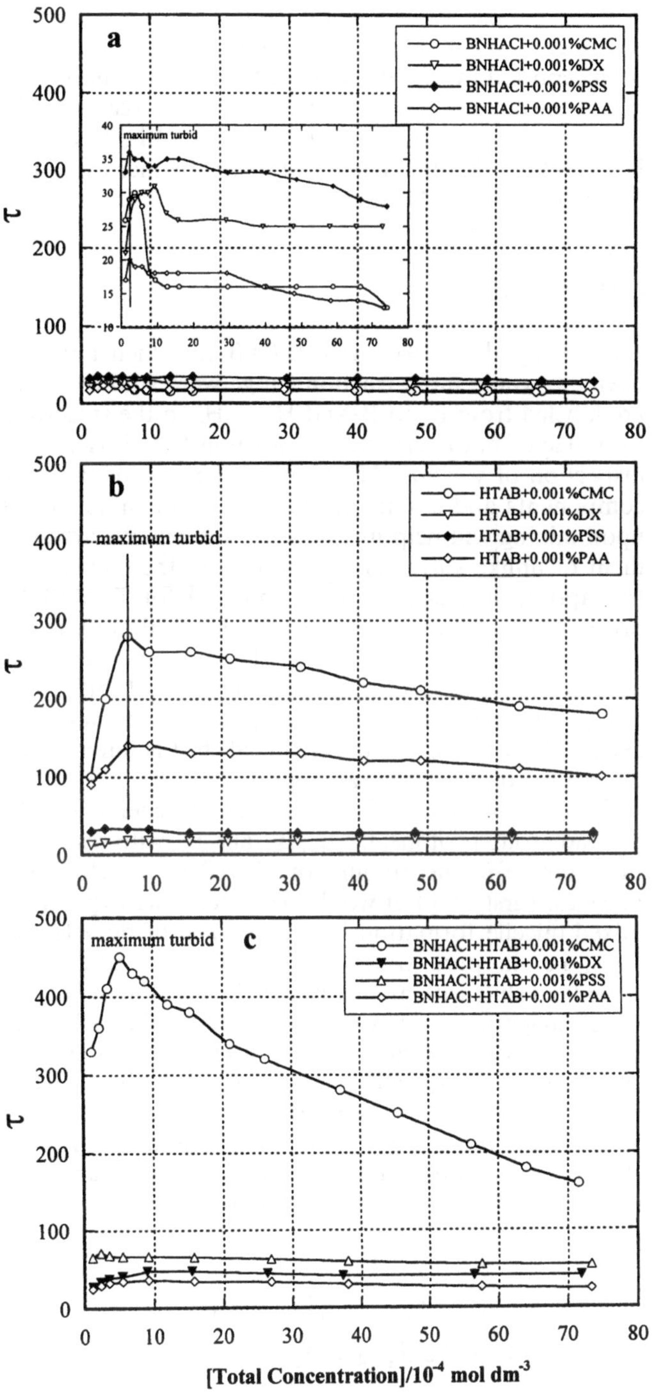

Fig. 4 τ versus total concentration of BHDACl in the presence of CMC, DX, PSS, and PAA. *Inset*: enlarged view with expanded τ values. **b** τ versus total concentration of HTAB in the presence of CMC, DX, PSS, and PAA. **c** τ versus total concentration of BHDACl + HTAB at equimolar proportions in the presence of CMC, DX, PSS, and PAA

that the pre-cac values are quite close to each other but are much lower than the cac. Similar results have already been reported in the literature [29,30]. Furthermore, the mixed surfactant–polyelectrolyte behavior seems to be the average contributions of both pure BHDACl and HTAB (Figs. 3c, 4c). These results are further explained on the basis of NMR measurements.

NMR behavior

The ^1H signals of BHDACl, HTAB, and their mixtures over the whole mole fraction range were studied. They are labeled from B_a to B_f and H_a to H_d in the structural formulae of both surfactants shown in Fig. 5. A careful inspection of various ^1H signals in the case of the pure components and a shift in the position of these signals upon mixing will help us to deduce the relative arrangement of unlike surfactant monomers in the mixed state (no appropriate signals were observed for B_d and H_c protons). Although the observed chemical shift is the weighted average of those of the monomer and the micellar molecules, these changes in chemical shift could not originate from the change in the cmc because the extent of the change in the chemical shift is different for different protons in the same molecule.

Figure 6 represents the variation in the chemical shifts of head group protons, i.e. B_a, B_b, and B_c, with respect to the increase in the amount of HTAB in the BHDACl micelles. First of all, it would be better if we explain the mixed-micelle formation in the absence of polyelectrolytes, i.e. in pure water. All the protons undergo a downfield shift with an increase in the fraction of HTAB, which suggests that they experience deshielding effects as soon as they realize the presence of HTAB monomers in their vicinity. This will push the aromatic rings further apart and will reduce the mutual effect of ring currents. This is also evident from the significant deshielding of the B_b proton in comparison to B_a and B_c. For instance, the magnitude of $\Delta\delta$ for B_a and B_c up to $\alpha_{HTAB} = 0.8$ is 0.065 and 0.045, respectively, in comparison to 0.1 for B_b. On the other hand, head group protons of HTAB, i.e. H_d, (Fig. 7) have undergone an upfield shift with the increase in the amount of BHDACl in the HTAB micelles. This behavior is complementary to that of head group protons of BHDACl, i.e. B_a, B_b, and B_c, (Fig. 6) since the deshielding effect of B_a in fact brings a shielding effect due to ring currents to H_d. Similar results have also been reported by Bhat and Gaikar [36] regarding the solubilization of butylbenzene sulfonates (BBS) by HTAB micelles. They observed that the intercalation of BBS between the HTAB head groups shields the latter owing to the ring current effects and results in the formation of loose micelles owing to steric incompatibility. Furthermore, Yuan et al. [37] observed that the trimethyl group attached to the polar head of HTAB locates between the first oxyethylene group next to the phenoxy ring of Triton X-100 (TX100).

Regarding the hydrocarbon tail protons, both B_e and B_f show an upfield shift which is very much pronounced in the latter case and can be explained on the basis of the shielding of these protons. The stronger shielding of the B_f proton upon incorporation of HTAB monomers suggests the greater compactness of the micellar core due to an increase in the hydrophobic interactions in the mixed state. In contrast, the H_a and H_b protons (Fig. 7) undergo downfield shifts, suggesting a dilution of HTAB micelles, also reported for HTAB–TX100 mixtures [37]. A relative comparison between the magnitude of $\Delta\delta$ for H_a and B_f, i.e. terminal methyl protons (Figs. 7, 6, respectively) suggests that the deshielding of H_a at $\alpha \approx 0.5$ ($\Delta\delta = 0.27$) is much more pronounced than the shielding of B_f at $\alpha \approx 0.5$ ($\Delta\delta = 0.027$) and hence the core of the mixed micelles should have much less hydrophobicity in comparison to that of pure HTAB micelles. Similarly, deshielding of B_a, B_b, and B_c upon induction of HTAB and shielding of H_d upon incorporation of BHDACl due to ring currents in fact brings about the compatibility in the head group region. Thus both factors in the mixed state will help in producing the synergism, which was earlier deduced from the cac plot (Fig. 1b).

The effect of various polyelectrolytes has also been shown in Figs. 6 and 7. Fig. 6 shows that B_a undergoes larger downfield and upfield shifts, in the BHDACl- and HTAB-rich regions, respectively, in the presence of CMC, PSS, and PAA than in pure water. The former effect is expected on the basis of the similar polarities of the aromatic group and the polyelectrolyte, which will bring deshielding effects to B_a. The latter effect is mainly governed by the shielding of the B_a proton owing to the predominance of strong electrostatic interactions between HTAB head groups and the polyelectrolyte. This can be explained on the basis of two facts; first, the mixed micelles are expected to be rich in HTAB monomers in this region and, second, the HTAB head group is free from any aromaticity and, hence, there would not be any

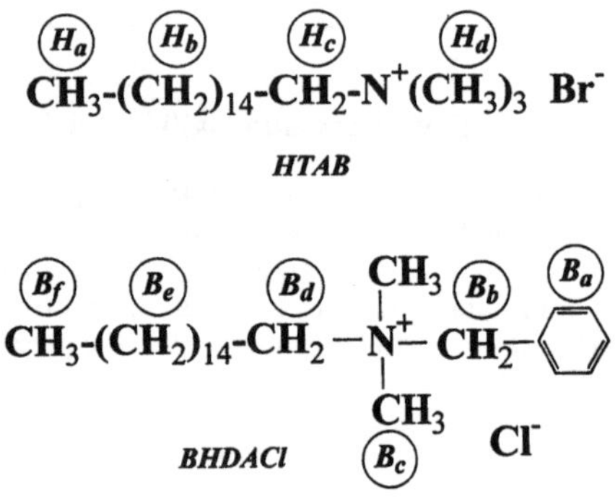

Fig. 5 Formulas and proton labeling of HTAB and BHDACl molecules

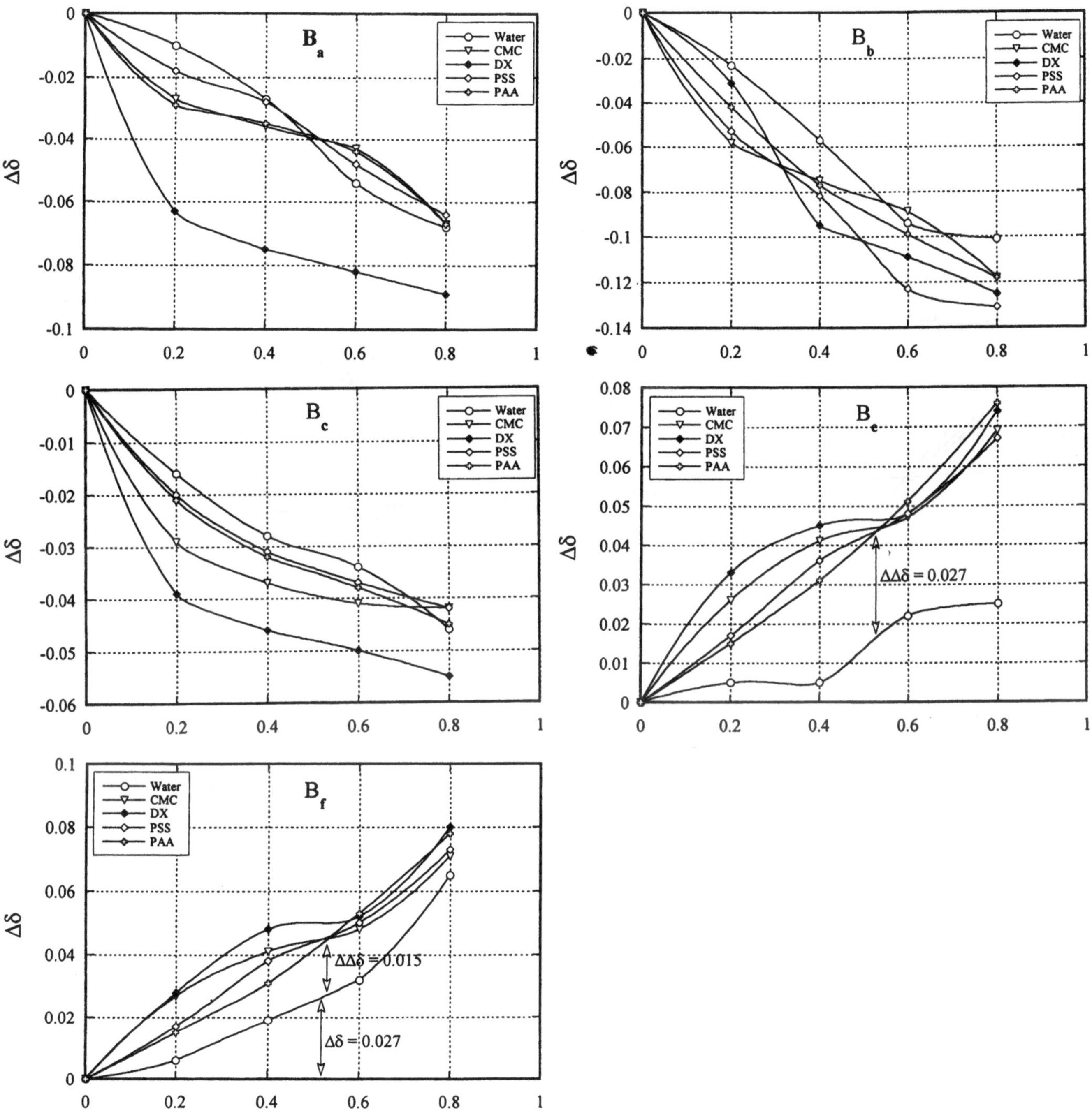

Fig. 6 ^{1}H chemical shifts (ppm) for BHDACl protons, i.e. B_a, B_b, B_c, B_e, and B_f

steric problems [36]. Similar behavior is also demonstrated by both B_b and B_c, though a relatively smaller downfield shift is observed over most of the mixing range. Regarding the effect of the polyelectrolyte on the head group protons of HTAB, i.e. H_d (Fig. 7), one can observe a systematic lesser upfield shift, being lowest in the case of DX in comparison to that in pure water. In other words, it can be said that the H_d protons have also undergone a downfield shift in comparison to pure water as observed for similar BHDACl protons, i.e. B_b and B_c. Thus, all the head group protons of BHDACl and HTAB undergo a downfield shift in the presence of all the polyelectrolytes [38] and the effect is more pronounced in the presence of DX. These results indicate that both surfactants in the mixed state interact with all the polyelectrolytes, though HTAB–polyelectrolyte

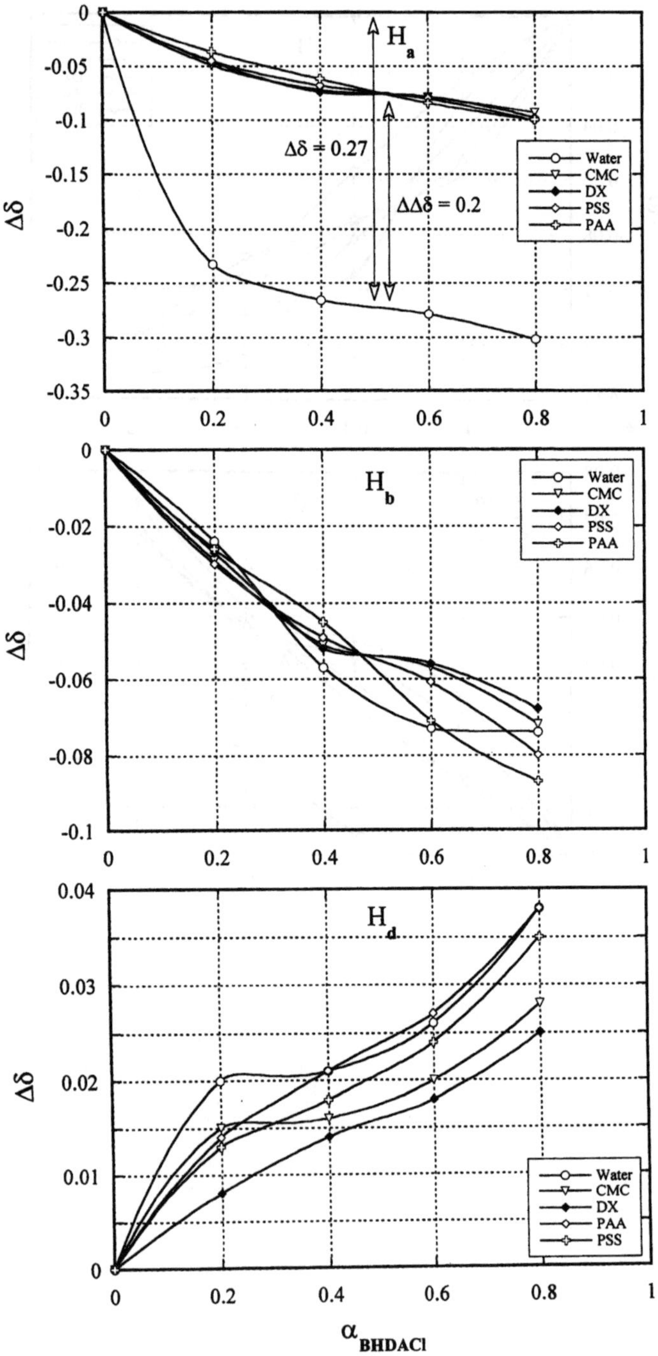

Fig. 7 ^{1}H chemical shifts (ppm) for HTAB protons, i.e. H_a, H_b, and H_d

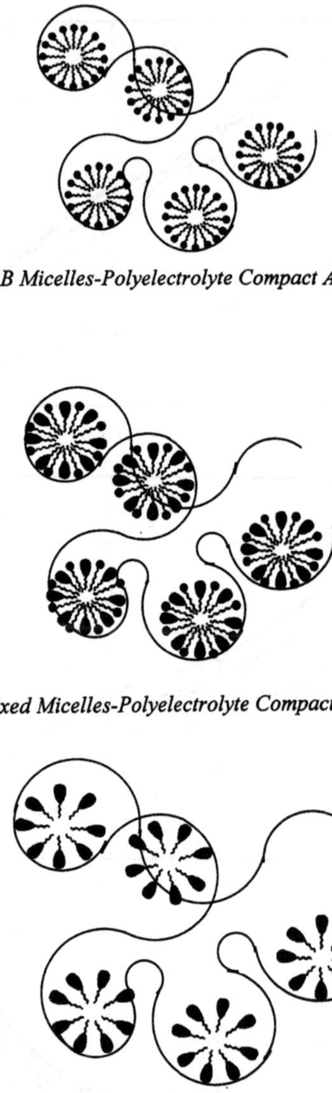

HTAB Micelles-Polyelectrolyte Compact Aggregates

Mixed Micelles-Polyelectrolyte Compact Aggregates

BHDACl Micelles-Polyelectrolyte Loose Aggregates

Fig. 8 Proposed HTAB micelle–polyelectrolyte, BHDACl micelle–polyelectrolyte, and HTAB + BHDACl mixed micelle–polyelectrolyte association

interactions are stronger than the BHDACl–polyelectrolyte ones since the latter interactions are screened by the steric effects of B_a protons [36]. This means that the lower η_r and higher τ of the HTAB–polyelectrolyte systems than the BHDACl–polyelectrolyte systems were mainly due to the presence of aromaticity in the BHDACl head group.

In contrast, the profiles of both B_e and B_f (Fig. 6) indicate a further upfield shift in the presence of each polymer in comparison to that in pure water over the whole mixing range. The magnitude of $\Delta\delta$ increases with the increase in the amount of HTAB, which is quite prominent in the case of B_e. For instance, $\Delta\Delta\delta = 0.027$ for B_e, whereas it is 0.015 for B_f at $\alpha = 0.5$. $\Delta\Delta\delta$ is the difference in the $\Delta\delta$ value in the absence and in the presence of polyelectrolyte (Fig. 6). This indicates that the B_e protons experience almost double the influence of polyelectrolyte than what the B_f protons experience. Consequently, the H_a protons experience pronounced shielding ($\Delta\Delta\delta = 0.2$) in the presence of polyelectrolyte in comparison to that in pure water, whereas the effect remains more or less the same on H_b (Fig. 7). Hence, all the protons of the hydrocarbon tails of BHDACl and HTAB undergo shielding in the presence of polyelectrolytes, though the effect is much stronger particularly in the case of H_a. The stronger shielding of the hydrophobic protons explains well the greater compactness of the mixed micelle–polyelectrolyte aggregates compared to that in pure water. Therefore, HTAB–polyelectrolyte association is thought to be much stronger with most of the polyelectrolytes than BHDACl–polyelectrolyte association. This can be explained on the basis of the wrapping of polyelectrolyte macromolecule around the spherical mixed micelles owing to the neutralization of oppositely charged polar sites, thus providing an additional nonpolar environment (Fig. 8). Among all the polyelectrolytes, again DX seems to have a greater effect on the shielding of the B_e, B_f, H_a, and H_b protons, especially in the BHDACl-rich region of the mixture and is a clear indication of stronger DX–BHDACl interactions than DX–HTAB interactions in this region. This could be the reason why η_r when DX is present (Fig. 2a) shows waterlike variation and is also the least turbid system (Fig. 2b). However, we could not find any evidence from the NMR study indicating stronger HTAB–CMC interactions, in contrast to the fact observed in Fig. 2. This could be accounted for by the low HTAB–CMC complex concentration in the solution phase in view of the insoluble salt formation as observed from τ measurements.

Conclusions

The present results indicate that even in the case of oppositely charged surfactant–polymer systems the extent of the electrostatic interactions is strongly influenced by the nature of the head group. Although both BHDACl and HTAB are 1:1 strong electrolytes, the weaker electrostatic interactions in the former case, as demonstrated by η_r and τ studies, were mainly attributed to the steric hindrance of the aromatic group in the BHDACl head group. NMR studies have also pinpointed this reason due to which the B_a proton resonance showed deshielding and shielding in the BHDACl- and HTAB-rich regions, respectively. The rest of the head group protons, i.e. either of BHDACl or HTAB, show deshielding owing to mixed micelle–polyelectrolyte interactions. All the hydrophobic protons undergo shielding owing to these interactions; however, the effect is quite significant on the terminal methyl protons of the HTAB hydrophobic tail, which suggests that the HTAB–polyelectrolyte system produces compact aggregates. η_r and τ showed stronger HTAB–CMC interactions in comparison to the rest of the polyelectrolytes, which were not observed from the NMR studies most probably owing to the insoluble salt formation in this case. NMR studies showed stronger BHDACl–DX interactions in the BHDACl-rich region in comparison to all the other polyelectrolytes. Thus, NMR has proved to be a quite useful technique in pinpointing the reasons behind the weak BHDACl–polyelectrolyte interactions, which could not be explained solely from κ, η, or τ measurements. It seems that NMR should be frequently applied for semiquantitative analysis of such kinds of complex systems.

Acknowledgements Research grants from the DST (SP/S1/H-22/2001) and CSIR [01(1777)/02/EMR-II], New Delhi, are gratefully acknowledged.

References

1. Nagarajan R, Kalpakci B (1985) In: Dubin P (ed) Microdomains in polymer solutions. Plenum, New York, pp 369–381
2. Jonsson B, Lindman B, Holmberg K, Kronberg B (1998) Surfactants and polymers in aqueous solutions. Wiley, New York
3. Kwak JCT (1998) Polymer–surfactant systems. Surfactant science series, vol 77. Dekker, New York
4. Tsujii K (1998) Surface activity (principles, phenomena, and applications). Academic, London
5. (a) Hayakawa K, Kwak JCT (1982) J Phys Chem 86:3866; (b) Hayakawa K, Kwak JCT (1983) J Phys Chem 87:506
6. Thalberg K, Lindman B (1990) Colloids Surf 47:147
7. Merta J, Stenius P (1995) Colloid Polymer Sci 273:974
8. (a) Goddard ED (1986) Colloids Surf 19:255; (b) Goddard ED (1986) Colloids Surf 19:301
9. Brackman JC, Engberts JBFN (1993) Chem Soc Rev 22:85
10. Liu J, Takisawa N, Shirahama K (1998) J Phys Chem B 102:6696
11. Saito S (1957) Kolloid Z 154:19
12. Saito S, Yukawa M (1969) J Colloid Interface Sci 30:211
13. Saito S, Yukawa M (1969) Kolloid Z Z Polym 234:1015
14. Hou Z, Li Z, Wang H (1999) Colloid Polymer Sci 277:1011
15. Bakshi MS, Kaur G (2000) J Surfactants Deterg 3:159
16. Bakshi MS (2000) Colloid Polymer Sci 278:524
17. Hoffman H (1994) In: Herb EA, Prud'homme RK (eds) Structure and flow in surfactant solutions. ACS symposium series 578. American Chemical Society, Washington, DC, chap 1
18. Hoff E, Nystrom B, Lindman B (2001) Langmuir 17:28
19. Dubin PL, The SS, Chew CH, Gan LM (1990) J Colloid Interface Sci 23:2500

20. Dubin PL, The SS, Gan LM, Chew CH (1990) Macromolecules 23:2500
21. Bakshi MS, Sachar S, Mahajan N, Kaur I, Kaur G, Singh H, Shegal P, Doe H (2002) Colloid Polymer Sci 280:990
22. Gao Z, Kwak JCT (1991) In: Mittal KL, Shah DO (eds) Surfactants in solution. Plenum, New York, p 261
23. Stilbs P (1998) In: Kwak JCT (ed) Polymer–surfactant systems. Surfactant science series, vol 77. Dekker, New York, pp 239–266
24. Ozeki S, Ikeda S (1980) J Colloid Interface Sci 77:219
25. (a) Bakshi MS (1993) J Chem Soc Faraday Trans 189:4323; (b) Bakshi MS (1997) J Chem Soc Faraday Trans 193:4005
26. Clint JH (1975) J Chem Soc Faraday Trans 171:1327
27. Macdonald PM, Staring D, Yue Y (1993) Langmuir 9:381
28. Bakshi MS, Kaur I (2002) Colloid Polymer Sci (in press)
29. Kogrej K, Skerjane J (1999) Langmuir 15:4251
30. Goddard ED, Ananthapadmanabhan KP (1998) In: Kwak JCT (ed) Polymer-surfactant systems. Surfactant science series, vol 77. Dekker, New York, pp 21–64
31. Isemura T, Imanishi A (1958) J Polym Sci 33:337
32. Saito S, Mizuta Y (1967) J Colloid Interface Sci 23:604
33. Jones MN (1967) J Colloid Interface Sci 23:36
34. Abin EB, Scaiano JC (1984) J Am Chem Soc 106:6274
35. Bocquenet Y, Siffert B (1984) Colloids Surf 9:147
36. Bhat M, Gaika VG (1999) Langmuir 15:4740
37. Yuan HZ, Zhao S, Cheng GZ, Zhang L, Miao XJ, Mao SZ, Yu JY, Shen LF, Du YR (2001) J Phys Chem 105:4611
38. Cabane B (1977) J Phys Chem 81:1639

Progr Colloid Polym Sci (2003) 122:47–55
DOI 10.1007/978-3-540-36114-5

Susan E. Burke
Rama M. Palepu
Samik K. Hait
Satya P. Moulik

Physicochemical investigations on the interaction of cationic cellulose ether derivatives with cationic amphiphiles in an aqueous environment

S. E. Burke · R. M. Palepu (✉)
Department of Chemistry,
St. Francis Xavier University,
Antigonish, NS B2G 2W5, Canada
e-mail: rpalepu@stfx.ca
Tel: +1-902-8673886
Fax: +1-902-8672414

S. K. Hait · S. P. Moulik (✉)
Center for Surface Science,
Department of Chemistry,
Jadavpur University,
Calcutta 700032, India
e-mail: cssju@yahoo.uk

Abstract The interactions of the cationic cellulose ether derivatives JR400 and LM200 with cationic surfactants differing in both the alkyl chain length and head group composition were investigated using microcalorimetric techniques. The presence of polymer-induced surfactant aggregation and polymer-bound aggregates was detected for the majority of the surfactants investigated. The thermodynamic properties of the premicellar aggregation, the binding, and the micellization of the surfactants in the presence of polymers were evaluated and analysed in terms of the strength of the interactions between the polymer and the surfactants. The nature of these aggregates was also evaluated using conductivity, fluorescence spectroscopy, and dye solubilization techniques.

Keywords Polymers · Surfactants · Microcalorimetry · Micellization · Thermodynamic properties

Introduction

Research in the area of polymer–surfactant complexes has received a considerable amount of attention over the last few decades owing to their wide range of technical applications and because of the fundamental interest in intermolecular interactions and hydrophobic aggregation phenomena. Polymer–surfactant interactions are important in areas such as colloid stability, food products, personal care products, pharmaceutical formulations, household and industrial detergents, paints, and enhanced oil recovery [1, 2, 3, 4, 5]. It is well established in the literature that nonionic polymers interact strongly with anionic surfactants, but very weakly with cationic surfactants [6, 7, 8]. These strong interactions of anionic surfactants with neutral polymers are attributed to favourable interactions between the anionic surfactants and the hydration shell of the polymer. In the case of cationic surfactants, the weak interactions are attributed to the bulkiness of the cation head groups and to an electrostatic repulsion between the polymer and the surfactant due to possible protonation of the polymer [9, 10, 11]. Hydrophobically modified, water-soluble polymers that contain a small number of strongly hydrophobic substituents are known to interact strongly with surfactants [12, 13, 14, 15]. Since contact between the hydrophobic groups and water is not favoured, these hydrophobically modified polymers have a tendency to self-aggregate and/or associate with the surfactants [14]. When surfactant is progressively added to a solution of hydrophobically modified polymer, the viscosity of the solution increases initially, followed by a decrease at higher surfactant concentrations. The initial increase in viscosity is ascribed to the formation of mixed micelles between the polymer alkyl chains and the surfactant molecules reinforcing the intermolecular cross-links in the polymer [16]. Upon further addition of surfactant molecules, the number of polymer alkyl chains in the mixed micelle decreases, thereby reducing the connectivity of the network and the solution viscosity [16].

Extensive studies involving polyelectrolytes with oppositely charged amphiphiles have been reported [17, 18, 19]. In the majority of these studies, the polymer and the surfactant molecules begin to interact at a particular concentration known as the critical aggregation concentration (cac), which usually occurs at a lower surfactant concentration than the critical micelle concentration (cmc) of the surfactant. An indication of the degree of interaction between surfactant and polymer can frequently be obtained from the ratio between the cmc and the cac values. The hydrophobic nature of these modified polymers provides them with the unique ability to interact with surfactants of the same charge; a phenomenon lacking in traditional polyion–surfactant systems [20].

By employing the I_1/I_3 ratio of the pyrene fluorescence spectrum, the existence of hydrophobic association between the hydrophohically modified cellulose ether Quatrisoft LM200 and alkyltrimethylammonium halide surfactants has been reported [21]. The evidence for polymer–surfactant interactions in this system was also established by other investigators employing other techniques, such as surface tension and phase equilibria [22, 23].

In our previous investigation [24], we established the existence of association between Quatrisoft LM200 and several cationic surfactants differing in the nature of the head group and chain length. Analysis of the results obtained through conductivity, steady-state fluorescence, and dye solubilization techniques suggested the presence of micellelike polymer–surfactant aggregate formation prior to the cmc. The degree of interaction depended on the chain length of the surfactant as well as on the nature and bulkiness of the head group.

The present investigation mainly focuses on the study of interactions of cationic surfactants with the cationic cellulose ether derivative JR400 and the hydrophobically modified derivative LM200, employing microcalorimetric techniques. Heat effects in polymer–surfactant solutions are relatively small, but modern calorimetric methods allow the measurement of the cac and the assessment of the relative strength of interactions between the polymer and the surfactants. It is well established in the literature that the physicochemical information obtained for a chemical process is often method-dependent [25]. The calorimetric results obtained in the present study were analysed along with the results obtained from JR400–surfactant systems employing techniques such as conductivity, fluorescence spectroscopy, and dye-solubilization methods, and were rationalized in terms of polymer–surfactant interactions.

Experimental

The cationic cellulose ether derivatives JR400 and Quatrisoft LM200 were gift samples from Amerchol and were used without further purification. The average molar masses were approximately 400,000 and 100,000 Da, respectively (Fig. 1). The degree of substitution was found to be 2.0×10^{-4} mol of hydrophobic chains

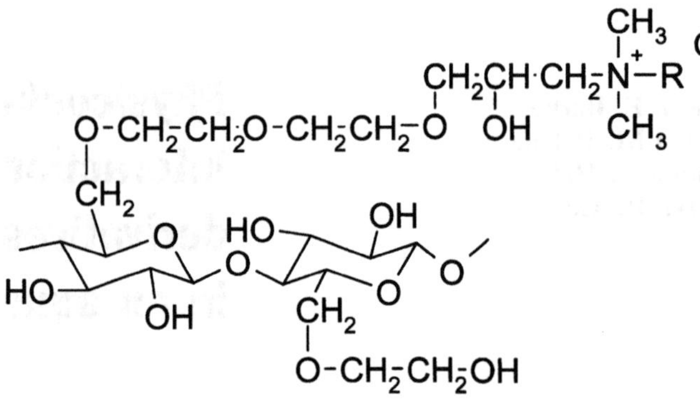

JR400 R = CH_3
LM200 R = $C_{12}H_{25}$

Fig. 1 Structure of a portion of modified cationic cellulose ether derivatives

per gram of polymer. This corresponds to approximately 3.4% by weight, or one substitution per every 19 glucose units [26]. The surfactants used were the same as those reported earlier [24]. Dodecylpyridinium chloride ($C_{12}PC$) and hexadecylpyridinium chloride ($C_{16}PC$) and the dodecyl-, tetradecyl- and hexadecyltrimethylammonium bromides (C_nTAB) were purchased from Aldrich. Decyltrimethylammonium bromide ($C_{10}TAB$) was obtained from TCI America. All these surfactants were purified by repeated recrystallization from acetone. Alkyldimethylammonium bromide (C_nDAB) and alkylammonium bromide (C_nAB) surfactants were synthesized by combining the corresponding amine with hydrogen bromide (48%). The amines were purchased from Aldrich, while the hydrogen bromide was received from Fischer Scientific. These surfactants were purified by repeated recrystallization from acetone and diethyl ether. The purity of these surfactants was tested by bromide ion analysis using the conductometric method. Silver nitrate solution was used as the standard. All the surfactants were found to be better than 99% pure. The two probes used in the fluorescence study were received from Aldrich. Pyrene (99% optical grade) was purified by repeated recrystallization followed by sublimation. The second probe, Auramine-O, was used as received. For the dye solubility studies Sudan III (from Sigma), an oil-soluble dye, was employed without further purification. All the solutions were made with deionized water, equivalent to triply distilled. The polymer stock solutions were made 24 h prior to use. All the polymer–surfactant solutions were prepared by employing the stock solutions and were sonicated to ensure homogeneity. The concentrations are either in percent (weight/volume) or in moles per litre.

Steady-state fluorescence measurements of the pyrene emission spectra were performed using a PerkinElmer MPF 66 fluorescence spectrophotometer. The pyrene concentration was maintained at 1×10^{-6} M, to prevent excimer formation. The fluorescence intensity of pyrene was followed over the range 360–500 nm. The excitation and emission slit widths were maintained at 1.5 and 2.0 nm, respectively. The experiments were carried out as a function of surfactant concentration in the presence of polymer.

The ratio of the first and third vibronic peaks, I_1/I_3, yields an estimate of the polarity of the pyrene molecules in their solubilization sites [27, 28, 29]. The aggregation numbers of the surfactants in the presence of polymers were determined using the static quenching method, employing both hexadecylpyridinium chloride and benzophenone as the quencher [30].

The Auramine-O studies were performed using solutions prepared from a probe stock solution as a function of surfactant concentration. The concentration of the probe was kept at 2.5×10^{-4} M in each sample. For the fluorescence studies with Auramine-O, an Aminco Bowmen series 2 luminescence spectrometer was employed. The slit width of 15 nm was maintained for both excitation and emission spectra. An excitation wavelength of 358 nm was employed and the spectrum was scanned over the range 450–600 nm. The dye solubilization experiments involved saturating each surfactant solution with the dye and then filtering off the excess dye after 24 h (equilibration at 298 K). The extent of solubulization of Sudan III was determined using a Hewlett-Packard 8452A diode-array spectrometer. The solution absorbance was monitored at 512 nm to determine the solubility of the dye.

The conductivity measurements were performed using an automatic conductivity bridge (Radiometer, Copenhagen) operating at 1,000 Hz, with a dip cell having a cell constant of 0.978 cm^{-1}. The specific conductivity at 298 K was measured as a function of surfactant concentration, and the values of the cac and the cmc were obtained from the break points of the plots of specific conductivity versus concentration. From the ratio of the slopes of the plots, the degree of counter ion binding, f, was estimated.

The microcalorimetric measurements were carried out employing an ITC instrument from Microcal (Northampton, USA), capable of recording heat flow as low as 10 μcal. In an experiment 1.325 ml aqueous polymer solution of the desired concentration was placed in the calorimeter cell. Under computer control, a preset number of 10–20-μl aliquots of temperature-equilibrated surfactant solutions were injected into the cell at 4-min intervals. The heat flow per injection and the corresponding enthalpy change per mole of added surfactant were obtained using Microcal software [31]. To determine the heat of dilution of the surfactants, control runs were carried out using water instead of polymer solution. All the runs were duplicated to check for reproducibility. All the measurements were carried out at 298 ± 0.001 K.

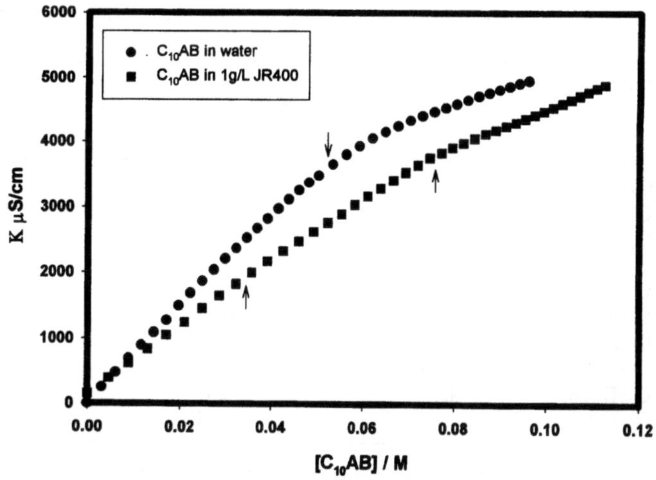

Fig. 2 Specific conductivity plot of decylammonium bromide ($C_{10}AB$) in water and in presence of JR400

Results and discussion

Conductivity

The deviation from linearity in the premicellar region of the specific conductivity versus surfactant concentration plots, in the presence of polymer, is taken as evidence for the polymer-induced aggregation (cac). A typical plot for $C_{10}AB$ interaction with JR400 is presented in Fig. 2. The interaction of the polymer with the surfactant is further reflected in the shift of the cmc of the surfactant to a higher value in the presence of the polymer. The cmc, the cac, f values, and the values obtained from calorimetry are presented in Tables 1 and 2. It is evident that the conductometric method is unable to provide evidence for polymer-induced aggregation in the premicellar region for several surfactants. However, more sensitive techniques, such as microcalorimetry, provide a better probe for determining the existence of such aggregates.

Calorimetry

The microcalorimetric method measures the heat changes associated with the interactions of the surfactants with the polymers [32, 33, 34]. Evidence for the existence of polymer–surfactant interactions is provided by this method for all the surfactants, except $C_{10}TAB$. The enthalpy of dilution plots of $C_{12}PC$ at 298 K in the absence and in the presence of LM200 as a function of surfactant concentration are presented in Fig. 3a. The differential enthalpy of the dilution curve obtained by subtracting the heat of dilution of $C_{12}PC$ in water from that in polymer solution is plotted in Fig. 3a. The results for the JR400–$C_{12}PC$ system are plotted in Fig. 3b. The pattern of the differential plots can be analysed as follows. The initial peaks in Fig. 3a and b are attributed to the enthalpy changes due to the amphiphilic aggregation induced by the presence of the polymer (cac). The inflection point in the initial curve is assigned to the cac value. From the crest, the enthalpy difference decreases slightly below zero and slowly approaches zero at higher concentrations of the surfactant. The enthalpy difference between points a and b in Fig. 3 is attributed to ΔH_{cac}. The enthalpy difference between points b and c represents the exothermic binding of aggregates with polymer, ΔH_b, along with other minor heat changes due to solvation–desolvation, conformational and organizational changes. Point c represents the value of the cmc, because the process of dilution of $C_{12}PC$ is associated with zero resultant heat change beyond this point. The enthalpy values are presented in Tables 3 and 4. In the case of the other surfactants (except for $C_{10}TAB$) with LM200, the enthalpy of dilution plots exhibited a third stage between points c and d and this is attributed to the formation of normal micelles. The representative plots for $C_{14}TAB$, $C_{14}DAB$, $C_{10}AB$ are presented in Figs. 4, 5, and 6. The majority of surfactants with JR400 also exhibited evidence for the existence of a cac. The calorimetric results of the noninteracting amphiphile $C_{10}TAB$ are presented in Fig. 7. The dilution curves with and without the polymer follow the same course and the transition points in the curves are also comparable. The strength of the influence

Table 1 The critical aggregation concentration (*cac*) with 1.0% (w/v) LM200 present, the critical micelle concentration (*cmc*) and *f* values of cationic surfactants, and the cmc/cac ratio

Surfactant	Cac (mM)[a]		Cmc (mM)[a]		f	cmc/cac
	Conductivity[b]	Calorimetry	Conductivity	Calorimetry		
$C_{12}PC$	5.3	11.0	17.0	19.7	0.6	1.79
$C_{16}PC$	0.2	0.16	0.92	1.13	0.57	7.06
$C_{10}TAB$	–	–	71.0	62.0	0.73	–
$C_{12}TAB$	–	6.62	18.0	18.1	0.7	2.73
$C_{14}TAB$	–	1.4	3.8	5.1	0.67	3.64
$C_{16}TAB$	–	0.15	0.94	1.2	0.7	8.0
$C_{10}AB$	14.8	17.7	74.4	66.1	0.6	3.73
$C_{12}AB$[c]	3.85	–	15.5	–	0.76	4.03
$C_{12}DAB$	4.7	6.4	13.4	17.1	0.75	2.67
$C_{14}DAB$	–	0.81	4.0	5.6	0.64	6.91

[a]The cmc and cac values are accurate to within 2%
[b]Conductivity results were taken from our previous publication [24]
[c]Calorimetric measurements could not be carried out owing to solubility problems

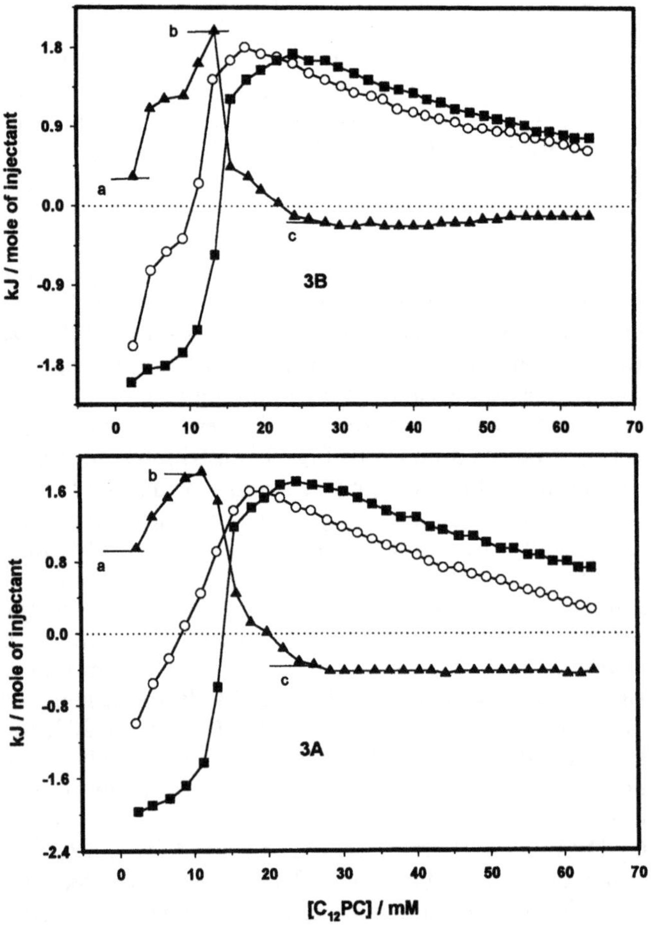

Fig. 3 Enthalpy of dilution of dodecylpyridinium chloride ($C_{12}PC$) in **a** LM200 and **b** JR400 solutions at 298 K: with no polymer (*squares*); with 1.0% (w/v) polymer (*circles*); and with 1.0% (w/v) polymer subtracting the $C_{10}PC$ heat of dilution in water (*triangles*)

of polymer on the amphiphile aggregation can be ascertained from the magnitude of the ratio of the cmc and the cac values. It is found that it follows the order $C_{16}PC \cong C_{16}TAB > C_{14}DAB > C_{14}TAB > C_{12}AB > C_{12}DAB > C_{10}AB > C_{12}TAB > C_{12}PC$. The enthalpy of the cac (ΔH_{cac}) and of binding (ΔH_b) exhibited similar behavior for both polymers, indicating that the introduction of a dodecyl chain in LM200 had very little influence on these parameters.

The free-energy and entropy changes at the cac and the cmc were evaluated by employing the following equations:

$$\Delta G_{cac} = RT \ln X_{cac}, \tag{1}$$

$$\Delta G_{cmc} = (1 + f)RT \ln X_{cmc}, \tag{2}$$

$$\Delta S_{cac} = \frac{\Delta H_{cac} - \Delta G_{cac}}{T}, \tag{3}$$

$$\Delta S_{cmc} = \frac{\Delta H_{cmc} - \Delta G_{cmc}}{T}, \tag{4}$$

where X_{cac} and X_{cmc} in Eqs. (1) and (2) are the values of the cac and cmc expressed on the unitary scale. The calculated values of ΔG and ΔS at the cac and the cmc are presented in the Tables 3 and 4 for LM200 and JR400, respectively. The values of ΔG_{cac} are found to be negative, with magnitudes dependent on the hydrophobic chain length and the nature of the head group of the surfactant. The values of ΔG_{cmc} are less negative when polymers are present, indicating that the micelle formation is less spontaneous in this environment. The enthalpies at the cac values are endothermic for all the surfactants. The values of the enthalpy of micellization, ΔH_{cmc}, are mostly positive, with several exceptions, primarily when polymer is absent. In most instances, the sign of ΔH_{cmc} changed when polymers were present. The

Table 2 The cac with 1.0% (w/v) JR400 present, the cmc and *f* values of cationic surfactants, and the cmc/cac ratio

Surfactant	cac (mM)[a]		cmc (mM)[a]		F	cmc/cac
	Conductivity	Calorimetry	Conductivity	Calorimetry		
C_{12}PC	–	13.3	17.0	21.0	0.7	1.61
C_{16}PC	–	0.1	0.91	1.1	0.74	8.61
C_{10}TAB	–	–	68.0	62.0	0.75	–
C_{12}TAB	–	6.6	16.0	18.1	0.65	2.74
C_{14}TAB	–	1.5	3.5	5.3	0.64	3.50
C_{16}TAB	–	0.1	0.9	1.2	0.64	8.00
C_{10}AB	28.0	22.5	72.0	66.8	0.60	2.97
C_{12}AB[b]	3.85	–	14.4	–	0.76	3.39
C_{12}DAB	4.70	6.7	13.0	22.8	0.75	3.42
C_{14}DAB	–	0.8	4.1	4.8	0.64	5.88

[a]The cmc and cac values are accurate to within 2%
[b]Calorimetric measurements could not be carried out owing to solubility problems

Table 3 The thermodynamic parameters for cac formation, aggregate binding, and micelle formation at 298 K. ΔG, ΔH, ΔS are accurate to within ±5%. The values in *parentheses* are the polymer concentrations in percent (w/v)

Surfactant + polymer	$-\Delta G_{cac}$ (kJ mol^{-1})	ΔH_{cac} (kJ mol^{-1})	ΔS_{cac} (J K^{-1} mol^{-1})	ΔH_B (kJ mol^{-1})	$-\Delta G_{cmc}$ (kJ mol^{-1})	ΔH_{cmc} (kJ mol^{-1})	ΔS_{cmc} (J K^{-1} mol^{-1})
C_{12}PC + no polymer	–	–	–	–	36.8	3.4	135.1
C_{12}PC + LM200 (1.0)	19.5	0.9	68.5	–	29.1	−1.8	91.6
C_{16}PC + no polymer	–	–	–	–	46.8	−6.0	136.8
C_{16}PC + LM200 (0.5)	30.9	1.5	108.7	−7.0	42.2	4.1	155.6
C_{16}PC + LM200 (1.0)	31.5	Too low	–	−7.8	42.2	4.6	157.2
C_{10}TAB + no polymer	–	–	–	–	29.1	2.3	105.8
C_{10}TAB + LM200 (1.0)	–	–	–	–	29.3	0.5	100.4
C_{12}TAB + no polymer	–	–	–	–	35.0	−2.4	109.6
C_{12}TAB + LM200 (1.0)	22.5	0.5	77.3	−1.5	34.0	1.2	118.3
C_{14}TAB + no polymer	–	–	–	–	41.7	−5.9	120.6
C_{14}TAB + LM200 (0.5)	26.8	0.2	90.5	−1.2	38.4	1.1	132.6
C_{14}TAB + LM200 (1.0)	26.9	0.2	90.5	−1.8	38.7	2.0	136.3
C_{16}TAB + no polymer	–	–	–	–	49.3	−10.7	129.2
C_{16}TAB + LM200 (0.5)	32.0	Too low	–	−5.2	45.6	4.5	167.7
C_{16}TAB + LM200 (1.0)	32.5	Too low	–	−5.4	45.6	4.8	169.4
C_{10}AB + no polymer	–	–	–	–	28.2	−1.6	89.3
C_{10}AB + LM200 (1.0)	19.9	0.5	68.4	−1.9	26.8	0.4	91.3
C_{12}AB	23.7	–	–	–	35.7	–	–
C_{12}DAB	–	–	–	–	35.0	−3.1	107.0
C_{12}DAB + LM200 (1.0)	22.5	0.60	77.5	−1.0	33.8	1.7	119.2
C_{14}DAB + no polymer	–	–	–	–	40.5	−9.0	105.7
C_{14}DAB + LM200 (1.0)	27.6	Too low	–	−3.7	38.1	5.0	131.2

positive entropy changes at both the cac and the cmc are attributed to the environmental disordering, resulting from the disruption of water structure during the amphiphile aggregation and the free movement of the surfactant tails in the hydrophobic environment.

Fluorescence spectroscopy

Micropolarity

The pyrene emission spectrum is used to determine the micropolarity of the medium [27]. The ratio I_1/I_3 of the (0,0) and (0,3) bands decreases sharply at the onset of micellization, as a result of the preference of pyrene to reside in the nonpolar regions of the aggregates. Fluorescence studies on the interactions of cationic surfactants with JR400 indicated evidence for the interaction of C_nAB and C_nDAB only. The existence of a cac and the shift in the cmc are evident from the plots of I_1/I_3 versus surfactant concentration (Fig. 8), indicating the presence of polymer–surfactant interactions. The ability of these surfactants to form hydrogen bonds with the cellulose ether polymer backbone allows interaction with the nonhydrophobic JR400 polymer molecules.

Table 4 The thermodynamic parameters for cac aggregate binding and micelle formation at 298 K. ΔG, ΔH, ΔS are accurate to within ±5%. The values in *parentheses* are the polymer concentrations in percent (w/v)

Surfactant + polymer	$-\Delta G_{cac}$ (kJ mol^{-1})	ΔH_{cac} (kJ mol^{-1})	ΔS_{cac} (J K^{-1} mol^{-1})	ΔH_B (kJ mol^{-1})	$-\Delta G_{cmc}$ (kJ mol^{-1})	ΔH_{cmc} (kJ mol^{-1})	ΔS_{cmc} (J K^{-1} mol^{-1})
$C_{12}PC$ + JR400 (1.0)	20.8	1.7	75.4	–	33.4	−2.0	105.4
$C_{16}PC$ + JR400 (0.5)	32.5	Too low	–	−6.6	46.8	4.6	172.7
$C_{16}PC$ + JR400 (1.0)	32.5	Too low	–	−7.5	46.8	5.1	174.4
$C_{10}TAB$ + JR400 (1.0)	–	–	–	–	29.7	0.6	101.6
$C_{12}TAB$ + JR400 (1.0)	22.5	0.3	76.7	Too low	32.8	0.5	111.6
$C_{14}TAB$ + JR400 (0.5)	26.3	0.4	89.5	−1.3	37.8	2.3	134.7
$C_{14}TAB$ + JR400 (1.0)	26.3	0.5	89.9	−1.5	37.9	2.9	137.2
$C_{16}TAB$ + JR400 (0.5)	31.5	Too low	–	−4.7	43.9	6.6	169.4
$C_{16}TAB$ + JR400 (1.0)	31.7	Too low	–	−4.9	43.9	7.4	172.3
$C_{10}AB$ + JR400 (1.0)	19.3	1.1	68.5	−3.4	26.6	0.7	91.7
$C_{12}AB$ + JR400 (1.0)	23.5	–	–	–	36.0	–	–
$C_{12}DAB$ + JR400 (1.0)	22.4	Too low	0	Too low	33.8	Too low	–
$C_{14}DAB$ + JR400 (1.0)	27.6	Too low	–	−2.1	38.3	5.0	145.3

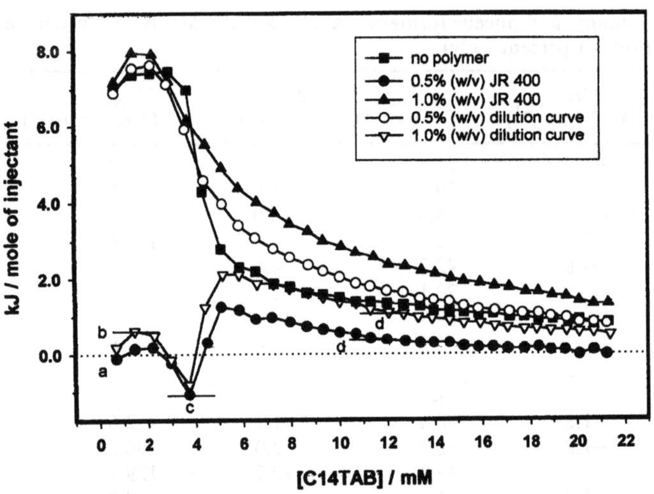

Fig. 4 Enthalpy of dilution of tetradecyltrimethylammonium bromide ($C_{14}TAB$) in JR400 solutions at 298 K

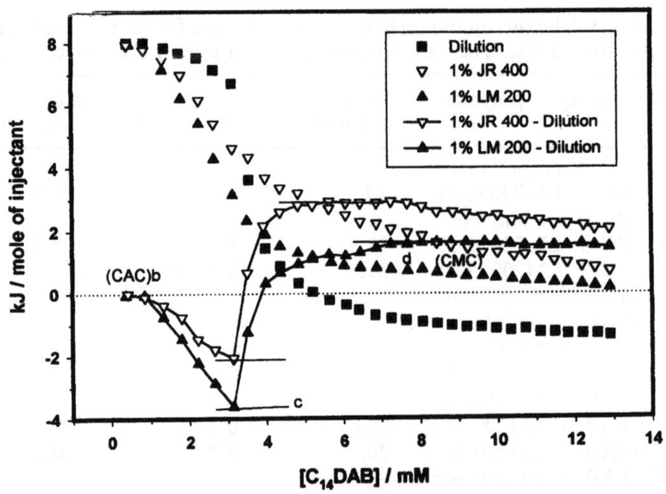

Fig. 5 Enthalpy of dilution of tetradecyldimethylammonium bromide ($C_{14}DAB$) in LM200 and JR400 solutions at 298 K

Aggregation numbers

It is well established in the literature that a comparison of the average size of the micellar aggregates in the presence and in the absence of the polymer can provide evidence for the polymer–surfactant interactions [35]. Average aggregation numbers of ammonium-based cationic surfactants in JR400 solution were determined by employing a static quenching method [36]. The values are presented in Table 5. Examination of the table reveals that for $C_{10}AB$, $C_{12}AB$, and $C_{12}DAB$, the aggregation numbers when JR400 was present decreased slightly from the values in water, indicting polymer–surfactant interactions. The aggregation numbers of C_nTABs in JR400 solution are found to be very similar to that in water.

Microviscosity

The microviscosity of the polymer–surfactant system was investigated by following the fluorescence of Auramine-O in these systems [37]. The intensity of Auramine-O fluorescence reached a maximum value at the cmc of the surfactant in water, and at the cmc when JR400 was present (Fig. 9). The increase in microviscosity when polymer was present along with the shift in the cmc value suggests that the probe senses a more viscous microenvironment created by the polymer–surfactant interaction [38]. Only $C_{10}AB$, $C_{12}AB$, and $C_{12}DAB$ exhibited an increase in microviscosity and a shift in cmc values. The intensity of the maximum microviscosity for each member of a given surfactant family is found to be independent of chain length, indicating that the probe molecules do not reside inside the aggregates but rather

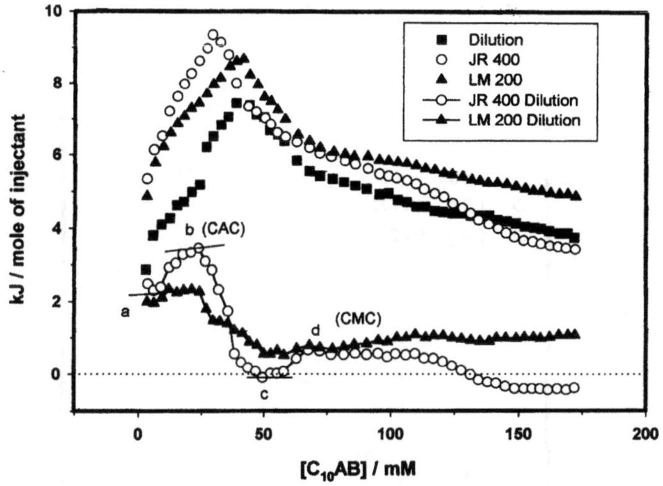

Fig. 6 Enthalpy of dilution of $C_{10}AB$ in LM200 and JR400 solutions at 298 K

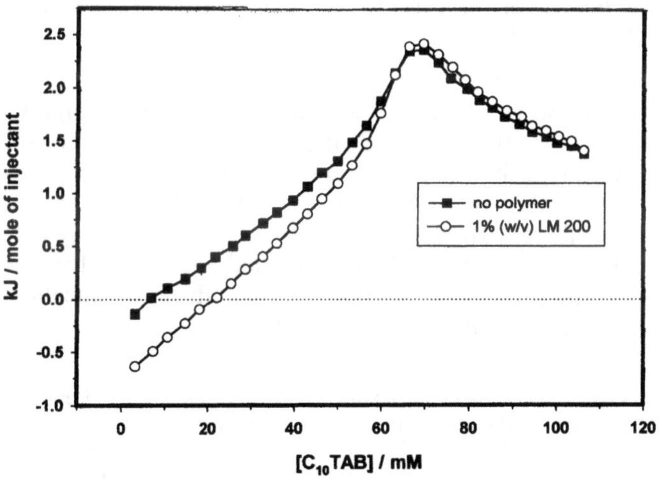

Fig. 7 Enthalpy of dilution of decyltrimethylammonium bromide ($C_{10}TAB$) in water and LM200 solution at 298 K

situate somewhere around the periphery of the aggregates.

Dye solubilization

Sudan III is a highly conjugated azo dye insoluble in water. In surfactant solutions, the dye is solubilized either deep in the palisade layer of the micelle or in the inner core of the micelle [39]. Dye solubilization studies can also reveal the presence of a premicellar aggregation of the surfactant molecules with the polymers, that is a cac [40]. The micellar aggregation creates relatively nonpolar regions to aid the solubilization of the dye. The dye solubilization results of Sudan III in an aqueous JR400 plus $C_{10}AB$ surfactant system along with a similar

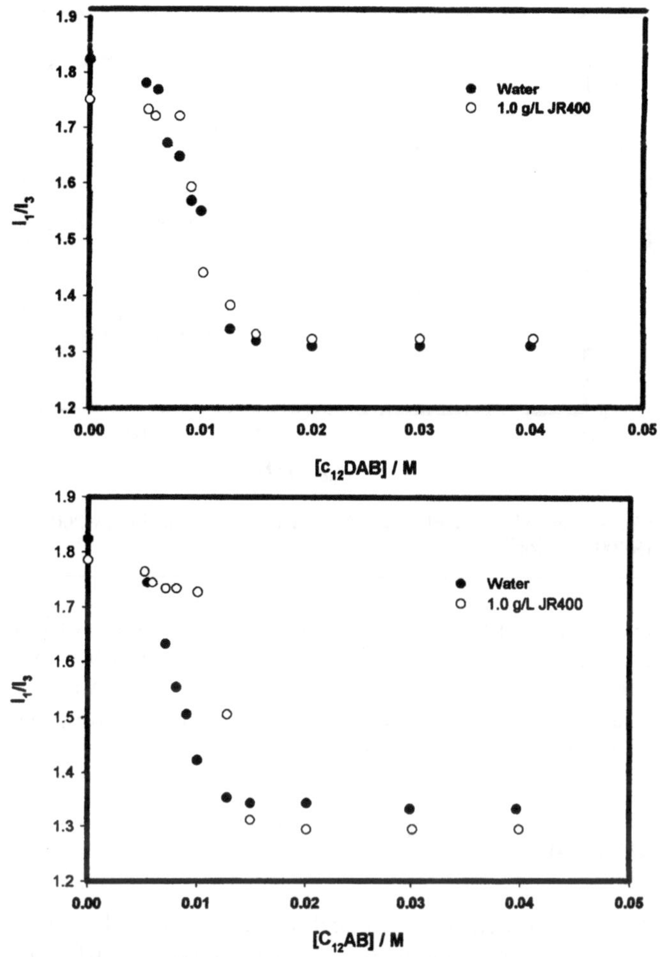

Fig. 8 Micropolarity plots for **a** dodecyldimethylammonium bromide ($C_{12}DAB$) and **b** dodecylammonium bromide ($C_{12}AB$) in water and JR400 solution

Table 5 Aggregation numbers of cationic surfactants at 298 K

Surfactant	Aggregation number	
	Water	JR400 1% (w/v)
$C_{10}TAB$	39	38
$C_{12}TAB$	58	57
$C_{14}TAB$	68	67
$C_{16}TAB$	88	87
$C_{10}AB$	39	32
$C_{12}AB$	56	52
$C_{12}DAB$	60	56

LM200 system are presented in Fig. 10. The formation of the cac and the solubilization of Sudan III in the LM200–$C_{10}AB$ system and a shift in the cmc value are evident from the figure. In the case of $C_{10}AB$ with JR400 there is no evidence for the solubilization of Sudan III in the premicellar region; however, the shift in the cmc is observed. With other C_nTABs, the solubilization data

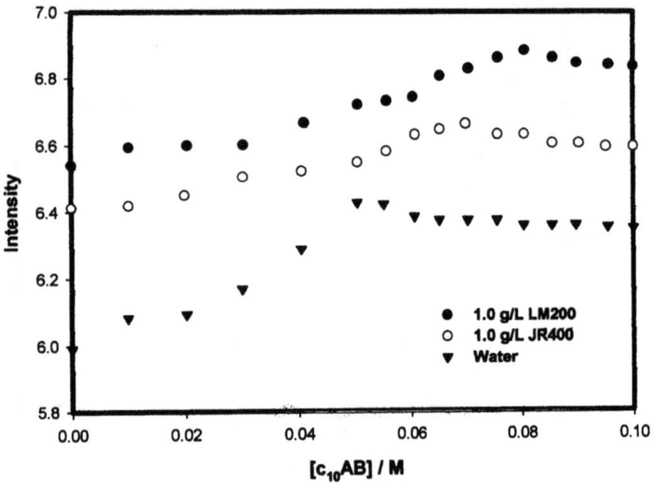

Fig. 9 Microviscosity plots for $C_{10}AB$ in water, and in LM200 and JR400 solutions

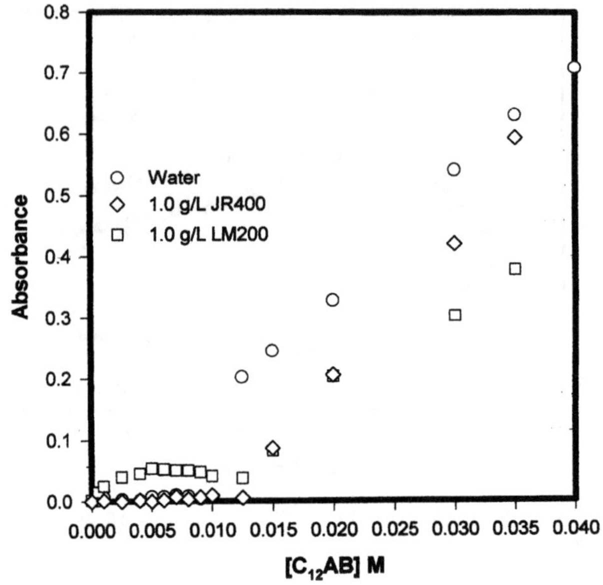

Fig. 10 Solubility of Sudan III in $C_{12}AB$ solutions and in $C_{12}AB$–polymer systems

revealed no interaction with JR400. From these studies, one can conclude that although the ammonium bromide surfactants interact with JR400, via the formation of hydrogen bonds, the resulting complex is not nonpolar enough to solubilize the dye.

Conclusions

1. The microcalorimetric technique proved to be a very sensitive method in the study of polymer–surfactant interactions and to derive thermodynamic properties.
2. Evidence for the existence of polymer-induced aggregation in the premicellar region is provided by the microcalorimetric technique. This is not clearly observed by conductometric and fluorescence methods.
3. The polymer JR400 was found to interact significantly with the C_nAB and C_nDAB surfactants and not with the C_nTAB compounds. This is attributed to the ability of C_nAB and C_nDAB to form hydrogen bonds to the backbone of the JR400 polymer. However, these hydrogen-bonded aggregates formed in the premicellar region are not sufficiently micellar-like to dissolve a nonpolar dye.

Acknowledgements R.P. acknowledges the Operating Research Grant from the Natural Sciences and Engineering Research Council (NSERC). S.K.H. thanks the University Grants Commission of India for a Junior Research Fellowship. The authors acknowledge J. Beck for his help in the preparation of the manuscript.

References

1. Breuer MM, Robb ID (1972) Chem Ind 13:530
2. Goddard ED (1986) Colloids Surf 19:255; 301
3. Goddard ED (1993) In: Goddard ED, Ananthapadmanabhan KP (eds) Interactions of surfactants with polymers and proteins. CRC, Boca Raton, pp 395–414
4. Brackman JC, Engberts JBFN (1993) Chem Soc Rev 22:85
5. Lindman B, Thalberg K (1993) In: Goddard ED, Ananthapadmanathan KP (eds) Interactions of surfactants with polymers and proteins. CRC, Boca Raton, pp 203–276
6. Evertsson H, Nilsson S (1997) Macromolecules 30:2377
7. Evertsson H, Nilsson S (1998) Carbohydr Polym 35:135
8. Majhi PR, Moulik SP, Burke SE, Rodgers MP, Palepu R (2001) J Colloid Interface Sci 235:227
9. Brackman JC, Engberts JBFN (1991) Langmuir 7:2097
10. Jones MN (1992) Chem Soc Rev 21:127
11. Garcia-Mateos I, Perez S, Velazquez MM (1997) J Colloid Interface Sci 194:356
12. Li Y, Dubin PL (1994) In: Herb CA, Prud'homme RH (eds) Structure and flow in surfactant solutions. ACS symposium series 578. American Chemical Society, Washington DC, p 587
13. Wei YC, Hudson SMJ (1995) J Macromol Sci Rev Macromol Chem Phys C 35:15
14. Regismond STA, Winnik FM, Goddard ED (1996) Colloids Surf A 119:221
15. Goldraich M, Schwartz JR, Burns JL, Talmon Y (1997) Colloids Surf 125:231

16. Rodenhiser AP, Kwak JCT (1998) In: Kwak JCT (ed) Polymer–surfactant systems. Dekker, New York, pp 1–20
17. Iliopoulos I, Wang K, Audebert R (1991) Langmuir 7:617
18. Holmberg C, Nilsson S, Singh SK, Sundelöf LO (1992) J Phys Chem 96:871
19. Guillement F, Piculell L (1995) J Phys Chem 99:920
20. Jönsson B, Lindman B, Holmberg K, Kronberg B (1998) Surfactants and polymers in aqueous solutions. Wiley, Chichester, p 219
21. Winnik FM, Regismond STA (1996) Colloids Surf A 118:1
22. Goddard ED, Leung PS (1992) Colloids Surf 65:221
23. Piculell L, Guillemet F, Thurresson K, Shubon V, Ericsson O (1996) Adv Colloid Interface Sci 63:1
24. Burke SE, Palepu R (2001) Carbohydr Polym 45:233
25. Majhi PR, Moulik SP, Rodgers MP, Burke SE, Palepu R (1999) J Surface Sci Technol 3:66
26. Thalberg K, Lindman B (1989) J Phys Chem 93:1478
27. Kalyanasundaran K, Thomas JK (1977) J Am Chem Soc 99:2039
28. Dong DC, Winnik MA (1985) Can J Chem 62:2560
29. Zana R (1987) Surfactant in solution: new method of investigation. Dekker, New York, p 244
30. Nilson S, Holmberg C, Sundelöf LO (1995) Colloid Polym Sci 273:83
31. Majhi PR, Moulik SP (1998) Langmuir 14:3986
32. Wang G, Olofsson G (1995) J Phys Chem 99:5588
33. Wang Y, Han H, Kwak JCT (1997) Langmuir 13:3119
34. Olofsson G, Wang G (1998) In: Kwak JCT (ed) Polymer–surfactant systems. Dekker, New York, pp 317–356
35. Zana R (1998) In: Kwak JCT (ed) Polymer–surfactant systems. Dekker, New York, pp 409–454
36. Turro NJ, Yekta A (1978) J Am Chem Soc 100:5951
37. Miyagishi S, Asakawa T, Nishida MJ (1987) J Colloid Interface Sci 115:199
38. Mwalupindi R, Rideau A, Agvaria R, Warner I (1994) Talanta 41:599
39. Rosen MJ (1978) Surfactants and interfacial phenomena. Wiley, New York, pp 125
40. Hayakawa K (1998) In: Kwak JCT (ed) Polymer–surfactant systems. Dekker, New York, pp 455–475

Progr Colloid Polym Sci (2003) 122: 56–66
DOI 10.1007/978-3-540-36114-5

K. Derek Berglund
A. E. Timko
T. M. Przybycien
R. D. Tilton

Use of nonionic ethylene oxide surfactants as phase-transfer catalysts for poly(acrylic acid) adsorption to silica against an electrostatic repulsion

Abstract Using a combination of optical reflectometry, atomic force microscopy, and streaming current measurements, we show that complexation with nonionic ethoxylated surfactants enables poly(acrylic acid) (PAA) to adsorb to negatively charged silica surfaces despite its unfavorable electrostatic interaction with the surface. In the absence of surfactant, PAA does not adsorb to silica to any extent that is measurable by our methods. We used optical reflectometry to measure the adsorption isotherm for *t*-octylphenoxypolyethoxyethanol (Triton X-100) surfactants on silica. When PAA is added to the surfactant solution, optical reflectometry reveals a significant enhancement of the total adsorbed mass below the surfactant critical micelle concentration. Atomic force microscopy confirms the presence of PAA in the mixed adsorbed layer. At lower surfactant concentrations, the presence of PAA in the adsorbed layer is manifested in the form of long-range bridging adhesion between opposing surfaces. Removing surfactant by thoroughly rinsing the mixed layer leaves an irreversibly adsorbed, "deposited" PAA layer. The force between silica surfaces that display the deposited PAA layers is dominated by a double-exponential repulsion that is of considerably longer range than the electrostatic double-layer repulsion, indicating a steric interaction between extended PAA chains. Streaming current measurements show that the magnitude of the ζ potential is less after rinsing a mixed adsorbed layer than it was before adsorption. This further supports the occurrence of surfactant-mediated deposition of PAA layers on silica. Our results suggest intriguing applications for surfactants as phase-transfer catalysts for polymers in surface-modification processes, enabling the attainment of single-component adsorbed polymer states that are inaccessible from single-component polymer solutions.

Keywords Polymer/surfactant binding · Adsorption hysteresis · Multicomponent adsorption

K.D. Berglund · A.E. Timko
T.M. Przybycien · R.D. Tilton (✉)
Department of Chemical Engineering,
Department of Biomedical Engineering,
and Center for Complex Fluids
Engineering, Carnegie Mellon University,
Pittsburgh, PA 15213, USA
e-mail: tilton@andrew.cmu.edu
Tel.: +1-412-2681159
Fax: +1-412-2687139

Introduction

A significant volume of research on the complexation of surfactants with polymers in solution may be traced to pioneering work conducted by Shuji Saito. This work has led to an increasing appreciation of the subtleties of multicomponent interactions that dictate the processing characteristics of industrial complex fluid formulations [1]. In the case of adsorption to solid or fluid interfaces, there are many opportunities to exploit polymer/surfactant interactions to engineer complex fluid behavior, including colloidal stability, suspension rheology, wetting, and electrokinetic phenomena. In this paper, we describe how polyelectrolyte/surfactant complexation can be used to deposit an adsorbed polyelectrolyte layer on an electrostatically repellant surface, and we describe how the structure of that layer responds when surfactant is removed from the system.

There is considerable experimental evidence that kinetic barriers often prevent equilibration in adsorbed polymer layers at solid/liquid interfaces, at least on practical timescales [2, 3, 4], and kinetic trapping may be the cause of heterogenities that have been observed in polymer layers [2, 5]. Physisorbed polymer layers are often hysteretic, i.e., some set of measured properties is path-dependent, and therefore the associated experimental observations do not represent a thermodynamic equilibrium state. Recognizing that hysteresis effects are magnified when polymers coadsorb with surfactants [6, 7, 8], it may be possible to design a surfactant system that directs polymer adsorption toward a persistent nonequilibrium state that displays desirable characteristics that might not be attained with a single-component polymer solution. Example design objectives would include creation of polymer steric layers with greater chain extension for colloid stabilization or for elimination of electroosmotic flow in capillary electrophoresis. Here we report on a coadsorbing polymer/surfactant system where the polymer is unable to adsorb in the absence of the surfactant. By complexing with the polymer in solution, the surfactant shuttles the polymer to the surface, and the polymer remains adsorbed even after removing surfactant from the solution. This is a case of extreme hysteresis.

Coadsorption from a polymer/surfactant mixture depends on the mutual attraction or repulsion between the solution species as well as the selectivity of the surface. By definition, a selective surface permits adsorption of one of the individual components, but not the other, from the respective single-component solution. Thus, we employ a very simple categorization system to describe polymer/surfactant/surface interactions:

1. Surfactant binds to polymer, and the surface is selective.
2. Surfactant binds to polymer, and the surface is nonselective.
3. Surfactant does not bind to polymer, and the surface is selective.
4. Surfactant does not bind to polymer, and the surface is nonselective.

For examples of each category, see Ref. [7] and references therein. Each of these categories may be further divided in the search for generalizable trends, for example, whether a selective surface favors the polymer or the surfactant or whether one or more components are ionic.

The current investigation concerns a type 1 system: a weakly charged polyelectrolyte poly(acrylic acid) (PAA), a nonionic surfactant *t*-octylphenoxypolyethoxyethanol, available commercially as Triton X-100, and a silica surface. This surface is negatively charged at a pH above its isoelectric point of approximately 1.8 to 2.8 [9, 10, 11]. Thus, silica is a selective surface at pH 3.2–3.4 where the coadsorption experiments were conducted: electrostatically repellant to PAA but attractive to ethoxylated surfactants.

Attractive interactions between PAA and nonionic surfactants of the poly(oxyethylene) type were reported as early as 1971 by Saito and Taniguchi [12]. These interactions have been studied extensively with a variety of experimental techniques, including pH titration [12, 13, 15], dye solubilization [13, 14], viscosity [12, 13, 15, 16], precipitation [17, 18, 19], and fluorescence [15, 19]. The attraction is attributed to hydrogen bonding between ether oxygens of the surfactant and protonated carboxylic acid residues of the polymer. Attraction is enhanced at a pH near 3.2–3.4 and weakens with increasing pH [12, 13, 14].

Adsorption of nonionic ethoxylated surfactants to silica, in the absence of polymers, has been studied with several experimental techniques, including atomic force microscopy (AFM) [20], optical reflectometry [21], ellipsometry [22–25], fluorescence spectroscopy [26, 27, 28] and calorimetry [29, 30, 31]. Hydrogen bonding between the ethoxylated chains and surface silanol residues is the dominant mechanism of adsorption at low surfactant concentrations. At higher concentrations, cooperative hydrophobic interactions between neighboring adsorbed surfactant molecules further promote adsorption. However, increasing the ethylene oxide headgroup to alkyl tail size ratio results in decreased adsorption owing to steric repulsions in the layer [24, 28]. At concentrations above the critical micelle concentration (cmc), the adsorbed layer for some nonionic surfactants comprises discrete aggregates [20, 24, 26, 28].

Using a combination of optical reflectometry, AFM, and streaming current measurements, we demonstrate that PAA does not adsorb to silica in the absence of surfactant, yet the nonionic surfactant shuttles PAA to the normally repellant surface. Subsequent removal of surfactant from solution leaves an irreversibly deposited PAA layer on the surface. PAA chains in the deposited layer are more highly extended than they were in the presence of the surfactant. We conclude with the suggestion that PAA is indeed surface-active on silica, but that a significant electrostatic barrier ordinarily inhibits this adsorption. Complexation with adsorbing ethoxylated surfactants perhaps weakens the electrostatic repulsion by favoring PAA protonation and/or provides an additional interfacial attraction that counteracts the electrostatic barrier. While the mechanism is uncertain, the result is that a fraction of PAA segments are deposited on the surface in an attractive interaction energy minimum. Thus, the surfactant plays the role of a phase-transfer catalyst, where the phase to which the PAA is transferred is the solid/liquid interface.

Experimental

Materials

We obtained PAA in a 25 wt% aqueous solution from Polysciences (Warrington, PA). According to the manufacturer, this PAA sample has a molecular weight of approximately 90,000. The pK_a of PAA is approximately 5.0 [13, 32]. In the pH 3.2–3.4 range encountered for PAA solutions in this study, less than 5% of the carboxyl groups are dissociated. We purchased Triton X-100 (peroxide and carbonyl free, stored under argon) from Sigma Chemical Company (St Louis, MO). This surfactant has an average of 9.5 ethylene oxide units per molecule, with an average molecular weight of 625. Pyrene (99% pure) was purchased from Aldrich Chemical (Milwaukee, WI). We purchased Chromerge cleaning solution, sodium chloride, hydrochloric acid, sulfuric acid, and toluene from Fisher Scientific (Pittsburgh, PA), sodium hydroxide pellets from EM Science (Gibbstown, NJ), and RBS 35 detergent from Pierce (Rockford, IL). All reagents were ACS grade and were used as received. Chromerge was prepared by mixing 25 ml Chromerge cleaning solution into 2.5 l 36 N sulfuric acid. We purified all water by reverse osmosis followed by treatment with the Milli-Q Plus system of ion-exchange and organic-adsorption cartridges from Millipore Corporation (Bedford, MA). The pH of solutions containing the surfactant and NaCl was unmodified from that of the air-saturated aqueous solution pH of 5.5–6.0, unless otherwise noted. No buffer salts were used.

Optical-grade silicon wafers purchased from Virginia Semiconductor (Fredericksburg, VA), with thermally grown silicon oxide surface films, served as adsorption substrates for optical reflectometry and AFM experiments. We used acid-treated soda lime glass microscope slides from Fisher Scientific in the streaming current experiments. Following Rebar and Santore [33], we first treated these glass slides with concentrated sulfuric acid (36 N) for 16 h to extract dopants and leave a surface whose composition matches that of pure silica. The cleaning procedure for all the adsorption substrates, detailed elsewhere [34], left them negatively charged and perfectly wetted by water. The substrates were stored in water and were rinsed profusely with water before being installed wet in the experimental apparatus. Only the AFM substrate was dried under nitrogen gas before being installed.

Fluorescence spectroscopy probes for polymer/surfactant interaction in solution

Pyrene is a fluorescent molecule that strongly partitions from aqueous solutions into surfactant aggregates. We exploited the well-known environmental sensitivity of pyrene fluorescence [35, 36, 37, 38] to determine the presence of surfactant aggregates in surfactant solutions both with and without PAA. Using the procedures detailed by Kim et al. [38], we detected the onset of surfactant aggregation as the surfactant concentration at which the intensity ratio of the first (371 nm) to third (382 nm) monomer vibronic emission bands, I_1/I_3, decreases sharply in an unsaturated pyrene solution. The effect is caused by the decrease in polarity of the pyrene microenvironment as it partitions from the aqueous pseudophase into surfactant aggregates. Also, we monitored the intensity ratio of the excimer emission band at 480 nm to the first monomer emission band at 371 nm, denoted as I_{ex}/I_{mon}, as it is closely related to the statistical distribution of pyrene among aggregates. We recorded fluorescence emission spectra with a SPEX Fluorolog 3 fluorescence spectrometer (Edison, NJ), using 238-nm excitation. Slit widths for excitation and emission were 5 and 3 nm, respectively. All fluorescence measurements were conducted at room temperature of 21–22 °C. The preparation of the samples for fluorescence spectroscopy has been described previously [34, 38]. For all experiments, we used a 0.5 μM pyrene concentration. This is below the pyrene aqueous solubility limit of 0.7 μM.

Adsorption measurements

We measured the amount of adsorption using optical reflectometry. The principles and applications of this technique are described elsewhere [8, 39, 40, 41, 42]. A detailed description of the scanning angle reflectometry technique that we used is available [42], and the particular instrument we used is the one described by Furst et al. [41] as modified by Velegol and Tilton [8]. All adsorption experiments were performed under laminar slit-flow conditions at a wall shear rate of approximately 3.1 s^{-1} and a constant temperature of 25 ± 0.1 °C. As a check against surface-active impurities, the reflectivity of each clean wafer was monitored as it soaked in 1.0 mM NaCl solution for at least 15 min before introducing polymer and/or surfactant solutions. We monitored the adsorption kinetics by recording the change in reflectivity of the p-polarized laser beam while the angle of incidence was fixed at the Brewster angle. The change in the square root of the reflection coefficient before and after adsorption under this condition is proportional to the effective optical thickness of the adsorbed layer, defined later. Steady-state surface excess concentrations were determined from a scanning angle regression of the reflection coefficient, $R_p(\theta)$, after attaining the adsorption plateau.

We evaluated the optical properties of the adsorbed layers using a homogeneous two-layer optical model, where the semi-infinite solution and bulk silicon are separated by an oxide layer of known refractive index and thickness and by an adsorbed layer of unknown refractive index and thickness. The refractive indices at the 632.8 nm He–Ne laser wavelength are $n_{Si} = 3.882 + 0.019i$ for bulk silicon and $n_{ox} = 1.46$ for the silicon oxide layer. Using the Abelès matrix method [43] to numerically evaluate the two-layer model, we fitted the measured reflectivity profile $R_p(\theta)$ via χ^2 minimization to obtain the optical average thickness, d_{av}, and the optical average refractive index, n_{av}, of the adsorbed layer. We know the oxide layer thickness because we measured it by scanning angle reflectometry prior to the start of adsorption in every experiment. Although the fitted d_{av} and n_{av} for the adsorbed layer are model-dependent and highly coupled, the effective optical thickness of the adsorbed layer, $d_{av}(n_{av}-n_0)$, where n_0 is the refractive index of the bulk solution, is model-independent.

For adsorption of an arbitrary number of independent components, the effective optical thickness is related to the surface excess concentration of each species, Γ_i, by

$$d_{av}(n_{av} - n_0) = \sum_i \Gamma_i \frac{dn}{dc_i}, \qquad (1)$$

where dn/dc_i is the refractive index increment of the adsorbing species i at the wavelength of interest. Using a Phoenix Precision Instrument Company differential refractometer, we determined that the refractive index increments of the surfactant and PAA at the He–Ne laser wavelength are 0.155 and 0.157 $cm^3\ g^{-1}$, respectively. Since reflectometry does not directly distinguish the relative amounts of each component in a mixed adsorbed layer, we used the average of the refractive index increments for the two components, 0.156 $cm^3\ g^{-1}$, and report the total surface excess concentration without regard to layer composition in this work [42].

Streaming current

We used an electrokinetic technique to independently test the adsorption and desorption of surfactant and polymer/surfactant mixtures by measuring their effect on the interfacial ζ potential. In our apparatus, we exploit the steady-state electrical current, i.e., the streaming current, that develops as a result of pressure-driven flow over a charged surface [34, 44, 45]. The ζ potential dictates the slope of the streaming current, I_s, versus pressure drop, ΔP, via

$$\zeta = -\frac{\mu L}{\varepsilon_0 \varepsilon w h}\left(\frac{I_s}{\Delta P}\right), \tag{2}$$

where μ is the viscosity of the solution, ε_0 is the permittivity of free space, ε is the dielectric constant of the solvent, and L, w, and h are the length, width, and height of the rectangular slit flowcell ($L > w >> h$). The upper and lower walls of the flowcell are formed by two glass slides. Instrumental and theoretical issues, particularly concerning the advantage of streaming current over streaming potential measurements, are described elsewhere [44]. The streaming current experiments were conducted at room temperature, approximately 22–23 °C.

Atomic force microscopy

Force measurements were performed with a Nanoscope IIIa atomic force microscope from Digital Instruments (now Veeco Metrology, Santa Barbara, CA), equipped with a fluid cell of approximate volume 1.0 ml. Gold-coated silicon nitride cantilevers used in the AFM experiments were obtained in wafer form from Thermomicroscopes (Sunnyvale, CA). All cantilevers used in this study came from the same line on the same wafer and had manufacturer-specified spring constants of 0.01 or 0.1 N m^{-1}. The cantilevers were first cleaned under a low-pressure mercury vapor grid lamp for 30 min in a UV–ozone cleaner, model 42 manufactured by Jelight Company (Irvine, CA). Following this the cantilevers were modified by attaching a 5-μm diameter colloidal silica sphere from Bangs Laboratories (Fisher, IN) at the apex of the cantilever using a UV curable adhesive, Norland optical adhesive UVE 61 (New Brunswick, NJ), following the procedure of Ducker et al. [46]. A detailed description of the procedure is available [47].

Before an experiment, the cantilevers were again cleaned in the UV–ozone chamber for a minimum of 30 min. After assembling the fluid cell with the cantilever and wafer in place, it was filled with a 10 mM NaOH solution for 30 min. This final surface preparation step was to ensure the hydrophilicity and negative charge. Next, the fluid cell was flushed with 1.0 mM NaCl and allowed to equilibrate for 1 h before force curves were recorded for the bare surfaces. Then, surfactant solutions or polymer/surfactant mixtures were allowed to adsorb for 3 h before measuring force curves. Whenever the solution composition was adjusted, we passed at least 100 fluid cell volumes of the new solution through the fluid cell to ensure complete exchange of solution contents. The experiments were conducted at room temperature, approximately 23 °C, with the temperature being recorded in all cases.

Detailed descriptions are available elsewhere for the use of AFM to measure surface forces [46, 48]. Briefly, a surface of interest is mounted to a piezoelectric ceramic. The piezo is translated with angstrom-level precision in the vertical direction until surface forces cause a deflection in the cantilever, as monitored by reflection of a laser from the cantilever onto a position-sensitive photodiode. Force measurements are made by maintaining the probe at one horizontal location adjacent to the surface and using the piezo to vertically translate the surface with an oscillatory motion. To facilitate data analysis, the amplitude of this oscillatory motion is chosen so that the probe samples both large probe-to-surface separation distances where the interaction force is so weak that there is no cantilever deflection, as well as very small distances over which the probe moves in constant compliance with the surface (i.e., where the cantilever deflection increases linearly with respect to the piezo motion). The force is calculated by measuring the cantilever deflection and applying Hooke's law with the cantilever spring constant.

Deflection versus piezo travel data profiles were converted to force (normalized by colloidal probe radius of curvature) versus separation distance profiles using a software algorithm developed in Mathematica 4.1 by Dagastine [47]. The zero-force baseline was calculated by applying a zero-slope line to the portion of the force curve that exhibited no deflection and subtracting the intercept from the data set. A linear fit to the constant compliance regime was calculated and the zero-separation distance was defined as the position corresponding to the onset of constant compliance. Displacement from this zero point is referred to as the apparent separation distance.

Results and discussion

Polymer/surfactant interaction in solution

The I_1/I_3 pyrene fluorescence intensity ratios are displayed in Fig. 1a as a function of the surfactant concentration for 1.0 mM NaCl solutions, both in the

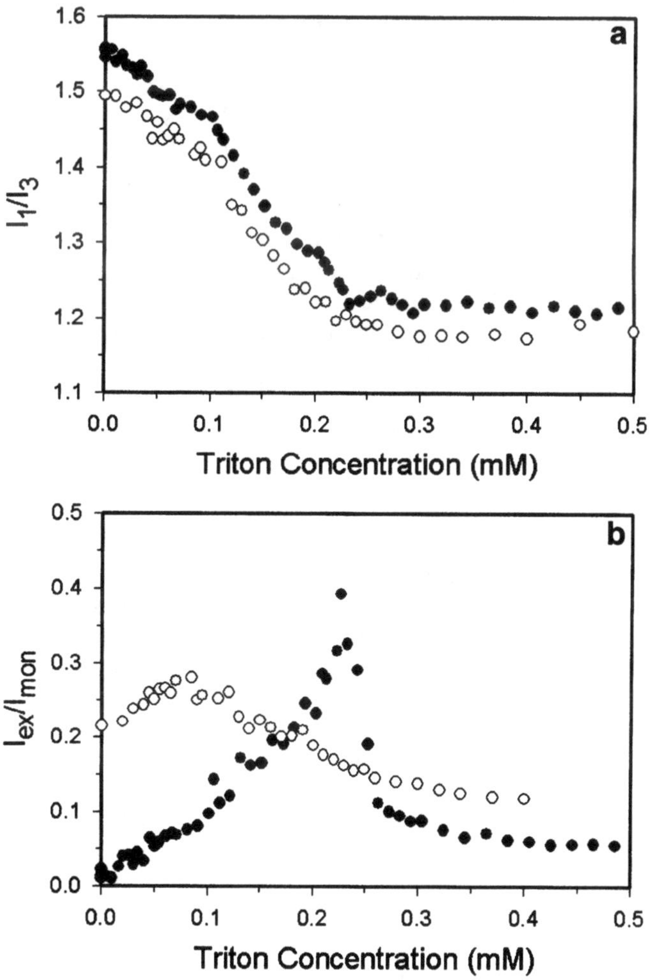

Fig. 1 **a** I_1/I_3 fluorescence ratios and **b** I_{ex}/I_{mon} fluorescence ratios for solutions containing Triton X-100 and 1.0 mM NaCl (*filled circles*) and Triton X-100, 0.1 wt% poly(acrylic acid) (*PAA*), and 1.0 mM NaCl (*open circles*)

absence of PAA and in the presence of 0.1 wt% PAA. Without surfactant, the I_1/I_3 ratio is 1.56 in the absence of PAA and 1.49 in the 0.1 wt% PAA solution. The lower I_1/I_3 ratio in the PAA solution indicates that there is already some interaction between the PAA and pyrene in the absence of any surfactant. At a surfactant concentration of approximately 0.02 mM, the I_1/I_3 ratio begins to decrease both in the presence and in the absence of PAA. This indicates that pyrene is partitioning into a less polar environment as surfactant is added to the solution. Above 0.02 mM, the I_1/I_3 ratios with and without PAA behave very similarly. In the absence of PAA, the I_1/I_3 ratio becomes fairly constant at 1.22 ± 0.01 above 0.23 mM surfactant. With PAA present, the I_1/I_3 ratio becomes fairly constant at 1.19 ± 0.01 above 0.22 mM surfactant. The pyrene microenvironmental polarity is therefore constant above these threshold surfactant concentrations.

The observation that the I_1/I_3 ratio decreased abruptly at 0.02 mM surfactant concentration both in the presence and in the absence of PAA was unexpected. The abrupt decrease in I_1/I_3 is usually taken as the cmc or critical aggregation concentration in surfactant solutions or polymer/surfactant mixtures [19, 34, 38, 49, 50], and I_1/I_3 usually remains constant at concentrations below the cmc. Yet, 0.02 mM is much lower than the cmc reported for this surfactant in the absence of polymers. The cmc is reported to be in the range of 0.22–0.24 mM [51, 52]. This is an indication of premicellar surfactant assembly. Prior theoretical calculations [53] and experimental results from sedimentation equilibrium [54], birefringence [55], NMR [56] and light or X-ray scattering [56, 57] studies indicate that Triton X-100 micelles are ellipsoidal and increase in size with increasing concentration. The I_1/I_3 behavior suggests that small premicellar clusters begin to form at 0.02 mM. Since a similar transition occurred both in the absence and in the presence of PAA, the I_1/I_3 ratio alone is insufficient to detect the onset of polymer/surfactant association.

The excimer fluorescence behavior provides stronger evidence for polymer/surfactant association below the cmc. The I_{ex}/I_{mon} ratios are displayed in Fig. 1b as a function of the surfactant concentration for the same solutions as just described. We note that excimer emission is promoted by PAA in the absence of surfactant, another indication that pyrene associates with PAA. A maximum in the I_{ex}/I_{mon} ratio occurs when all of the available pyrene in solution has partitioned into surfactant or polymer/surfactant aggregates. In the absence of PAA, the I_{ex}/I_{mon} maximum occurs at 0.23 mM surfactant concentration. With PAA present, the maximum occurs at a much lower surfactant concentration, approximately 0.08 mM.

Since the pyrene concentration is the same in both sets of data, the earlier occurrence of the I_{ex}/I_{mon} maximum suggests that the aggregates that form at low surfactant concentrations have significantly larger solubilizing capacities for pyrene in the presence of PAA than they do in its absence. Solubilizing capacity refers to the maximum number of solubilizates that can be borne by one aggregate. Thus, the polymer-associated aggregates are probably larger than the premicellar aggregates that form in the absence of polymer.

Adsorption on silica from single-component solutions

Electrostatically driven PAA adsorption has been studied on positively charged surfaces such as alumina [58, 59] and kaolinite platelet edges [60]. PAA has also been found to be weakly surface active at the air/water interface [15, 61, 62]. Using optical reflectometry we detected no adsorption from 0.1 wt% PAA in 1.0 mM NaCl solutions onto negatively charged silica over the course of 1 h. Displacing a PAA solution from the flowcell by a 1.0 mM NaCl rinse left no deposit behind, to within the 0.05 mg m^{-2} resolution limit of our reflectometer. Our AFM experiments also produced no evidence of adsorption on silica from 1.0 mM NaCl solutions containing 0.1 wt% PAA, i.e., the force curves measured in 1.0 mM NaCl before incubation, after a 1-h incubation in the PAA solution, and after a 1.0 mM NaCl rinse were identical. Thus, our results are consistent with previously published studies using mica or charged polystyrene latex: PAA does not adsorb to negatively charged surfaces [63, 64].

Optical reflectometry measurements with surfactant solutions containing 1.0 mM NaCl produced the adsorption isotherm on silica, plotted as filled circles in Fig. 2. Surfactant adsorption was in all cases reversible, and 2 h was sufficient to attain equilibrium for all surfactant concentrations that we examined below 0.2 mM (near the commonly reported cmc). Above 0.2 mM, adsorption required less than 5 min to establish equilibrium. The isotherm plateau, with a maximum average surface excess concentration of 2.30 mg m^{-2} or 2.2 molecules nm^{-2}, begins just above 0.2 mM. This suggests that a concentration of approximately 0.2 mM satisfies the definition of the cmc as the surfactant concentration above which the surfactant chemical potential remains constant, although the fluorescence results reveal the existence of premicellar aggregation in this system. This adsorption isotherm is consistent with that previously reported by Levitz et al. [26] and is characteristic of surfactants with weak interaction energies between the headgroups at the interface [24]. Since the pH of solutions containing PAA ranged from 3.2 to 3.4, we repeated several surfactant adsorption measurements for low surfactant concentrations at pH 3.2 by adding a small amount of concentrated HCl to the surfactant solutions. These results, plotted as triangles in Fig. 2, are indistinguishable from the measurements made at pH 5.5. Thus, the surfactant adsorption is not sensitive to pH changes between the natural pH

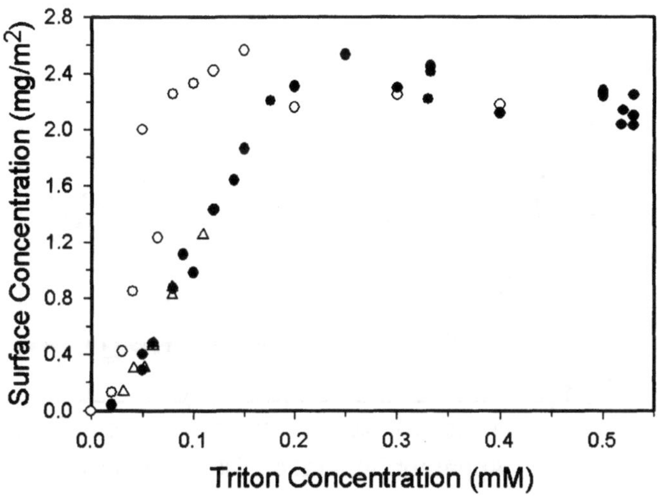

Fig. 2 Adsorption isotherm on the silica surface for Triton X-100 solutions in 1.0 mM NaCl with an unmodified pH of 5.5 (*filled circles*), Triton X-100 solutions in 1.0 mM NaCl acidified to pH 3.2 (*triangles*), and solutions containing Triton X-100, 0.1 wt% PAA in 1.0 mM NaCl (*open circles*)

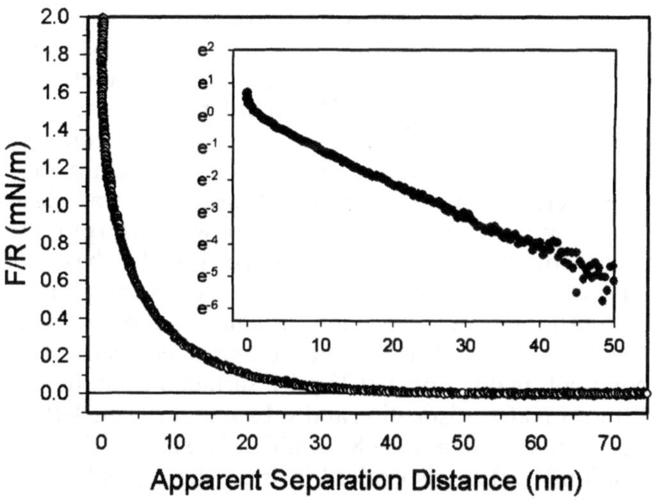

Fig. 3 Advancing (*filled circles*) and retracting (*open circles*) forces normalized by colloidal probe radius measured between a silica surface and a colloidal silica probe in 1.0 mM NaCl after adsorption of 0.21 mM Triton X-100. The nominal spring constant is 0.1 N m^{-1}. *Inset*: semilog plot of the advancing force curve displaying a decay length of 9.2 nm

of the surfactant solution in unbuffered, air-saturated water and the pH of 0.1 wt% PAA solutions. This will be important when interpreting the coadsorption results in the next section.

The forces between the colloidal silica probe and the silica surface attained via AFM after exposure to a pH 5.5 solution containing 0.21 mM surfactant and 1.0 mM NaCl are displayed in Fig. 3. A monotonically decaying repulsive force dominates both the advancing and retracting force curves. After desorbing the adsorbed surfactant layer by multiple exchanges with 1.0 mM NaCl solutions, the approaching and retracting force curves were identical to the force curves shown in Fig. 3. As illustrated in the inset, a semilog plot of the advancing force curves measured in the presence of the surfactant after adsorption reveals a hard wall repulsion followed by single-exponential decay. Averaging over multiple independent force curves yields a decay length of 9.1 ± 0.5 nm. This is consistent with the 9.6 nm Debye length expected for electrostatic double-layer repulsion in a 1.0 mM NaCl solution. Before surfactant adsorption, the measured exponential decay length was 9.4 ± 1 nm for bare silica interactions. Thus, double-layer repulsion is the dominant long-range force between the surfaces before adsorption, in the presence of the adsorbed surfactant layer, and after desorption of the layer.

Coadsorption on silica from polymer/surfactant mixtures

The coadsorption isotherm for solutions containing 0.1 wt% PAA and 1.0 mM NaCl with varying surfactant concentrations is plotted as open circles in Fig. 2. Adsorption from the polymer/surfactant mixtures was reversible to within the detection limit of our reflectometer. There was also no difference between the equilibration times in the absence or presence of PAA. Between surfactant concentrations of 0.04 and 0.15 mM, the total adsorbed mass is much greater in the presence of PAA than in its absence. The total coadsorbed mass plateaus at 2.23 mg m^{-2} for surfactant concentrations exceeding approximately 0.2 mM, and there is a slight overshoot in adsorption at approximately 0.1 mM.

Although the increase in total adsorbed mass below 0.2 mM probably indicates that PAA is present in the adsorbed layer, optical reflectometry does not provide composition information for mixed layers. For independent verification that PAA is indeed shuttled to the surface by the surfactant, we first report the results of streaming current measurements. The streaming current responds to any surfactants or polymers, either ionic or nonionic, adsorbed at the silica–solution interface. These molecules shift the hydrodynamic shear plane away from the surface and/or add charge to the interfacial region, causing the ζ potential to change. Although we previously have used the measurement of the ζ potential before and after nonionic polymer adsorption to determine the adsorbed layer thickness [3, 34, 44, 45] the presence of ionic molecules such as PAA at the interface complicates the interpretation of the ζ potential in terms of layer thickness. Therefore, we utilized streaming current measurements primarily to compare the adsorption reversibility for surfactant solutions and polymer/surfactant mixtures.

As a check against surface-active impurities, we first passed 500 ml (4,500 flowcell volumes) 1.0 mM NaCl solution through the streaming current flowcell. We measured I_s as a function of ΔP in order to calculate the ζ potential of the bare surface in 1.0 mM NaCl solution. Then, surfactant solutions or polymer/surfactant mixtures were pumped through the flowcell at a wall shear rate of approximately 110 s^{-1}. We allowed adsorption to occur for more than 3 h, judged on the basis of reflectometry experiments to be more than adequate to establish plateau adsorption. Finally, we rinsed with 500 ml 1.0 mM NaCl solution and measured the ζ potential again. For a 0.22 mM surfactant solution in 1.0 mM NaCl, adsorption and subsequent rinsing left the ζ potential unchanged: −88.3 mV before adsorption and −90.3 mV after adsorption/rinsing. The typical reproducibility in measuring the ζ potential was approximately ± 5 mV, leading us to conclude that the surfactant adsorption on silica is completely reversible. For comparison, we examined a polymer/surfactant mixture that contained 0.1 wt% PAA and 0.1 mM surfactant in 1.0 mM NaCl, since this mixture produces the same total adsorbed amount as the 0.22 mM surfactant solution. The streaming current indicates that adsorption from this polymer/surfactant mixture, followed by rinsing, leaves a deposit on the surface. The process of mixed coadsorption followed by rinsing changed the ζ potential from −90.3 to −56.8 mV in the presence of 1.0 mM NaCl. There is clearly a remnant, and since the surfactant adsorbs reversibly, the remnant must be PAA. The fact that the ζ potential becomes less negative in the presence of the PAA indicates that the effect of adsorption on the ζ potential is dominated by the outward shift in the hydrodynamic shear plane rather than the addition of negatively charged PAA segments to the interface.

AFM measurements support the conclusion drawn from the streaming current experiments that the surfactant deposits PAA on the silica surface. The advancing and retracting force curves between the silica surfaces after adsorption from a 1.0 mM NaCl solution containing 0.1 wt% PAA and 0.1 mM surfactant are displayed in Fig. 4. This surfactant concentration is just below the plateau on the coadsorption isotherm. The force curves were measured with both the polymer and the surfactant present in solution. A detectable repulsion begins at an apparent separation distance of approximately 30 nm on the advance. Plotting the advancing force curve on a semilog scale (inset) reveals a hard wall repulsion followed by a single-exponential decay. Averaging multiple force curves yields a decay length of 6.6 ± 0.6 nm. Given that the solution is approximately pH 3 and contains 1.0 mM NaCl, providing a total ionic strength of 2 mM, the expected Debye length would be 6.7 nm. The measured long-range force is therefore consistent with electrostatic double-layer repulsion.

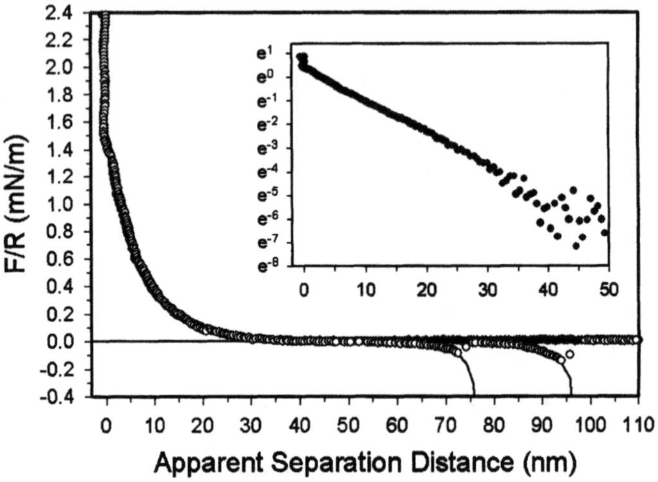

Fig. 4 Advancing (*filled circles*) and retracting (*open circles*) forces normalized by colloidal probe radius measured between a silica surface and a colloidal silica probe after adsorption of 0.1 wt% PAA and 0.1 mM Triton X-100 in a pH 3 1.0 mM NaCl solution. The nominal spring constant is 0.1 N m^{-1}. Attractive bridging forces occurred in 15% of the retracting force curves. The wormlike chain fits for the adhesive wells are also shown. *Inset*: semilog plot of the advancing force curve displaying a decay length of 7.4 nm

Those advancing force curves contain no evidence for an extended PAA layer in the presence of the surfactant. Nevertheless, the force curves measured on retraction do indicate the presence of PAA in the layers. Significant hysteresis existed between the advancing and retracting force curves, with one or more attractive minima appearing in approximately 15% of the retraction force curves. These attractive forces are probably due to the bridging of individual polymer molecules between the two interfaces. These adhesive wells were observed to occur anywhere from 40 to 150 nm in apparent separation distance. Instances of multiple adhesive wells in a single force curve may be caused by multiple polymer chains spanning the two interfaces, or they could be caused by a single bridging chain that has more than one tethering point on one or both of the interfaces. The form of the adhesive wells is quantitatively consistent with polymeric bridging forces.

We compared the adhesive forces with two entropy-based statistical models of polymer stretching: the freely jointed chain (FJC) model [65] and the wormlike chain (WLC) model [66]. In the FJC model, the polymer comprises independent segments of length l joined in linear succession and connected by freely rotating pivots with equal probability for rotation in all directions. The force, F_{st}, needed to stretch a FJC to a length D is given as

$$F_{st}(D) = \frac{k_b T}{l} L^{-1}\left(\frac{D}{L_c}\right), \tag{3}$$

where k_b is the Boltzmann constant, T is temperature, L_c is the polymer contour length and L^{-1} is the inverse Langevin function [67].

The WLC model considers the polymer chain to be continuously curved with a trajectory that varies smoothly at any point along the chain, with a stiffness characterized by a persistence length b. The force required to stretch a WLC of contour length L_c to a given length D is

$$F_{st}(D) = \frac{k_b T}{b}\left(\frac{D}{L_c} + \frac{1}{4\left(1-\frac{D}{L_c}\right)^2} - \frac{1}{4}\right). \quad (4)$$

We regressed adhesive wells against both the FJC and WLC models using the contour length and either the segment length (FJC) or persistence length (WLC) as fitting parameters. In order for the FJC model to adequately reproduce the form of the adhesive well, the predicted contour length was less than the measured detachment distance (release from the bridging adhesion) in all cases. Thus, the contour lengths from the FJC model will not be discussed since the contour length cannot be less than the detachment distance.

The WLC model consistently gave a satisfactory fit to the adhesive well, and the measured detachment occurred in all cases at 91 ± 2% of the regressed contour length. For all the adhesive wells observed, the regressed contour lengths were greater than 40 nm, with approximately 70% less than 80 nm. The persistence length was of the order of 5 Å, but we note that according to Eq. (4), the persistence length is sensitive to the magnitude of the force, whereas the contour length is sensitive to the decay of the force. The distance decays of the reported force curves are accurate, but since we used nominal spring constants and nominal colloidal probe radii to interpret force data, the magnitudes could easily be in error. Therefore, our regressed contour lengths, but not the persistence lengths, have physical meaning in terms of polymer bridging. Other studies of polymer bridging have also found the distance of detachment to be approximately 90% of the WLC-regressed contour length [67, 68, 69].

This analysis supports the claim that PAA is responsible for bridging. What is the nature of the PAA interaction with a surfactant-covered surface that produces bridging? Increasing the surfactant concentration to a value well into the coadsorption isotherm plateau eliminates the bridging attraction. The advancing and retracting force curves after adsorption from a 1.0 mM NaCl solution containing 0.1 wt% PAA and 0.55 mM surfactant are displayed in Fig. 5. As earlier, the PAA and surfactant remained in solution during the force measurements. A repulsive single-exponential decay was again the dominant feature of the force curve. By averaging multiple force curves, the decay length is 6.4 ± 0.4 nm, again consistent with electrostatic double-layer repulsion as the dominant long-range force. No hysteresis was observed in any of the force curves

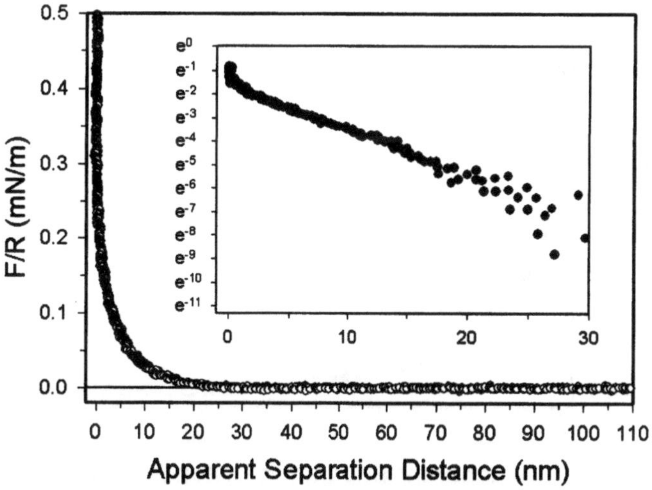

Fig. 5 Advancing (*filled circles*) and retracting (*open circles*) forces normalized by colloidal probe radius measured between a silica surface and a colloidal silica probe in 1.0 mM NaCl after adsorption of 0.1 wt% PAA and 0.55 mM Triton X-100 in a pH 3 1.0 mM NaCl solution. The nominal spring constant is 0.01 N m^{-1}. *Inset*: semilog plot of the advancing force curve displaying a decay length of 6.9 nm

collected at this surfactant concentration, indicating that polymer bridging did not occur.

A result to be shown later indicates that an adsorbed PAA chain cannot attach directly to an opposing silica surface, indicating that PAA/surfactant interactions produce bridging. Nevertheless, bridging does occur below the coadsorption plateau but not in the plateau. We therefore hypothesize that PAA bridging is promoted by defects in the adsorbed surfactant layer. A fluorescence decay study by Levitz et al.[26] indicated that at concentrations between 0.2cmc and 0.65cmc a Triton X-100 adsorbed layer on silica consists of isolated surface aggregates. Above the cmc (and thus above the adsorption isotherm plateau), they found that the adsorbed layer could be described as an assembly of close-packed adsorbed micelles. For the present system, the 0.1 mM surfactant concentration is 0.43cmc and is below the coadsorption plateau. If discrete surface aggregates exist, PAA may preferentially interact with the edges of the adsorbed surfactant aggregates. A PAA tail extending from one surface may access the aggregate edges on the opposing surface. The 0.55 mM surfactant concentration is 2.4cmc and is well within the coadsorption plateau. The close packing of adsorbed surfactant aggregates would decrease the probability of a PAA tail locating a defect on the opposing surface and thus would decrease the probability of bridging.

These results confirm that the surfactant shuttles PAA to the interface, yet the most remarkable effects are observed after we rinse the mixed adsorbed layers with a 1.0 mM NaCl solution that contains no surfactant or PAA. Two typical advancing force curves measured after rinsing a coadsorbed layer are displayed in Fig. 6a.

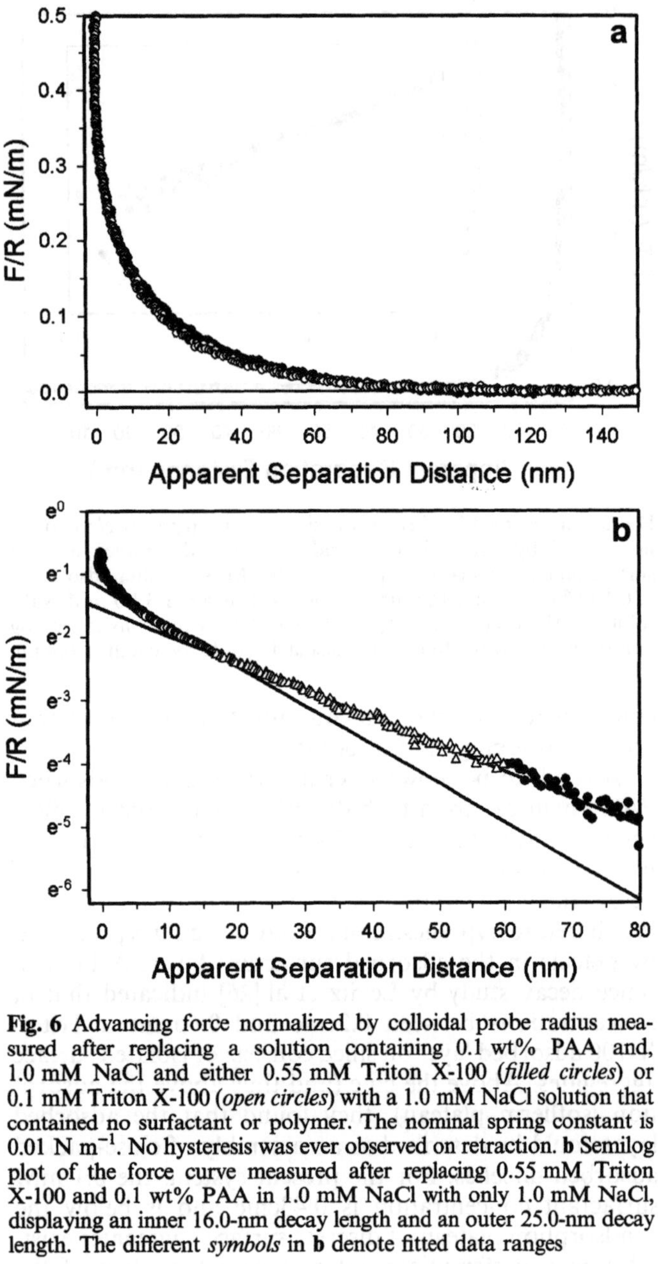

Fig. 6 Advancing force normalized by colloidal probe radius measured after replacing a solution containing 0.1 wt% PAA and, 1.0 mM NaCl and either 0.55 mM Triton X-100 (*filled circles*) or 0.1 mM Triton X-100 (*open circles*) with a 1.0 mM NaCl solution that contained no surfactant or polymer. The nominal spring constant is 0.01 N m^{-1}. No hysteresis was ever observed on retraction. **b** Semilog plot of the force curve measured after replacing 0.55 mM Triton X-100 and 0.1 wt% PAA in 1.0 mM NaCl with only 1.0 mM NaCl, displaying an inner 16.0-nm decay length and an outer 25.0-nm decay length. The different *symbols* in **b** denote fitted data ranges

The adsorbed layers were initially formed from a 1.0 mM NaCl solution containing 0.1 wt% PAA and either 0.1 or 0.55 mM surfactant. The force curves were measured in 1.0 mM NaCl solution, with no polymer or surfactant present in solution. Advancing force curves were hardly distinguishable for the two surfactant concentrations and neither case displayed any hysteresis. There was no bridging attraction after rinsing, so evidently a deposited PAA chain on one silica surface could not adsorb to the opposing silica surface in the absence of surfactant.

Compared with the force curves measured with PAA and surfactant present in solution shown in Fig. 5, a considerably longer range repulsion is evident after rinsing. The onset of repulsion varied from approximately 80 to 90 nm (depending on the experiment) and no trend with initial surfactant concentration was observed. Given that the "onset of repulsion" depends on the atomic force microscope sensitivity and therefore on the spring constant, one must examine the force decay to better compare the range of the repulsion.

The semilog plot of the advancing force curve of a rinsing experiment when the coadsorbed layer was originally formed with 0.55 mM surfactant is displayed in Fig. 6b. Regressing the linear regimes of the semilog plots consistently indicates a double-exponential decay. Averaging the regression results from multiple experiments at both surfactant concentrations yields an inner decay length of 16.7 ± 1.1 nm and an outer decay length of 24.6 ± 0.8 nm. Therefore, the dominant long-range repulsion in these 1.0 mM NaCl solutions is clearly not electrostatic in origin, and it may be attributed to polymer steric forces.

The Alexander–de Gennes [70] scaling model for end-grafted nonionic polymer brushes is often successful in describing steric forces in terms of the layer thickness, even for physisorbed homopolymers that do not form brushes. The appropriateness of the Alexander-de Gennes force model predictions for non-brush-forming homopolymers was verified by Luckham [71]. This model predicts that the repulsive force curve should have a power law exponent of −2.25 in the polymer chain overlap (osmotic pressure) regime and −0.75 in the polymer compression (elastic deformation) regime. When plotted on log–log scales, the PAA force curves display more than one power-law regime. Log–log plots of multiple postrinsing force curves, for either surfactant concentration case, yield power-law exponents of −0.5 and −1.2 for the inner and outer regimes, respectively. So, it would be inappropriate to fit force curves to the Alexander–de Gennes model to extract layer thicknesses, even though the PAA is very weakly charged. The polyelectrolyte character of PAA must be taken into account.

Zhulina et al. [72] have modeled the static forces in confined layers of weak polyelectrolytes. PAA falls into the category of "annealing" polyelectrolytes, where the ionization equilibrium for the adsorbed polyelectrolyte is a function of the separation distance. The model identifies three distinct force regimes for annealing polyelectrolytes. Upon approach of the surfaces, a quasi-neutral regime should display a −3 power-law exponent for force versus separation. In the salt-dominance regime, interaction of the counterion clouds of the adsorbed polyelectrolytes should produce a −2 power-law exponent. Lastly, highly compressed layers enter the osmotic regime where the force should scale with distance to the −0.5 power. This matches the −0.5 power in the inner regime of our force measurements. Our outer-regime power-law scaling does not match the model predictions. We believe this is due to the sparse nature of the deposited PAA layer.

The annealing polyelectrolyte force model was strictly developed for brushes, but the authors noted that the force law in the osmotic regime, where chains are highly compressed, should also apply to nonbrush layers. The model conditions for the outer regimes are certainly not satisfied by the sparse deposited PAA layers in our experiments. Although deposition was demonstrated by the AFM and streaming current measurements, optical reflectometry did not detect a measurable PAA remnant surface concentration. The detection limit of our optical reflectometer is approximately 0.05 mg m^{-2}, or at least 3,000 nm^2 per PAA molecule. Since our PAA has a radius of gyration of approximately 35 nm [62], this minimum value for the area per molecule would indicate little if any of the interchain packing that normally leads to stretching in a conventional brush.

Finally, we must address the difference in the long-range advancing forces between mixed layers measured before rinsing (electrostatic double-layer repulsion) and those measured after rinsing (rather long range steric repulsion). Since the PAA is attracted to the adsorbed surfactant layer, it is probably in a relatively flat conformation in the presence of surfactant, intimately mingling polymer segments with adsorbed surfactant aggregates. When the surfactant is removed from the system and the additional attraction of the PAA to the interfacial region disappears, electrostatic repulsion of charged monomers from the surface probably stretches the adsorbed chains. The anchorage to the surface is almost certainly due to protonated PAA segments. Protonation would be enhanced by the negative electrostatic potential of the silica surface.

Conclusions

PAA is electrostatically repelled from the negatively charged silica surface and cannot adsorb in the absence of surfactant. Triton X-100 surfactant adsorbs readily and reversibly to silica. The favorable interaction between this ethoxylated surfactant and PAA, observable in bulk solution, allows the surfactant to shuttle PAA to the interface. This shuttling of a polyelectrolyte to an otherwise repellant surface was independently confirmed by optical reflectometry, streaming current, and AFM techniques, each of which is based on entirely distinct physics.

Observations of polymer bridging adhesion in the presence of mixed layers, and the dependence on surfactant concentration, suggest that PAA adsorbs preferentially at defects in the adsorbed surfactant layer. PAA molecules that are deposited on silica by coadsorption with the surfactant remain on the surface after surfactant is removed from the system. In the absence of the surfactant, these deposited PAA chains introduce a long-range steric repulsion between the surfaces, and they are unable to form bridging contacts with an opposing silica surface.

When considered together, all of our results suggest that PAA is indeed surface-active on silica, even though one cannot observe direct adsorption from a single-component PAA solution. In a surfactant-free solution, the long-range electrostatic repulsion presents an effectively insurmountable energy barrier against adsorption. In the presence of the ethoxylated surfactant, PAA adsorbs, perhaps owing to the addition of an attractive force that compensates for the electrostatic repulsion, or by a surfactant-induced decrease in the degree of ionization of PAA segments. The persistence of the PAA deposit after exhaustive rinsing indicates that at least some of the PAA is irreversibly adsorbed. Yet the inability of the irreversibly adsorbed, postrinse PAA deposit to make bridging contact with an opposing silica surface in the absence of surfactant indicates the strength of the electrostatic barrier against "unshuttled" PAA contact with silica. Thus, the surfactant functions as a phase-transfer catalyst, delivering PAA to the solid/liquid interface. This use of surfactants to deliver polymers to normally nonadsorptive surfaces suggests new types of polymeric surface modification strategies.

Acknowledgements Acknowledgement is made to the donors of The Petroleum Research Fund, administered by the American Chemical Society, for partial support of this research. This material is also based in part on work supported by National Science Foundation grants CTS-9623849 and 9871110 and by the Pennsylvania Infrastructure Technology Alliance. We thank Ijeoma Nnebe and Ray Dagastine of the Carnegie Mellon Department of Chemical Engineering for help with AFM.

References

1. Goddard ED, Ananthapadmanabhan KP (1993) Interactions of surfactants with polymers and proteins. CRC, Boca Raton
2. Schneider HM, Frantz P, Granick S (1996) Langmuir 12:994
3. Pagac ES, Prieve DC, Solomentsev Y, Tilton RD (1997) Langmuir 13:2993
4. Sukhishvili SA, Dhinojwala A, Granick S (1999) Langmuir 15:8474
5. Kelley TW, Schorr PA, Johnson KD, Tirrell M, Frisbie CD (1998) Macromolecules 31:4297
6. Dedinaite A, Claesson PM, Bergström M (2000) Langmuir 16:5257
7. Braem AD, Prieve DC, Tilton RD (2001) Langmuir 17:883
8. Velegol SB, Tilton RD (2001) Langmuir 17:219

9. Iler RK (1979) The chemistry of silica: solubility, polymerization, colloid and surface properties and biochemistry. Wiley, New York
10. Scales PJ, Grieser F, Healy TW, White LR, Chan DYC (1992) Langmuir 8:965
11. Hartley PG, Larson I, Scales PJ (1997) Langmuir 13:2207
12. Saito S, Taniguchi T (1971) Kolloid Z Z Polym 248:1039
13. Saito S, Taniguchi T (1973) J Colloid Interface Sci 44:114
14. Saito S (1989) J Am Oil Chem Soc 66:987
15. Anghel DF, Winnik FM, Galatanu N (1999) Colloids Surf A 149:339
16. Anghel DF, Saito S, Baran A, Iovescu A (1998) Langmuir 14:5342
17. Saito S, Taniguchi T, Matsuyama H (1976) Colloid Polym Sci 254:882
18. Saito S (1977) Tenside 14:113
19. Vasilescu M, Anghel DF, Almgren M, Hansson P, Saito S (1997) Langmuir 13:6951
20. Grant LM, Tiberg F, Ducker WA (1998) J Phys Chem B 102:4288
21. Luokkala BB, Garoff S, Tilton RD, Suter RM (2001) Langmuir 17:5917
22. Tiberg F, Joesson B, Lindman B (1994) Langmuir 10:3714
23. Tiberg F, Joensson B, Tang J-A, Lindman B (1994) Langmuir 10:2294
24. Tiberg F (1996) J Chem Soc Faraday Trans 92:531
25. Brinck J, Joensson B, Tiberg F (1998) Langmuir 14:1058
26. Levitz P, Van Damme H, Keravis D (1984) J Phys Chem 88:2228
27. Levitz P, Van Damme H (1986) J Phys Chem 90:1302
28. Somasundaran P, Krishnakumar S (1997) Colloids Surf A 123:491
29. Gu T, Zhu B (1990) Colloids Surf 44:81–87
30. Lindheimer M, Keh E, Zaini S, Partyka S (1990) J Colloid Interface Sci 138:83
31. Narkiewicz-Michalek J, Rudzinski W, Keh E, Partyka S (1992) Colloids Surf 62:273
32. Kono N, Ikegami A (1966) Biopolymers 4:823
33. Rebar VA, Santore MM (1996) J Colloid Interface Sci 178:29
34. Berglund KD, Przybycien TM, Tilton RD (2002) (manuscript in preparation)
35. Turro NJ, Baretz BH, Kuo PL (1984) Macromolecules 17:1321
36. Zana R, Lianos P, Lang JJ (1985) J Phys Chem 89:41
37. Ananthapadmanabhan KP, Goddard ED, Turro NJ, Kuo PL (1985) Langmuir 1:352
38. Kim J-H, Domach MM, Tilton RD (1999) J Phys Chem B 103:10582
39. Dijt JC, Cohen Stuart MA, Hofman JE, Fleer GJ (1990) Colloids Surf 51:141
40. Dijt JC, Cohen Stuart MA, Fleer GJ (1994) Adv Colloid Interface Sci 50:79
41. Furst EM, Pagac ES, Tilton RD (1996) Ind Eng Chem Res 35:1566
42. Tilton RD (1999) In: Farinato RS, Dubin PL (eds) Colloid–polymer interactions: from fundamentals to practice. Wiley, New York, p 331
43. Azzam RMA, Bashara NM (1977) Ellipsometry and polarized light. North Holland, Amsterdam
44. Braem AD(2001) PhD dissertation. Carnegie Mellon University, Pittsburgh
45. Braem AD, Prieve DC, Tilton RD (2002) (manuscript in preparation)
46. Ducker WA, Senden TJ, Pashley RM (1992) Langmuir 8:1831
47. Dagastine R (2002) PhD dissertation. Carnegie Mellon University, Pittsburgh
48. Butt H-J (1991) J Biophys 60:1438
49. Kim J-H, Domach MM, Tilton RD (1999) Colloids Surf A 150:55
50. Kim J-H, Domach MM, Tilton RD (2000) Langmuir 16:10037
51. Tiller GE, Mueller TJ, Dockter ME, Struve WG (1984) Anal Biochem 141:262
52. Sigma Chemical Company (1999) Triton X-100 product information
53. Robson RJ, Dennis EA (1977) J Phys Chem 81:1075
54. Tanford C, Nozaki Y, Rohde MF (1977) J Phys Chem 81:1555
55. Wright AK (1976) J Colloid Interface Sci 55:109
56. Brown W, Rymden R, Van Stam J, Almgren M, Svensk G (1989) J Phys Chem 93:2512
57. Paradies HH (1980) J Phys Chem 84:599
58. Dupont L, Foissy A, Mercier R, Mottet B (1993) J Colloid Interface Sci 161:455
59. Das KK, Somasundaran P (2001) Colloids Surf A 182:25
60. Sastry NV, Sequaris JM, Schwuger MJ (1995) J Colloid Interface Sci 171:224
61. Ishimuro Y, Ueberreiter K (1980) Colloid Polym Sci 258:928
62. Maloney C, Huber K (1994) J Colloid Interface Sci 164:463
63. Berg JM, Claesson PM, Neuman RD (1993) J Colloid Interface Sci 161:182
64. Burns JL, Yan Y-d, Jameson GJ, Biggs S (2002) J Colloid Interface Sci 247:24
65. Kuhn W, Gruhn F (1942) Kolloid Z 101:248
66. Marko JF, Siggia ED (1995) Macromolecules 28:8759
67. Butt H-J, Kappl M, Mueller H, Raiteri R, Meyer W, Ruehe J (1999) Langmuir 15:2559
68. Ortiz C, Hadziioannou G (1999) Macromolecules 32:780
69. Yamamoto S, Tsujii Y, Fukuda T (2000) Macromolecules 33:5995.
70. Alexander S (1977) J Phys (Paris) 38:983
71. Luckham PF (1991) Adv Colloid Interface Sci 34:191
72. Zhulina EB, Borisov OV, Birshtein TM (2000) Macromolecules 33:3488

Progr Colloid Polym Sci (2003) 122: 67–72
DOI 10.1007/978-3-540-36114-5

L. E. Vîjan
E. Volanschi
Mihaela Hillebrand

Molecular modeling of anthracycline–DNA interaction

Paper presented in part at the 34th Midwest Theoretical Chemistry Conference, Minneapolis, Minnesota, USA, 5–6 October 2001

L.E. Vîjan · E. Volanschi
M. Hillebrand (✉)
Department of Physical Chemistry,
Faculty of Chemistry,
University of Bucharest,
Bd. Regina Elisabeta 4–12,
70346 Bucharest, Romania
e-mail: mihh@gw-chimie.math.unibuc.ro
Fax: +4021-3159249

Abstract A series of anthracycline antibiotics, daunomycin, doxorubycin and epirubycin, was investigated by both molecular mechanics and quantum chemistry (semiempirical AM1) methods. The charge distributions used to evaluate the electrostatic contribution to the binding energy were calculated by the semiempirical AM1 method. In order to evidence the sequence specificity of the drugs, some model single- and double-stranded DNA containing the bases adenine (A), thymine (T), cytosine (C) and guanine (G) in AAAA, TTTT, CCCC, GGGG, ATAT and CGCG sequences were used. In the drug–nucleic acid model complexes the quinone chromophore is intercalated between the base pairs of the DNA helix with the glycoside extended into the minor groove. The results outline the differences in the relative contributions of the electrostatic and van der Waals interactions to the total binding energy.

Keywords Molecular mechanics · Semiempirical AM1 method · DNA · Anthracycline antibiotics

Introduction

A number of cancers, especially leukemia, have their growth preferentially reduced by a variety of intercalating compounds, as a consequence of inhibition of DNA functions. The binding constants of the drug–DNA association have been determined using absorption and fluorescence spectral data, viscometry and equilibrium dialysis experiments [1, 2, 3, 4, 5]. Among these drugs, some anthracycline antibiotics, daunomycin, doxorubycin and epirubycin (Fig. 1), are widely used in cancer chemotherapy. The structural parts of these compounds are a four-cycle moiety (A, B, C, D) called the aglycon chromophore attached to a daunosamine fragment.

The drugs differ by the position of the hydroxyl group attached to C4′ of the daunosamine fragment and the nature of the substituent *R* attached to ring A. The hydroxyl group attached to C4′ of the daunosamine fragment is in an axial position in daunomycin and doxorubycin and in an equatorial position in epirubycin.

Previous experimental studies [1, 2, 3] have pointed out the complex nature of the binding process, evidencing besides intercalation an external cooperative interaction. The dependence of the binding constants on the ionic strength of the medium allowed the dissection of the binding free energy into electrostatic and nonelectrostatic contributions. It was found that the nonelectrostatic contribution prevails.

The purpose of this paper is to perform theoretical modeling of the interaction of these drugs with some model DNAs. We focus on the following aspects.

1. The study of the electronic structure of the isolated drugs. This implies the analysis of the possible conformers in order to find the optimal conformations for the interaction with the nucleic acids, the calculation of the charge distributions and the electrostatic potentials.
2. The theoretical modeling of the drug–DNA complexes in order to estimate the relative contributions of the van der Waals and electrostatic terms to the

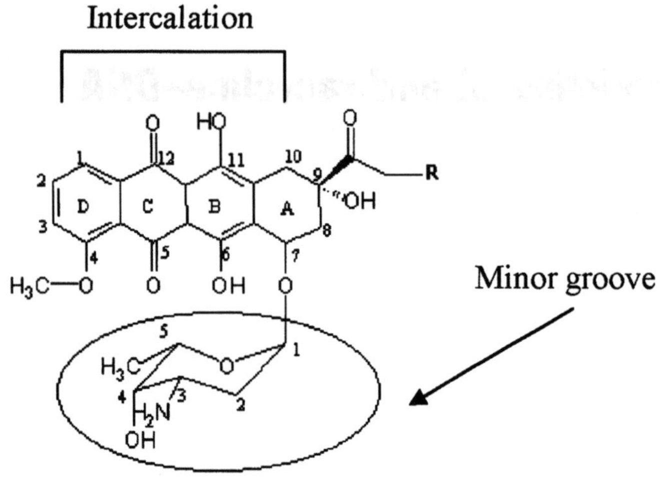

Fig. 1 The molecular structure of anthracycline antibiotics: daunomycin (*R*–H), doxorubycin and epirubycin (*R*–OH)

interaction energy and to get an insight into the sequence selectivity of the drugs.

Computational details

The structures of the drugs and the sequences of the nucleic acids were built within the HyperChem (version 5.02) program and optimized by the semiempirical AM1 method (parameters: self-consistent-field control of 0.01, restricted Hartree–Fock spin pairing, Polak–Ribiere optimizer, root-mean-square (RMS) gradient 0.01 kcal/mol Å for drugs and RMS gradient 0.1 kcal/mol Å for sequences of nucleic acids). The electronic structures of the isolated drugs were also optimized by the ab initio method using the STO-3G basis (Gamess program [6]). The calculations on the drug–nucleic acid complexes, were performed in vacuo by both the molecular mechanics (MM+ force field) and the AM1 methods. The optimization criteria were 0.1 kcal/mol Å for the MM method, 0.3 kcal/mol Å for complexes of the drug with sequences of single-stranded DNA and 1 kcal/mol Å for complexes of the drug with sequences of double-stranded DNA.

Results and discussion

Isolated ligands

Circular dichroism spectroscopy of a series of anthracyclines and anthracycline–DNA complexes in aqueous solution indicated the drastic modifications of the binding constants of anthracyclines with opposite chirality at C7 and C9 chiral centers. It was found that strong drug–DNA interaction and large cytotoxicity result when the absolute configuration at C7 and C9 is 7S, 9S. In contrast, the absolute configuration 7R, 9R implies weak drug–DNA interaction and low cytotoxicity.

We started with two conformers (1, 2), which differ by the orientation of the hydrogen atom of the hydroxyl group attached to C6 of the aromatic ring B and have S chirality at C7 and C9. The optimized torsion angles (Fig. 1) describing the position of the sugar linkage, C8–C7–O7–C1′ (ϕ_1) and C7–O7–C1′–O5′ (ϕ_2) are listed in Table 1 and the ab initio optimized structures of the doxorubycin conformers are presented in Fig. 2.

Conformer 2 is more stable than conformer 1, although the differences between them are small. It can be seen that different hydrogen bonds are formed depending on the orientation of the two hydrogen atoms from the hydroxyl groups of the aromatic ring B. The data in Table 1 point out the poor agreement with the experimental data for ϕ_1 for both conformers and both methods. For ϕ_2, the AM1 method predicts a planar conformation ($\phi_2 \sim 180°$), while the ab initio method predicts a quasi orthogonal conformation ($\phi_2 \sim 280°$), except for conformer 2 of epirubycin, in better agreement with experiment. However, the calculations performed considering the experimental value [7, 8] for ϕ_1 led to higher energies, and therefore these conformers were not considered further.

The charge distributions in the two conformers of each drug are similar, except for the atoms directly implied in the hydrogen bonds, i.e. the oxygen atoms bonded to C5, C6 and C7. In conformer 1, the smallest charge density was found at the oxygen atom on C7, while in conformer 2, the oxygen atom on C5 bears the smallest charge density.

Table 1 Values of the torsion angles obtained by AM1 and ab initio methods (conformers 1 and 2)

Torsion angle	Method	Daunomycin		Doxorubycin		Epirubycin		Experiment [7]
		1	2	1	2	1	2	
ϕ_1 (degrees)	AM1	39.99	52.45	41.83	54.61	40.13	56.71	125
	Ab initio	–	–	77.65	77.79	77.05	65.91	
ϕ_2 (degrees)	AM1	182.37	171.49	182.76	171.80	183.54	172.06	292
	Ab initio	–	–	278.08	276.31	278.37	188.65	

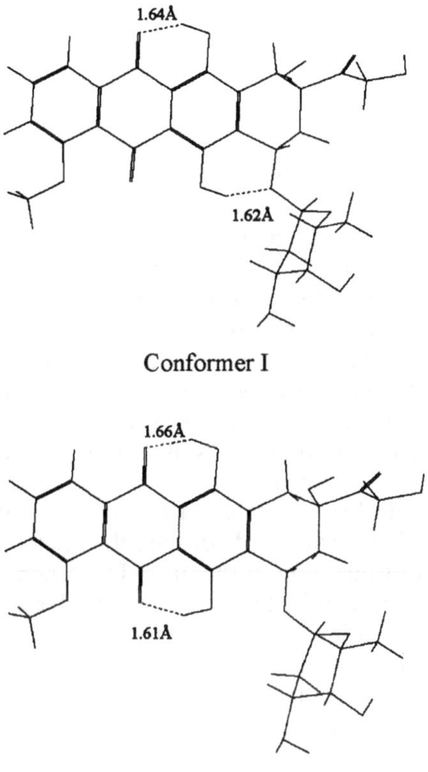

Fig. 2 The molecular structures of the doxorubycin conformers obtained by ab initio method

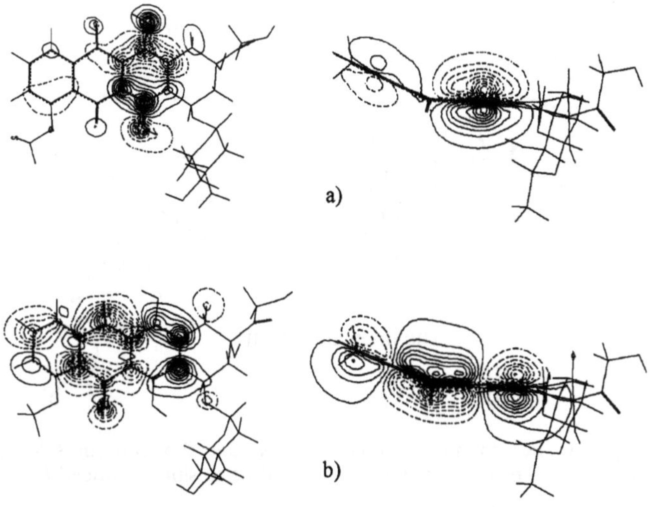

Fig. 3 a The highest occupied molecular orbitals and **b** the lowest unoccupied molecular orbitals of epirubycin

The frontier molecular orbitals (Fig. 3) are π orbitals. The highest occupied molecular orbital ($\varepsilon_{homo}\sim-9.1$ eV) is preferentially localized on the aromatic ring B, ensuring a high superposition with the π system of the DNA base pairs, and thus explaining the strong intercalating tendency of these drugs. The first vacant orbital ($\varepsilon_{lemo}\sim-1.35$ eV) is delocalized on the entire aromatic moiety. The nonimplication of the semisaturated ring A and of the daunosamine fragment on the frontier orbitals can also be noticed.

The electrostatic potential has emerged in the last 2 decades as a useful analytical tool in the analysis of chemical reactive behavior involving both electrophilic and nucleophilic processes, as well as recognition and hydrogen-bonding interactions. An important feature of the electrostatic potential is that it is a real physical property that can be determined experimentally by diffraction methods as well as computationally. The electrostatic potential was calculated for each compound and the minimum and maximum values for the two conformers are presented in Table 2.

Owing to the presence of the lone pairs, the oxygen atoms are characterized by negative regions of the electrostatic potential, while the aromatic rings and hydroxyl groups are characterized by a positive electrostatic potential. There are no significant differences considering the extension of the positive and negative regions of the electrostatic potential either between the conformers or between the drugs.

Isolated DNA

As the literature data point out that these intercalating drugs have a DNA sequence selectivity that does not extend beyond the two or three base pair level [4, 9, 10, 11, 12, 13, 14, 15, 16] we have chosen some model single- and double-stranded DNA containing the AAAA, TTTT, CCCC, GGGG, ATAT and CGCG sequences. Two further simplifying hypotheses were

Table 2 Results of AM1 calculations

	Daunomycin		Doxorubycin		Epirubycin	
	1	2	1	2	1	2
ε_{homo} (eV)	−9.087	−9.055	−9.097	−9.145	−9.159	−9.136
ε_{lumo} (eV)	−1.365	−1.555	−1.260	−1.634	−1.439	−1.614
V_{min} (kcal/mol)	−61.89	−42.37	−42.37	−39.06	−32.44	−47.01
V_{max} (kcal/mol)	332.99	260.49	267.12	329.35	638.17	233.36

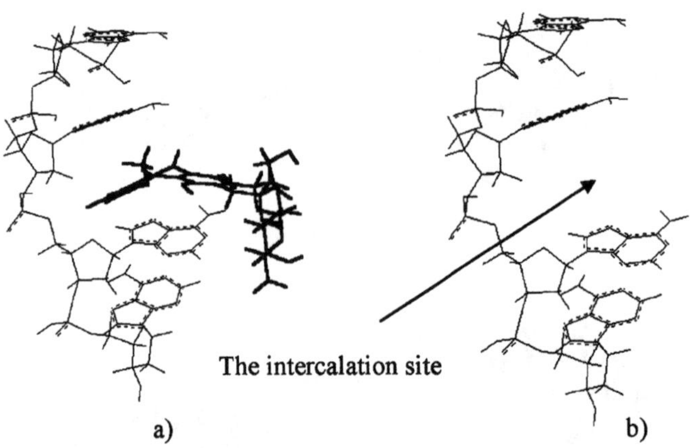

Fig. 4 **a** The optimized geometries of the doxorubycin–AAAA complex. **b** The intercalation site of the drug to single-stranded DNA

used: the solvent effect was not considered and the charge of the phosphate groups was neutralized with hydrogen atoms. The nucleic acids were optimized by both MM and AM1 methods. The values of the DNA energies calculated by the MM method were used to calculate the drug–nucleic acid interaction energy, according to Eq. (1). The results obtained by the AM1 method indicate a low net charge on the nitrogen atoms and negative area for the electrostatic potential on the oxygen and nitrogen atoms, which have lone pair electrons.

Antracycline–DNA complexes

The optimized conformers of the drugs and the DNAs were further utilized in the optimizations of the drug–nucleic acid complexes. The starting structures of the drug–nucleic acid complexes were built by the docking procedure. Initially, several restraints were imposed, so that the aglycon fragment of the drug intercalates between the base pairs of the DNA.

We considered that rings B and C are stacked in the duplex, ring D protrudes out into the major groove of the double helix, whereas ring A and daunosamine lie on the minor groove side of the DNA helix [15, 16, 17, 18]. After optimization of the drug–DNA complexes, the restraints were eliminated and the complexes optimized again. The optimized geometries of the drug–DNA complexes are presented in Figs. 4 and 5.

The energies of the drug–nucleic acid complexes were used for the evaluation of the following quantities [19, 20]:

Fig. 5 **a** The optimized geometries of the doxorubycin-CCCC–GGGG complex. **b** The intercalation site of the drug to double-stranded DNA

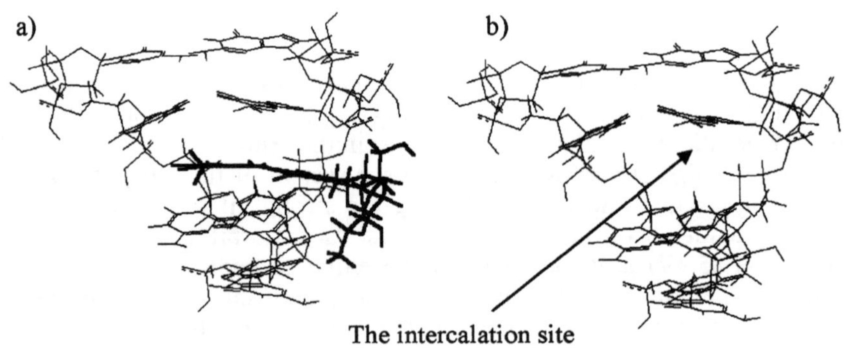

Table 3 Molecular mechanics results of drug–nucleic acid interaction

Nucleic acid	Daunomycin		Doxorubycin		Epirubycin	
	$E_{binding}$ (kcal/mol)	% van der Waals	$E_{binding}$ (kcal/mol)	% van der Waals	$E_{binding}$ (kcal/mol)	% van der Waals
AAAA	−12.09	87.68	−18.91	83.22	−27.76	68.37
TTTT	−19.89	95.46	−19.46	60.02	−11.32	92.40
ATAT	−16.96	95.46	−19.13	55.88	−16.60	87.47
CCCC	−9.08	90.97	−12.20	96.97	−12.68	77.44
GGGG	−20.15	70.67	−14.92	70.38	−10.69	94.76
CGCG	−23.56	72.41	−21.11	73.19	−14.25	93.54
AAAA–TTTT	−53.28	75.83	−46.81	86.35	−61.43	67.15
ATAT–TATA	−37.42	72.47	−45.65	76.36	−44.24	87.03
CCCC–GGGG	−61.12	63.07	−59.91	51.58	−60.37	66.82
CGCG–GCGC	−3.82	79.84	−6.50	74.15	−13.83	54.95

Table 4 Results of drug–nucleic acid interaction obtained by the AM1 method

Nucleic acid	$E_{binding}$ (kcal/mol)		
	Daunomycin	Epirubycin	Doxorubycin
AAAA	0.35	1.20	−1.24
TTTT	4.80	8.50	2.00
ATAT	4.16	1.30	0.90
CCCC	−2.40	−1.21	0.40
GGGG	−8.95	−2.33	−6.17
CGCG	−1.90	1.90	1.50
AAAA–TTTT	6.98	7.78	–
CCCC–GGGG	13.18	4.56	6.87
CGCG–GCGC	13.42	10.23	5.00

1. The interaction energy

$$E_{\text{interaction}} = E_{\text{complex}} - \left(E_{\text{drug}} + E_{\text{DNA}}\right)_{\text{optimized}} . \quad (1)$$

2. The binding energy

$$E_{\text{binding}} = E_{\text{complex}} - \left(E_{\text{drug}} + E_{\text{DNA}}\right)_{\text{frozen in complex}} . \quad (2)$$

3. The perturbation energy

$$E_{\text{perturbation}} = E_{\text{interaction}} - E_{\text{binding}} . \quad (3)$$

The values of the binding energies calculated by MM are presented in Table 3. In all cases, the binding energies have negative values, reflecting the drug–DNA interaction. A slight preference for the CG sequences can be noticed for daunomycin and doxorubycin for both single- and double-stranded DNA. The van der Waals contribution to the binding energy is also included in Table 3.

The results underline the significant van der Waals contribution to the binding energy and, consequently, the low percentage of the electrostatic interactions, in agreement with our previous experimental data [1, 2, 3].

The binding energies obtained by the AM1 method are presented in Table 4. The results are ambiguous. For the single-stranded DNA, negative values of the binding energies were obtained for the cytosine- and guanine-containing sequences, in agreement with the experimental observations. For double-stranded DNA, the binding energies are positive, probably owing to the rough approximations used.

The geometries of the complexes reveal a perturbation of the DNA structure. Some relevant geometric parameters for the isolated DNA and the drug–nucleic acid complexes, reflecting these interactions, are given in Table 5 for the CCCC sequence. Four distances between certain atoms of two neighboring base pairs at the intercalation site were considered: the distance between the N1 atoms (d_1), the distance between the C2 atoms (d_2), the distance between the C5 atoms (d_3) and the distance between the N4 atoms (d_4).

The calculations predict an increase in the distance between bases at the intercalation sites (cytosine 2 and cytosine 3). The turning of the polynucleotidic helix is produced and an "accordion-type" motion (Figs. 4, 5) takes place and determines a break in the hydrogen bonds between base pairs.

Conclusions

Our molecular modeling points out that the anthracycline–DNA complexes are stabilized mainly by van der Waals forces involving the aromatic chromophore and the DNA bases and that the electrostatic term brings only a minimal contribution (less than 20%) to the binding energy. A slight preference for cytosine and guanine bases was found. As a result of the drug–nucleic acid interaction, only the DNA structure is significantly perturbed, the structure of the drug being practically unchanged.

Table 5 Characteristics of the geometry of the drug–nucleic acid complexes

		CCCC optimized	Drug–CCCC complex		
			Daunomycin	Epirubycin	Doxorubycin
Cytosine 1 + cytosine 2	d_1 (Å)	5.15	4.87	5.15	4.87
	d_2 (Å)	5.41	5.43	5.83	5.43
	d_3 (Å)	5.15	4.22	3.92	4.22
	d_4 (Å)	5.81	5.37	4.79	5.37
Cytosine 2 + cytosine 3	d_1 (Å)	4.99	7.03	6.66	7.03
	d_2 (Å)	5.05	7.23	6.85	7.23
	d_3 (Å)	4.82	8.38	8.60	8.38
	d_4 (Å)	4.93	9.59	9.88	9.59
Cytosine 3 + cytosine 4	d_1 (Å)	5.87	4.50	4.83	4.50
	d_2 (Å)	6.04	4.53	4.77	4.53
	d_3 (Å)	5.01	4.43	4.21	4.43
	d_4 (Å)	4.70	4.99	4.14	4.99

References

1. Rizzo V, Penco S, Menozzi M, Geroni C, Vigevani A, Arcamone F (1988) Anti-Cancer Drug Des 3:103–115
2. Remeta DP, Mudd CP, Berger RL, Breslauer KJ (1993) Biochemistry 32:5064–5673
3. Volanschi E, Vîjan LE (2000) Rom J Biophys 10:1–15
4. Volanschi E, Vîjan LE (2001) Rev Roum Chim 46:163–173
5. Vîjan LE, Volanschi E, Hillebrand M (2001) Ann West Univ Timisoara 10:711–720
6. Schmidt MW, Baldridge KK, Boatz JA, Elbert ST, Gordon MS, Jensen J.J, Koseki S, Matsunaga N, Nguyen KA, Su S, Windus TL, Dupuis M, Mongomery JA (1993) J Comput Chem 14:1347
7. Neidle S, Taylor G (1977) Biochim Biophys Acta 479:450–459
8. Neidle S, Taylor G (1979) FEBS Lett 107:348–354
9. Ward D, Reich E, Goldberg IH (1965) Science 149:1259–1263
10. Pigram WJ, Fuller W, Hamilton LD (1972) Nature New Biol 235:17–19
11. Gabbay EJ, Grier D, Fingerle RE, Reimer R, Levy R, Pearce SW, Wilson WD (1976) Biochemistry 15:4209–4215
12. Wilson WD, Jones RL (1981) Adv Pharmacol Chemother 18:177–222
13. Wang AHJ, Ughetto G, Quigley GJ, Rich A (1987) Biochemistry 26:1152–1163
14. Chaires JB, Herrera JE, Waring M, (1990) Biochemistry 29:6145–6153
15. Wang AHJ (1992) Opin Struct Biol 2:361–368
16. Chaires JB (1998) Curr Opin Struct Biol 8:314–320
17. Frederick CA, Williams LD, Ughetto G, van der Marel GA, van Boom JH, Rich A, Wang AHJ (1990) Biochemistry 29:2538–2549
18. Nunn CM, Meervelt L van, Zhang SD, Moore MH, Kennard O (1991) J Mol Biol 222:167–177
19. Madrid JM, Villfruella M, Serrano R, Mendicuti F (1999) J Phys Chem B 103:4847
20. Cervero M, Mendicuti F (2000) J Phys Chem B 104:1572

Progr Colloid Polym Sci (2003) 122: 73–81
DOI 10.1007/978-3-540-36114-5

Alejandro J. Müller
Yvanova Garcés
Miguel Torres
Benjamin Scharifker
A. Eduardo Sáez

Interactions between high-molecular-weight poly(ethylene oxide) and sodium dodecyl sulfate

A.J. Müller (✉) · Y. Garcés · M. Torres
Grupo de Polímeros USB,
Departamento de Ciencia de los Materiales,
Universidad Simón Bolívar,
Caracas 1080-A, Venezuela
e-mail: amuller@usb.ve

B. Scharifker
Departamento de Química,
Universidad Simón Bolívar,
Caracas 1080-A, Venezuela

A.E. Sáez (✉)
Department of Chemical and
Environmental Engineering,
University of Arizona, Tucson,
AZ 85721, USA
e-mail: esaez@engr.arizona.edu
Tel.: +1-520-6215369
Fax: +1-520-6216048

Abstract In this work, we study interactions between poly(ethylene oxide) (PEO) and sodium dodecyl sulfate (SDS) in aqueous solutions, using surface tension, electrical conductivity and shear rheometry. The main distinguishing feature of this investigation with respect to previous works is the use of high-molecular-weight PEOs (up to 8×10^6 g/mol). The aggregation pattern of the surfactant in the form of micellar aggregates attached to the polymer chain causes specific changes in surface tension, electrical conductivity and zero-shear-rate viscosity that qualitatively resemble those found previously for lower-molecular-weight PEOs when no additional electrolytes are present in the solution. In the presence of 0.1 M NaCl and for PEOs with molecular weights of 600,000 g/mol and more, the electrical conductivity and surface tension of the solutions remain constant when surfactant is added beyond the critical aggregation concentration, until saturation of the polymer with micellar aggregates is achieved. This behavior indicates complete complexation of the dodecyl sulfate ions into micellar aggregates attached to the polymer, as well as a strong binding of the counterions to the PEO/SDS complex.

Keywords Poly(ethylene oxide) · Sodium dodecyl sulfate · Polymer/surfactant complexes

Introduction

Solutions of water-soluble polymers and surfactants are widely employed in practice to achieve effects such as colloidal stability, emulsification, viscosity and texture control, suspension of solids, flocculation, and in many cases combinations of more than one of such effects. Applications include paints, petroleum recovery, cosmetics, detergents, food products, polymer synthesis and controlled drug release [1, 2, 3].

Specific combinations of polymers and surfactants might form complexes in solution, depending on their chemical structure. In particular, certain nonionic polymers, like poly(ethylene oxide) (PEO), form intermolecular complexes with anionic surfactants, like sodium dodecyl sulfate (SDS). The PEO/SDS system is the most widely studied polymer/surfactant interacting pair [1, 2, 3]. Nevertheless, most studies have been performed with relatively low molecular weight PEO, typically less than 1×10^6 g/mol. It is widely acknowledged that the PEO/SDS complexes consist of micellar aggregates of SDS attached to the PEO chain.

In this work, we apply the techniques of surface tension, electrical conductivity and shear rheology to study the intermolecular complex formation between SDS and high-molecular-weight PEO in aqueous solutions. Although most of the results are generally consistent with trends reported in previous works on PEO/SDS interactions in the absence of electrolytes, we report a significantly different aggregation behavior when high-molecular-weight PEO is used in combination with SDS in the presence of an electrolyte (NaCl) at relatively high concentrations.

Experimental

PEO samples of different molecular weights were obtained from Polysciences. Approximate molecular weights were provided by the manufacturer. These values usually correspond to peak average values determined by size exclusion chromatography. The polymer samples are expected to have high polydispersity. The PEO samples used had reported molecular weights of 1×10^5, 6×10^5, 1×10^6, 2×10^6, 4×10^6, 5×10^6 and 8×10^6 g/mol. The SDS was supplied by Riedel de Haën with 90% purity. It was further purified by the recrystallization procedure reported by Okubo [4].

Electrical conductivity measurements were carried out with a CDM 3 conductimeter from Radiometer, Copenhagen, using a model CDC304 immersion cell. Surface tension was determined by the du Noüy ring method in a Fisher tensiometer equipped with a platinum ring. All surface tension and conductivity measurements were performed at 25 °C, except where stated.

For the simple shear experiments, a Rheometrics ARES shear rheometer with a double-wall Couette geometry was used to measure shear viscosity as a function of shear rate. All measurements were performed at 25 °C. The zero-shear-rate viscosity, η_0, was determined from either the measured Newtonian plateau at low shear rates or by fitting the Ellis model to the data [5]. Values of η_0 were determined for PEO solutions, PEO with selected SDS concentrations and for PEO/SDS and 0.1 M NaCl solutions.

The standard Huggins and Kramer extrapolation procedures were employed to determine intrinsic viscosities [5, 6]. The viscosity-average molecular weight was calculated from the Mark–Houwink–Sakurada equation [6]:

$$[\eta] = K\bar{M}_v^a \ , \quad (1)$$

where $[\eta]$ is the intrinsic viscosity, $\bar{M}_v$ is the viscosity-average molecular weight and K and a are constants whose values depend on the solvent and temperature. For unperturbed PEO coils in water at 25 °C the reported values are $a = 0.67$ and $K = 49.9 \times 10^{-3}$ ml/g [6].

Results and discussion

PEO/SDS interactions in the absence of electrolyte

The results of surface tension measurements for aqueous solutions of SDS and PEO/SDS are shown in Fig. 1. These results are fairly typical of the PEO/SDS system, and other polymer/surfactant pairs [1, 2, 3, 7]. The SDS solution exhibits constant surface tension at SDS concentrations higher than the critical micelle concentration (cmc) since further addition of SDS beyond the cmc goes into micelle formation, and the free micelles have no detectable surface activity. Solutions with 100 ppm 2×10^6 g/mol PEO at low SDS concentrations show a slightly lower surface tension than SDS solutions, which indicates a slight surface activity of the PEO. In this case, the most interesting changes in the surface tension curve occur at higher SDS concentrations, and they are characterized by three notable concentrations, denoted as T_1, T_2' and T_2 in Fig. 1. The concentration T_1 at which there is a distinct change of slope in the surface tension curve is the critical aggregation concentration (cac), which represents the onset of association of SDS to PEO by means of micellar aggregates attached to the polymer chain. The binding of the surfactant to the polymer in the form of micellar aggregates is driven by hydrophobic interactions between them [8, 9, 10, 11]. The fact that the surface tension does not remain precisely constant as the concentration of SDS is further increased might be taken as an indication that there is an increase in free surfactant unimer concentration. A second distinct change in slope occurs at T_2', indicating the point at which the PEO is saturated with micellar aggregates, termed the polymer saturation point (PSP). After this point, the surface tension drops at a rate similar to that observed before the cac, which indicates that all of the new surfactant added dissolves as free unimers, until the unimer concentration reaches the cmc at T_2. Beyond the cmc, free SDS micelles begin to form. This interpretation of surface tension measurements identifies T_2' with the PSP, a view that is widely held [1, 2, 3, 7, 12]. However, a certain number of previous works have taken T_2 to be the PSP, including the original work of Jones [8], and even other recent works [13].

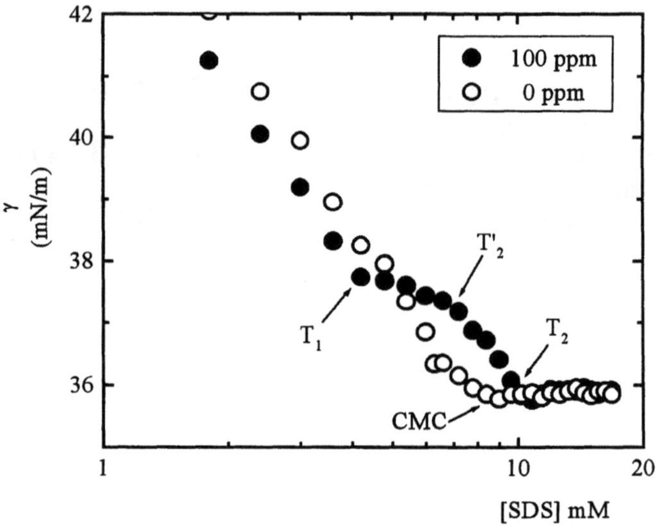

Fig. 1 Surface tension of sodium dodecyl sulfate (*SDS*) and poly(ethylene oxide) (PEO)/SDS solutions for a PEO of 2×10^6 g/mol molecular weight. The *inset* shows PEO concentration

Another parameter commonly used to study the binding of ionic surfactants to nonionic polymers is electrical conductivity. Changes in the slope of a plot of conductivity versus surfactant concentration are thought to relate to the critical interaction concentrations previously defined. However, in most conductivity plots that are available in the literature for the PEO/SDS system, only two critical concentrations are reported and they are usually identified as the cac and the PSP [1]. One exception can be found in the work of François et al. [14], where three changes in slope were observed. In terms of increasing SDS concentration, the first critical concentration was identified as the cac, the last one as the PSP/cmc, and the intermediate one was assigned to a transitional saturation concentration, where the PEO chains were saturated with micelles with a low aggregation number. At this point, according to François et al., further addition of SDS would contribute to increase the aggregation number of attached SDS micelles and also to increase the concentration of unimers in solution. Nevertheless, fluorescence measurements have indicated a negligible change of the aggregation number of micellar aggregates throughout the attachment process [14].

The idea that at concentrations lower than T_2 no free micelles are formed in the PEO/SDS solutions has been challenged by Minatti and Zanette [15]. They found a nonlinear region between T_1 and T_2 in their conductivity plots at low PEO concentrations in the absence of added electrolytes. They interpreted this result as evidence that the true saturation point occurs below T_2, and that it is possible that micellar aggregates and free micelles coexist in solution below T_2. There is evidence that micellar aggregates and free micelles might form at the same time upon addition of SDS at some point after the cac [16].

We have found that surface tension measurements provide the most accurate determination of T_1, T_2' and T_2. Once these values are known, corresponding changes of slope of the conductivity curves can be detected at approximately the same concentrations. In some cases, these changes of slope are slight and difficult to detect independently in conductivity curves, especially those occurring at T_2'. An example of the identification of T_1, T_2' and T_2, employing surface tension and conductivity measurements, is presented in Fig. 2. The slope of the conductivity curve decreases at each transition for various reasons. At T_1, the aggregation of dodecyl sulfate ions into micelles attached to the polymer chain reduces their mobility. Similarly, at T_2, the formation of free micelles reduces the charge-carrying capacity of the solution. At T_2', however, one would expect an increase of the slope owing to the fact that more free dodecyl sulfate ions would be generated once the polymer is saturated with micellar aggregates. However, the reduction in the mobility of the free sodium ions, which tend to screen the net negative charge of the polymer/surfactant complex, might be the determining factor.

Experiments conducted with various molecular weights and polymer concentrations (Table 1) confirm the main trends reported in the literature; namely:

1. T_1, T_2' and T_2 are independent of molecular weight, including results at 8×10^6 g/mol.
2. T_1 is approximately independent of polymer concentration.

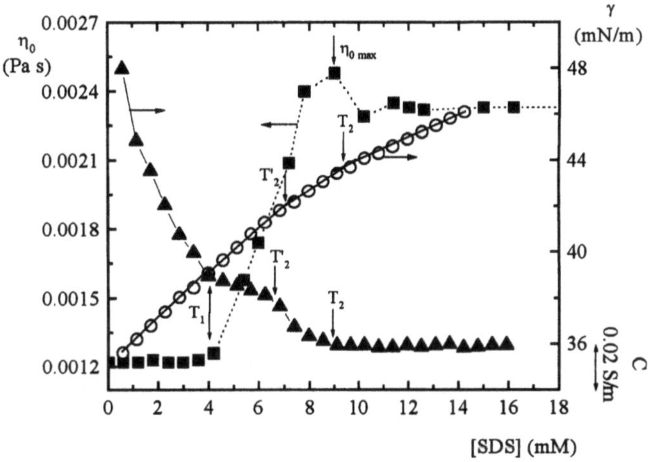

Fig. 2 Surface tension, γ, electrical conductivity, C and zero-shear-rate viscosity, η_0, for 100 ppm PEO/SDS solutions. The PEO molecular weight is 8×10^6 g/mol. *Lines* have been drawn between data points for the surface tension and viscosity curves. *Three straight lines* of different slopes, joined at the critical concentrations, have been drawn through the conductivity data points

3. T_2' and T_2 vary linearly with polymer concentration, with T_2 reaching the cmc of SDS in pure water as the polymer concentration tends to zero.

The values of the cac obtained in this work lie in the range 4.1–5 mM. They fall within the wider range that has been reported in the literature for samples with lower-molecular-weight PEO: 3–5.8 mM [15, 17, 18, 19, 20, 21].

Increases in the viscosity of solutions of certain polymers upon addition of SDS were reported in the pioneering work of Saito [22]. This effect is attributed to changes in polymer conformation due to the formation of polymer/surfactant complexes. Typical results showing the dependence of η_0 on the SDS concentration are shown in Fig. 2. At an SDS concentration below the cac (T_1), the SDS and PEO molecules are dissolved without interactions and no effect of SDS on the viscosity is observed. Once micellar aggregates form along the polymer chain, the viscosity increases with surfactant concentration. This viscosity increase has been attributed to the coil expansion resulting from electrostatic repulsion between micellar aggregates [18]. It is relevant to note that two other effects might contribute, in general, to a viscosity increase: the coil expansion due to the change in size of the polymer coil because of the attached micellar aggregates, which would increase the coil's excluded volume, and the increase in interchain interactions due to the possibility that a micellar aggregate might interact with two polymer coils simultaneously, producing temporal cross-links. Upon further increase of the SDS concentration, a viscosity maximum is reached. The subsequent reduction in viscosity has been attributed to the screening of the electrostatic repulsion between micellar aggregates along

Table 1 Critical transition concentrations for the poly(ethylene oxide) (PEO)/sodium dodecyl sulfate (SDS) system, as determined by surface tension measurements (ST), electrical conductivity measurements (C) and shear rheology (R). $[SDS]_{max}$ is the surfactant concentration at which the maximum zero-shear rate-viscosity was achieved; when no absolute maxima were obtained (as for 100 and 250 ppm PEO in Fig. 3), $[SDS]_{max}$ is the surfactant concentration at which the viscosity plateau is reached

$\bar{M} \times 10^{-6}$ (g/mol)	[PEO] (ppm)	T_1 (mM)	T_2' (mM)	T_2 (mM)	$[SDS]_{max}$ (mM)	Technique
2	50	4.2	7.2	9.3	–	ST
2	100	4.1	7.9	10.0	–	C
2	100	4.1	7.6	10.0	–	ST
8	25	4.1	6.4	8.3	–	C
8	50	4.2	7.2	9.1	–	ST
8	100	4.2	7.4	10.0	–	C
8	100	4.3	7.5	9.7	–	ST
8	150	4.2	8.0	12.0	–	C
8	200	4.5	7.6	12.1	–	ST
0.6	3,000	4.8	–	–	38.4	R
0.6	4,000	5.0	–	–	44.0	R
1	700	4.8	–	–	13.2	R
1	1,000	4.8	–	–	16.8	R
2	100	4.8	–	–	8.6	R
2	250	4.8	–	–	10.1	R
2	400	4.8	–	–	11.3	R
2	700	4.8	–	–	15.6	R
4	100	4.8	–	–	8.6	R
5	100	4.8	–	–	8.4	R
5	250	4.8	–	–	11.0	R
8	50	4.8	–	–	7.2	R
8	100	4.6	–	–	9.1	R

the polymer chain due to the increased concentration of sodium ions. Notice that, for the case presented in Fig. 2, this screening effect only partly compensates the viscosity increase achieved at the maximum point.

It has been argued that the point of maximum viscosity corresponds to the PSP [18]. Our results do not indicate that the exact concentration at which the viscosity is a maximum coincides either with T_2' (i.e., the PSP) or with T_2. In fact, the maximum viscosity is nearly always reached between T_2' and T_2. The fact that the viscosity keeps on increasing after T_2' suggests that more micellar aggregates are being attached to the polymer chain and the coil keeps expanding. However, this could be related to an enhancement in the coil's ability to take on micellar aggregates in shear-flow experiments owing to the chain deformation that naturally occurs in the flow.

The observation of a maximum in η_0 followed by a slight decrease and a region of constant viscosity was the general case for all the PEO/SDS systems examined in this work, in the absence of added electrolytes; however, the maximum was usually less noticeable at low PEO concentrations and in some cases only a rise in viscosity followed by a plateau was observed, as shown in Fig. 3. In these cases, the screening of micellar charges might be occurring gradually as more micelles are attached to the polymer coil.

The values of the intrinsic viscosity determined for solutions of PEO/SDS at fixed SDS concentrations are shown in Table 2. For PEO solutions, the viscosity-

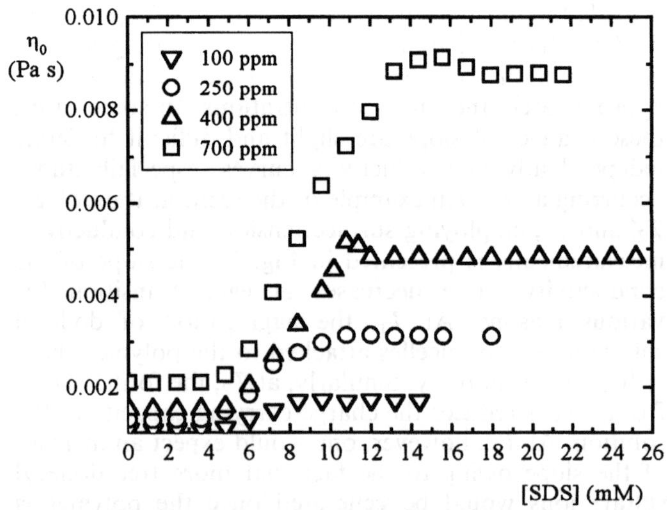

Fig. 3 η_0 of PEO/SDS solutions for a PEO with 2×10^6 g/mol molecular weight. The *inset* shows PEO concentration

average molecular weight, $\bar{M}_v$, calculated from Eq. (1) is reported. Note that $\bar{M}_v$ is consistent with the manufacturer's value for the lower-molecular-weight samples. At higher molecular weights, the high dilution levels that need to be achieved in order to remain within the dilute regime impose restrictions on the sensitivity of the rheological measurements. For PEO/SDS mixtures, the SDS concentration employed for the data in Table 2 was

Table 2 Intrinsic viscosity and derived parameters for the PEO/SDS system. The intrinsic viscosity was calculated from experimental data using Huggins and Kramer extrapolation. The viscosity-average molecular weight was calculated from Eq. (1). The coil expansion factor was determined from Eq. (3), as explained in the text

$\bar{M} \times 10^{-6}$ (g/mol)	[SDS] (mM)	$[\eta]$ (ml/g)	$\bar{M}_v \times 10^{-6}$ (g/mol)	α
0.6	0	394	0.65	–
0.6	38.8	1,448	–	1.5
1	0	624	1.3	–
1	13.2	3,205	–	1.7
1	15.6	3,352	–	1.7
2	0	1,011	2.7	–
2	12.0	5,234	–	1.7
2	13.2	5,290	–	1.7
2	15.6	5,395	–	1.7
4	0	1,365	4.2	–
4	8.4	7,630	–	1.8
4	12.0	8,307	–	1.8
5	0	1,490	4.8	–
5	9.6	7,331	–	1.7
5	12.0	7,634	–	1.7
8	0	2,668	11.4	–
8	8.4	17,477	–	1.9
8	10.8	30,250	–	2.2

chosen so that it was always above T_2' for all PEO concentrations to ensure saturation of the coil with micellar aggregates.

In order to quantify changes in the coil conformation in the presence of micellar aggregates, we calculated a coil expansion factor, α, analogous to that determined for a pure polymer from the intrinsic viscosity values under theta conditions, $[\eta]_\theta$, and its value under good solvent conditions by [23]

$$\alpha^3 = \frac{[\eta]}{[\eta]_\theta} \ . \tag{2}$$

We employed a modified relation based on Eq. (2) to determine a coil expansion factor induced by the presence of micellar aggregates on the PEO coils:

$$\alpha^3 = \frac{[\eta]_{SDS}}{[\eta]} \ , \tag{3}$$

where $[\eta]_{SDS}$ is the intrinsic viscosity of PEO in the presence of excess SDS and $[\eta]$ is the intrinsic viscosity of the solution in the absence of SDS. The results are presented in Table 2. The calculated coil expansion factor varies in the range 1.5–2.2. Note that α tends to increase with molecular weight, as could be anticipated from the increase in coil size.

PEO/SDS interactions in the presence of NaCl

The formation of micellar aggregates along the PEO chain confers the polymer a polyelectrolytic nature. The micellization process is also affected by electrostatic interactions; therefore, changes in the ionic strength of the solution play a very important role in the behavior of PEO/SDS complexes. In this section we explore PEO/SDS interactions for high-molecular-weight PEO in 0.1 M NaCl solutions.

The effect of NaCl addition on a pure SDS aqueous solution is to lower its cmc from 8.2 to 2.0 mM (as confirmed in this work with conductivity and surface tension measurements), while it raises the micelle aggregation number from 56 to 93 [24]. These changes are due to electrostatic effects of the counterions on the electrical double layer of the micelles that have been described quantitatively through solutions of the Poisson–Boltzmann equation [24].

Conductivity plots for several PEO concentrations of PEO/SDS/NaCl solutions are shown in Fig. 4. For analogous reasons to the behavior of the cmc of SDS solutions, when NaCl is added to a PEO/SDS solution, the cac decreases. In addition, more surfactant can be attached to a given polymer coil in the presence of NaCl in view of the higher micelle aggregation number and the higher number of micellar aggregates that can attach owing to the screened electrostatic interactions. This is reflected by the fact that the cac is lowered much more than T_2' and T_2 when NaCl is added: our results indicate that the cac is approximately lowered to half its original value upon addition of 0.1 M NaCl, while T_2' and T_2 only decrease by 15–20%.

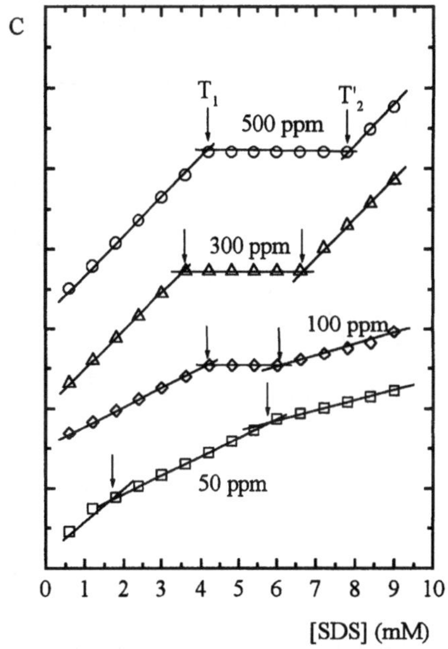

Fig. 4 Electrical conductivity of PEO/SDS solutions in 0.1 M NaCl for the PEO concentrations indicated. The PEO molecular weight is 6×10^5 g/mol. The transition concentrations T_1 and T_2' are indicated for each curve. The conductivity scale is different for each curve to allow separation. *Solid lines* have been drawn through straight portions of each curve

The results in Fig. 4 show the T_1 and T_2' transitions. The T_2 transition occurs at higher SDS concentrations than the range used in Fig. 4 (see Fig. 7 for an example showing the three transitions). The 50 ppm PEO solution in Fig. 4 exhibits the typical behavior that has been previously reported for PEO samples of lower molecular weight [1, 2, 3], which is qualitatively the same as solutions in the absence of electrolyte (Fig. 2). However, when the concentration of PEO is increased, the conductivity curves exhibit a region in which the conductivity becomes insensitive to SDS concentration. This remarkable result implies that added ions that are produced by the dissociation of the surfactant added in this range do not contribute to changes in the measured conductivity. Another unusual effect is that the cac fluctuates with the PEO concentration for this PEO molecular weight.

In order to further explore the effect just described, a PEO concentration of 100 ppm was selected and the molecular weight was varied, keeping the salt concentration constant at 0.1 M NaCl. The results are shown in Fig. 5. The plateau region is not present for the sample with the lowest molecular weight employed, but is present in all the other samples examined.

The plateau effect on the conductivity measurements was not observed for PEO/SDS/NaCl solutions when samples of PEO below a molecular weight of 6×10^5 g/mol were used. The results obtained for this transitional molecular weight exhibited cac values that fluctuated with PEO concentration. Nevertheless, once the PEO molecular weight was equal to or greater than 10^6 g/mol, the plateau effect was always present for all the PEO concentrations examined and the cac was remarkably constant (between 2.0 and 2.1 mM). As an example, the effect of PEO concentration for a molecular weight of 8×10^6 g/mol is presented in Fig. 6. The width of the region where the conductivity is constant increases with the amount of PEO in the solution, as a logical consequence of the constancy of the cac (T_1) and the increase of T_2' with polymer concentration.

The existence of a plateau in electrical conductivity variations with SDS concentrations for PEO/SDS/NaCl solutions has not been reported previously in the literature, to our knowledge. Many studies [14, 15, 19] have been performed in the absence of added electrolyte (where the effect is not present) and those that use electrolytes have employed PEO with lower molecular weights (at least those that have measured conductivity) [15].

An example of the correspondence between surface tension and conductivity measurements for the highest-molecular-weight PEO employed in this work is given in Fig. 7. This was generally the situation in all the cases explored, as can be seen in Table 3. In the region between T_1 and T_2', the conductivity is constant and the surface tension of the solution is also constant. The fact that the surface tension is constant indicates that all of the dodecyl sulfate ions added to the solution in this range form micellar aggregates, since they do not contribute to the surfactant's surface activity. The fact that the conductivity does not change means that the added Na^+ counterions are not participating in the conduction

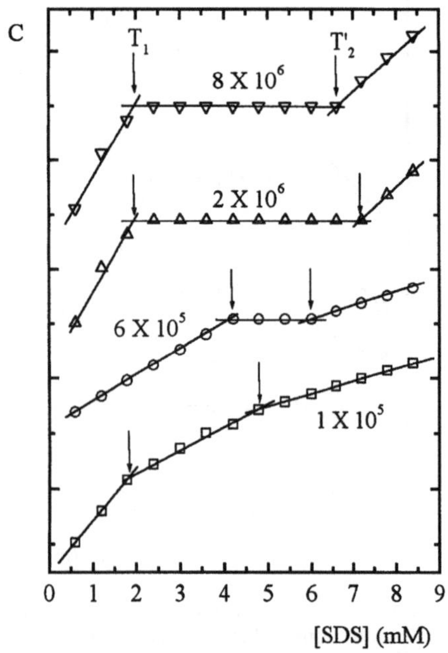

Fig. 5 Electrical conductivity of PEO/SDS solutions in 0.1 M NaCl for PEOs of the indicated molecular weight at a PEO concentration of 100 ppm

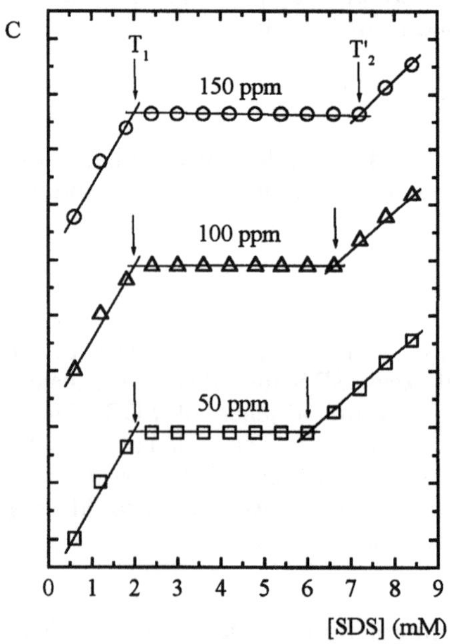

Fig. 6 Electrical conductivity of PEO/SDS solutions in 0.1 M NaCl for the PEO concentrations indicated. The PEO molecular weight is 8×10^6 g/mol

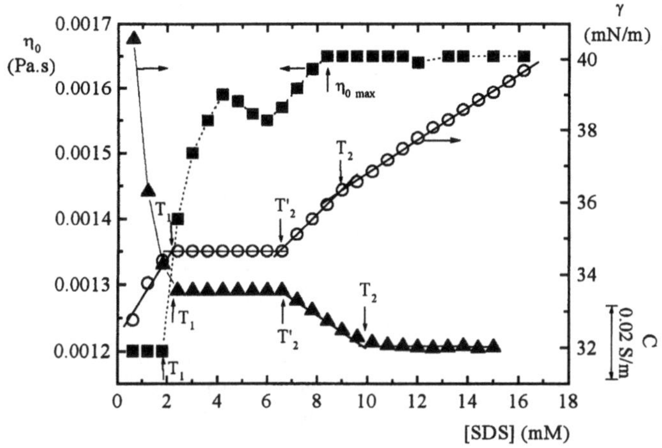

Fig. 7 γ, C and η_0 for 100 ppm PEO/SDS/ 0.1 M NaCl solutions. The PEO molecular weight is 8×10^6 g/mol

process either, which suggests that the mobility of added counterions is restricted by their interaction with the PEO/SDS complex.

Figure 7 also includes the corresponding behavior of the zero-shear-rate viscosity. At the cac (T_1), there is a sharp increase in viscosity as the concentration of SDS is increased. As in the case when no salt was added, good correspondence was always found between the cac determined by conductivity or surface tension and a sharp rise in viscosity also occurred, indicating coil expansion. The rather weak maximum followed by a second rise in viscosity was encountered only in the case of the 8×10^6 g/mol PEO sample and is unexpected. For all the other molecular weights examined the viscosity versus SDS concentration curves followed an approximately sigmoidal curve with the viscosity increasing to a plateau value without local maximum or minimum points. This is shown in Fig. 8, where zero-shear-rate viscosities of solutions with and without salt are compared. The most evident difference between the two curves is the substantial reduction in viscosity caused by electrolyte addition. This is expected since the excess counterions shield the electrostatic repulsion between micelles attached to the polymer molecules. Nevertheless, the PEO coils are still expanded in the case where they are saturated with SDS micelles when compared to the pure PEO solution. We were able to corroborate this by experimentally determining the coil expansion coefficients, which are reported in Table 4 (compare with the values in solutions without salt, Table 2). The values of α in the presence of NaCl and relatively high SDS concentrations oscillate between 1.1 and 1.5 and no trend was found with the molecular weight, even though an increase with molecular weight was expected (note, however, that the highest values of α were found with the highest-molecular-weight sample). In the cases for which several concentrations of SDS were employed, a slight increase in α was obtained upon increasing the SDS concentration, although sometimes the increase was almost negligible, indicating that saturation of the PEO coils with SDS micelles had already been achieved. When the same PEO molecular weight is compared with and without salt, the coil contraction effect caused by the electrolyte is obvious. For instance, for the case of the 2×10^6 g/mol molecular-weight sample of Fig. 8, α values for PEO/SDS at 15.6 mM SDS are listed in Table 2 and for PEO/SDS/NaCl at 12 mM SDS in Table 4. In both cases, their respective zero-shear-rate viscosities are close to their maximum values and α changes from 1.7 in the case without salt to 1.1 when excess salt was added.

Table 3 Critical transition concentrations for the PEO/SDS/0.1 M NaCl system, as determined by surface tension measurements (ST), electrical conductivity measurements (C) and shear rheology (R). The concentration $[SDS]_{max}$ is the surfactant concentration at which the maximum zero-shear-rate viscosity was achieved; when no absolute maxima were obtained (as for the PEO/NaCl curve in Fig. 8), $[SDS]_{max}$ is the surfactant concentration at which the viscosity plateau is reached

$\bar{M} \times 10^{-6}$ (g/mol)	[PEO] (ppm)	T_1 (mM)	T'_2 (mM)	T_2 (mM)	$[SDS]_{max}$ (mM)	Technique
0.1	100	2.0	4.8	–	–	C
0.6	50	1.8	5.8	–	–	C
0.6	100	4.1	6.0	–	–	C
0.6	150	3.6	6.7	–	–	C
0.6	200	4.2	7.8	–	–	C
0.6	50	2.0	6.6	9.8	–	ST
0.6	100	–	7.0	10.6	–	ST
0.6	150	–	7.5	12.0	–	ST
1	50	2.6	6.1	9.7	–	ST
1	100	2.3	6.6	10.7	–	ST
1	200	2.3	7.8	12.7	–	ST
2	100	2.0	7.1	–	–	C
2	50	2.3	6.6	10.3	–	ST
2	100	2.3	7.0	11.0	–	ST
2	150	2.3	7.4	11.9	–	ST
2	200	2.0	7.8	12.6	–	ST
4	50	2.3	5.8	9.4	–	ST
4	100	2.6	6.5	9.9	–	ST
4	150	2.3	7.4	12.3	–	ST
8	50	2.1	6.0	–	–	C
8	100	2.1	6.6	–	–	C
8	150	2.1	7.2	–	–	C
8	50	2.3	6.0	9.6	–	ST
8	100	2.3	7.0	9.8	–	ST
8	150	2.1	7.3	11.7	–	ST
8	200	2.3	8.4	12.4	–	ST
0.6	3,000	1.8	–	–	12.0	R
1	400	1.8	–	–	10.8	R
1	1,000	1.9	–	–	24.0	R
2	100	1.8	–	–	4.8	R
2	250	1.8	–	–	8.4	R
2	400	1.8	–	–	10.8	R
2	700	1.8	–	–	13.0	R
5	100	1.8	–	–	4.8	R
5	250	1.8	–	–	7.3	R
8	50	2.2	–	–	4.2	R
8	100	2.2	–	–	8.4	R

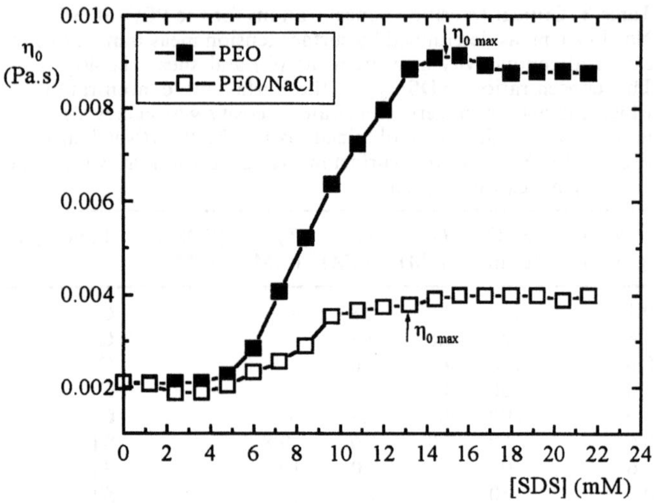

Fig. 8 Comparison of η_0 of PEO/SDS solutions with 0.1 M NaCl and without salt. The PEO concentration is 700 ppm, and it molecular weight is 2×10^6

Table 4 Intrinsic viscosity and coil expansion factor for the PEO/SDS/0.1 M NaCl system. The intrinsic viscosity was calculated from experimental data using Huggins and Kramer extrapolation. The coil expansion factor was determined from Eq. (3)

$\bar{M} \times 10^{-6}$ (g/mol)	[SDS] (mM)	$[\eta]$ (ml/g)	α
1	10.8	1,341	1.3
1	12.0	1,364	1.3
1	15.6	1,673	1.4
2	10.8	1,237	1.1
2	12.0	1,349	1.1
4	3.6	4,080	1.4
4	7.2	4,309	1.5
5	4.8	2,791	1.2
5	7.2	3,209	1.3
8	4.8	9,356	1.5

Another difference encountered in Fig. 8 and caused by the electrolyte addition is that the PEO solution without salt exhibits a viscosity maximum for the PEO concentration employed at an SDS concentration of approximately 13 mM. Upon salt addition, the general trend for most of the samples examined (with the exception of 8×10^6 g/mol PEO, Fig. 7) was that there was no maximum in viscosity, but the viscosity increased until a plateau was reached. This is consistent with the explanation given earlier for the origin of the viscosity maximum in the case where no electrolyte was added, since no maximum would be expected if there is already an excess of counterions in solution.

One final observation that is noteworthy in Fig. 8 is the behavior of the viscosity at low SDS content just above the cac. A small but noticeable reduction in zero-shear-rate viscosity was obtained just above the cac for the solution with salt, indicating that the coil size was reduced compared to the situation with PEO coils in solution without SDS. This effect is a result of intramolecular interactions induced in the PEO coils when only very few SDS micelles are attached to each polymer coil. A similar effect on shear viscosity measurements was reported by François et al. [14]. These intramolecular effects in PEO/SDS/NaCl solutions at SDS concentrations above but close to the cac are more evident in elongational flows [25, 26].

The plateau in the solution conductivity between T_1 and T_2' was also present when the chemical nature of the electrolyte employed was changed. Experiments were performed for PEO/SDS solutions using 0.1 M LiCl, NaCl, NH_4Cl, KCl, CuCl, $MgCl_2$ and $CoCl_2$, and in all cases the constant conductivity range was clearly observed.

As stated before, the fact that the conductivity remains constant between T_1 and T_2' means that both the surfactant anion and the added sodium counterion are immobilized by the PEO/SDS complex formed. What is different from the results in which the plateau is not observed (without salt, and with salt at relatively low molecular weights and/or polymer concentrations) is the immobilization of the counterion. This type of behavior has been observed in certain solutions of polyelectrolytes complexing with oppositely charge surfactants. Fielden et al. [27] observed that polymer/surfactant association slowed considerably the rate of increase of conductivity with surfactant concentration for mixtures between a cationic copolymer and SDS because of the immobilization of sodium ions. The PEO/SDS complex has a polyelectrolytic nature owing to the negatively charged micellar aggregates, but it is the particular structure of the complex formed with high-molecular-weight PEO and salt that leads to the conductivity plateau.

One possible explanation for the existence of the plateau could be the presence of two phases. All the solutions were carefully inspected optically and they were transparent with no apparent changes to the naked eye. This only rules out macroscopic phase segregation and precipitation. Therefore, it is still be possible that micro-phase separation could be taking place and other experimental techniques (such as light scattering) must be employed to rule out this possibility. Nevertheless, we performed some conductivity experiments at different temperatures and the results do not seem to favor the micro-phase segregation hypothesis. The experiments were conducted at three different temperatures, 25, 35 and 40 °C, with the 8×10^6 g/mol PEO at 150 ppm and 0.1 M NaCl (the curves are not shown since they are similar to those in Fig. 6). Upon increasing the temperature, T_1 did not change appreciably, while T_2' decreased from 7.2 mM at 25 °C to 6.2 mM at 35 °C and finally to 5.2 mM at 40 °C. Therefore, the T_1–T_2' range of the constant conductivity plateau decreases with increasing temperature. High-molecular-weight PEO exhibits lower critical solution temperature behavior with a critical temperature that is very close to 100 °C. An increase in temperature therefore reduces the solubility of PEO in water and would help phase separation. However, the extent of the plateau decreases upon heating, indicating that the effect may not be necessarily related to microphase segregation. In fact, as the temperature is lowered, the solubility of PEO increases and the coils expand. In the absence of SDS, this would lower the coil overlap concentration, c^*, and favor interchain aggregation.

The experimental evidence presented in this section indicates that the state of aggregation of PEO and SDS suffers an important change when electrolyte is added to the solution above certain values of the PEO molecular weight and concentration. The criticality in concentration and molecular weight is typical of phenomena that involve intermolecular interactions. At molecular weights of 10^5 g/mol or lower, addition of NaCl to PEO/SDS solutions only affects values of the critical concentrations, especially the cac; the aggregation number and the number of micelles attached to each coil increase and the coil is made more compact in view of the excess of counterions [1, 2]. The use of high-molecular-weight PEO (6×10^5 g/mol and higher) implies a reduction in c^*. This critical concentration will be reduced even more by any factor that induces a coil expansion (or increases in α). The addition of electrolytes causes a coil contraction in PEO coils that are charged with bound SDS micelles. If the coils are entangled with one another (as they would be at concentrations above c^*) they might form interpenetrated, multichain structures that might form independent microenvironments within them. This concept is very similar to the "swollen cage" model proposed by Chari et al. [18] to describe the PEO coil macroconformation when it is loaded with SDS micelles, as opposed to the pearl-necklace model previously suggested [9]. In our case, the "swollen cages" conformed by PEO and SDS would have the ability to trap ions in the range T_1–T_2' when an excess of NaCl is present in the solution with such effectiveness that all added SDS unimers and their counterions would be trapped within these multichain structures and, consequently, they would not contribute to the conductivity or the surface tension of the solution. The presence of electrolyte would favor the formation of these structures by allowing the polymer chains to interact among themselves, partly because of the relatively high number of large micelles.

Concluding remarks

Surface tension and conductivity measurements have revealed a state of aggregation between SDS and PEO that is particular for high-molecular-weight PEO in the presence of relatively high concentrations of an electrolyte (NaCl). In this state of aggregation, the addition of surfactant does not change the conductivity or surface tension of the solution at all, until the polymer becomes saturated with surfactant micelles. We speculate that the PEO/SDS complexes formed under these conditions consist of multichain associations that form independent microenvironments within them, which are capable of absorbing added surfactant (as part of micellar aggregates), and the added counterion.

Acknowledgements A.J.M. is grateful to the Venezuelan Science and Research Council (grant 97-003590) for financial support.

References

1. Goddard ED, Ananthapadmanabhan KP (1993) Interactions of surfactants with polymers and proteins. CRC, Boca Raton
2. Jönsson B, Lindman B, Holmberg K, Kronberg B (1998) Surfactants and polymers in aqueous solutions. Wiley, Chichester
3. Kwak JCP (1998) Polymer–surfactant systems. Dekker, New York
4. Okubo T (1979) Proc R Soc Lond Ser A 81:366
5. Macosko CW (1994) Rheology. Wiley-VCH, New York
6. Brandup J, Immergut EH, Grulke EA (1999) Polymer handbook, 4th edn. Wiley-Interscience, New York
7. Goddard ED (2002) J Colloid Interface Sci 256:228
8. Jones MN (1973) J Colloid Interface Sci 44:475
9. Shirahama K, Ide N (1976) J Colloid Interface Sci 54:450
10. Gao Z, Wasylishen RE, Kwak JCT (1991) J Phys Chem 95:462
11. Gjerde MI, Nerdal W, Høiland H (1996) J Colloid Interface Sci 183:285
12. Cabane B, Duplessix R (1982) J Phys (Paris) 43:1529
13. Li F, Li GZ, Xu GY, Wang HQ, Wang M (1998) Colloid Polym Sci 276:1
14. François J, Dayantis J, Sabbadin J (1985) Eur Polym J 21:165
15. Minatti E, Zanette D (1996) Colloids Surf A 113:237
16. Gjerde MI, Nerdal W, Høiland H (1998) J Colloid Interface Sci 197:191
17. Turro NJ, Baretz BH, Kuo P (1984) Macromolecules 17:1321
18. Chari K, Antalek B, Lin MY, Sinha SK (1994) J Phys Chem 100:5294
19. Gjerde MI, Nerdal W, Høiland H (1998) Colloid Polym Sci 276:503
20. Froehner SJ, Belarmino A, Zanette D (1998) Colloids Surf A 137:131
21. Zanette D, Lima CF, Ruzza A, Belarmino AT, Santos S, Frescura VL, Marconi D, Froehner J (1999) Colloids Surf A 147:89
22. Saito S (1957) Kolloid Z 154:19
23. Billingham NC (1977) Molar mass measurements in polymer science. Kogan Page, London
24. Evans DF, Wennerström H (1999) The colloidal domain, 2nd edn. Wiley-VCH, New York
25. DaRocha CM, Patruyo LG, Ramírez NE, Müller AJ, Sáez AE (1999) Polym Bull 42:109
26. Smitter LM, Guédez JF, Müller AJ, Sáez AE (2001) J Colloid Interface Sci 236:243
27. Fielden ML, Claesson PM, Schillen K (1998) Langmuir 14:5336

Progr Colloid Polym Sci (2003) 122: 82–86
DOI 10.1007/978-3-540-36114-5

Keishiro Shirahama
Hideyuki Koga
Noboru Takisawa

Diverse actions of added alkanols on the binding of dibucaine cation to an anionic polymer

K. Shirahama · H. Koga · N. Takisawa
Department of Chemistry and Applied Chemistry, Saga University, Saga 840-8502, Japan

K. Shirahama (✉)
Wajiro 5-1-20-910, Higashi-ku, Fukuoka City 811-0202, Japan
e-mail: shirakey@wa-sun.gr.jp
Tel.: +81-92-6070063

Abstract The binding of dibucaine cation to a polyanion [poly(acrylamide 2-methylpropane sulfonate)] in the presence of various alkanols(C_1–C_3, C_5, and C_7) is measured by using an electrode responsive to dibucaine cation. The binding is characteristic of cooperative interaction. The binding affinity is enhanced by added pentanol and heptanol, and is well correlated with thermodynamic activities of these alkanols. In contrast, methanol and ethanol reduce the binding affinity, which is well correlated with the free energy of transfer of dibucaine cation from water to mixed solvent. The effect of added propanol lies between the medium- and the small-chain alkanols. The significance of the present results is briefly referred to in relation with the solution state of dibucaine in a body fluid.

Keywords Dibucaine · Selective electrode · Anionic polymer · Binding isotherm · Alkanols

Introduction

At the earliest stage of a bioprocess, there should be binding of a biomolecule (ion) to some site where a series of biological events starts. So it is very interesting and informative to study binding of biomolecules to macromolecules as a model system [1]. A biomolecule in vivo may experience various molecular circumstances which are ionic as well as hydrophobic. In the present work, the binding of cationic dibucaine (Fig. 1a) to a synthetic anionic polyelectrolyte [poly(acrylamide 2-metheylpropane sulfonate)] (PAMPS) (Fig. 1b) was measured in the presence of various alkanols(C_1–C_3, C_5, and C_7). It is considered that these systems may mimic the actual situations in body fluid. Dibucaine employed as a model compound is a typical local anesthetic in surgery and dental clinical fields, and is cationic in a physiological environment (pK 9). The binding of dibucaine cation was measured by using an electrode selective to the organic cation.

Experimental

Materials

The dibucaine hydrogen chloride was commercially available (Sigma, D0638) and was used without further purification. The polyelectrolyte (PAMPS) was prepared by radical polymerization of acrylamide 2-methylpropane sulfonate and was purified by dialysis [2].

Electrode

The electrode sensing dibucaine cation is essentially a concentration cell:

Ag/AgCl|KCl bridge|C_o|M|C|KCl bridge|AgCl/Ag (cell A)

where M is the electrode membrane consisting of a slightly negatively charged copolymer of poly(vinyl chloride) and sodium acrylamide propane sulfonate and polymeric plasticizer (Elvaloy L-742, Du Pont) in a mass ratio of 1:2. There is no worry about dissolution of membrane materials even in aqueous alkanol solution, because of the polymer materials [3]; thus, it is tolerable for the present systems. This electrode was used in two ways in the

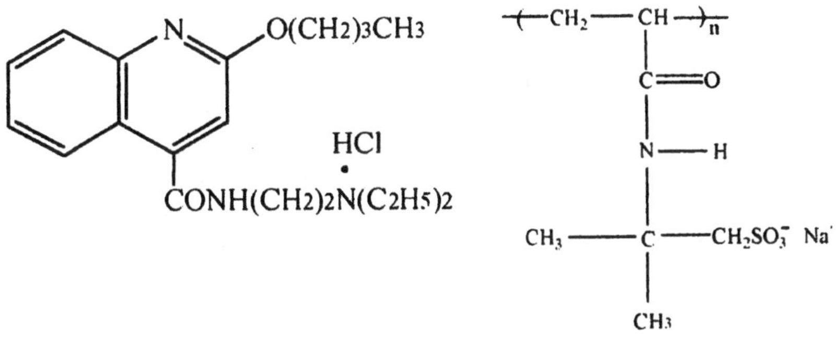

Fig. 1a,b Chemical structures

present study. Firstly, it was applied to the measurement of the binding of dibucaine cation to the polymer, when the same solvent was used in both the reference (inner) and the sample (outer) compartment of the electrode cell A. Secondly, it was utilized to evaluate the free energy of transfer for dibucaine cation from water to water–alkanol mixed solvents where the water was a solvent in the inner compartment and a mixed solvent was placed in the outer compartment. All measurements were carried out at 25 °C.

Results and discussion

Examples of the electrode response are shown in Fig. 2, where the electromotive force, E, is plotted against the logarithm of the dibucaine concentration. A linear (nearly Nernstian) response was observed in the absence of the polyion (calibration), which means that the electrode is sensing dibucaine cations.

However there is a deviation from the linear response in the presence of the polyion. Some of the dibucaine added must be taken up by the polyion. To see the interactions in more detail, a binding isotherm is very convenient. The differential response in Fig. 2 gives the equilibrium concentration of dibucaine cation, c_f, and the amount of binding, c_t-c_f. By plotting $\beta=(c_t-c_f)/c_p$ versus c_f, c_p being the polyion concentration (sulfonate group), the biding isotherms are built up as seen in Fig. 3, where binding of dodecylpyridinium is also shown for comparison. There is no sign of binding at low c_f, but at a certain concentration binding suddenly occurs, and soon tends to be saturated around $\beta=1$. This is characteristic of cooperative binding. The binding motivation is relatively little at a low equilibrium concentration of dibucaine cation. Once a dibucaine cation is attached at a binding site on the polymer, another dibucaine cation is bound with much more ease through hydrophobic interaction with a dibucaine molecule already bound to a site adjacent to the binding site in question. This kind of cooperative binding has often been observed for ionic surfactant–polyion systems with opposite electrical charges [4]. The bulky hydrophobic group of dibucaine allows efficient interaction when bound in sequential binding sites just as observed with surfactant molecules.

It would be interesting to see how added alkanols affect the binding affinity of dibucaine when the alkanol chain length is varied.

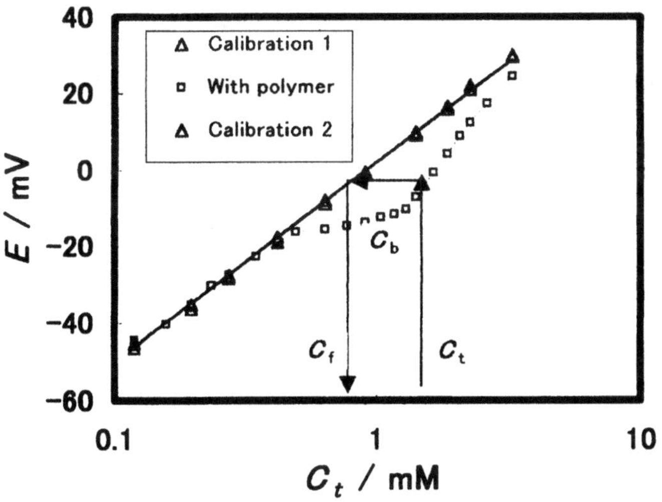

Fig. 2 Potentiogram of dibucaine hydrochloride

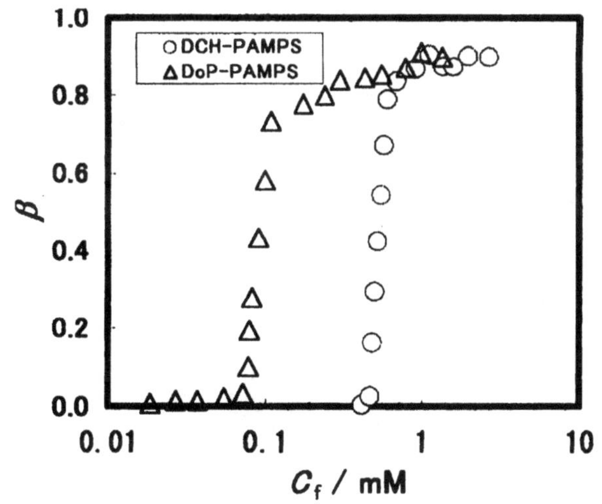

Fig. 3 Binding isotherms

Effects of medium-chain alkanols

The binding isotherms obtained with pentanol and heptanol added in various concentrations are shown in Fig. 4. It is seen that the addition of the alkanols causes the binding isotherm to shift to the left, i.e., to the lower equilibrium concentration of dibucaine cation, and this tendency becomes more marked with increasing alkanol concentration. This result shows that added alkanol enhances binding. The origin of the synergetic effect should be ascribed to the hydrophobicity of added alkanol, since more hydrophobic heptanol more effectively increases the dibucaine binding. The bound dibucaine molecules are necessarily separated by binding sites equidistantly spaced along a polymer strand. Hydrophobic interaction may be more effective at portions of bound dibucaine molecules far from the binding sites, since it is easier for them to be in contact with each other, while portions close to the binding sites are forced to be separated. There are exactly these portions where added alkanol molecules are partitioned to relieve hydrophobic interaction.

The binding sites on the polymer may be regarded as a one-dimensional adsorbent, for which the binding isotherm is expressed by the following equation in accordance with the statistical thermodynamic theory for a one-dimensional cooperative system [5]:

$$\beta = \frac{1}{2}\left[1 - (1-s)/\sqrt{(1-s)^2 + 4s}\right] \qquad (1)$$

where s is the reduced concentration ($s = uKc_f$, with K an intrinsic binding constant) and u is a cooperativity constant. The quantity, uK is a useful parameter describing interaction including the cooperative effect. $\log uK$ is plotted against c_A, the alkanol concentration in Fig. 5, where it is clearly seen that more hydrophobic heptanol has a greater enhancement effect.

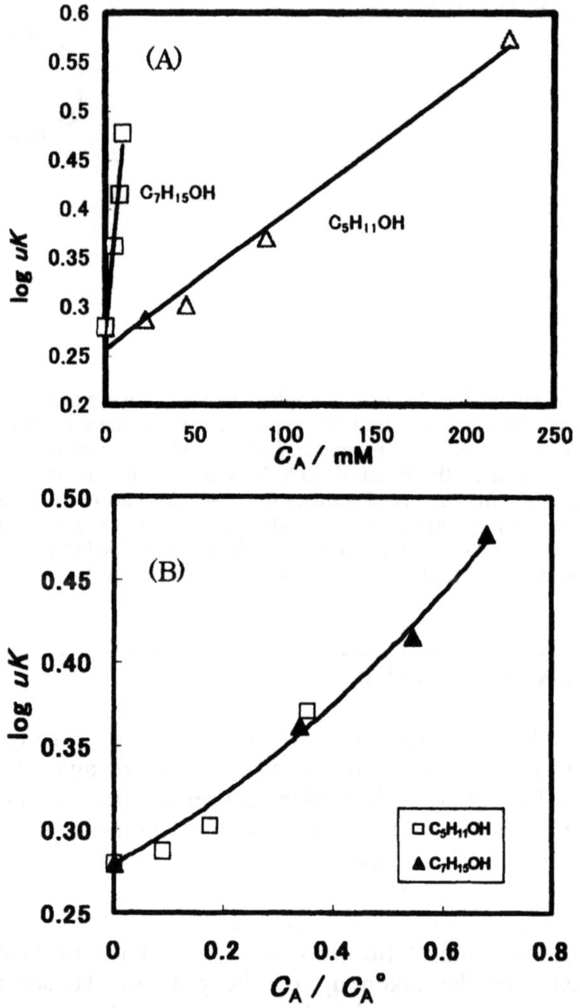

Fig. 5 Binding affinities plotted against **a** alkanol concentration and **b** against activity

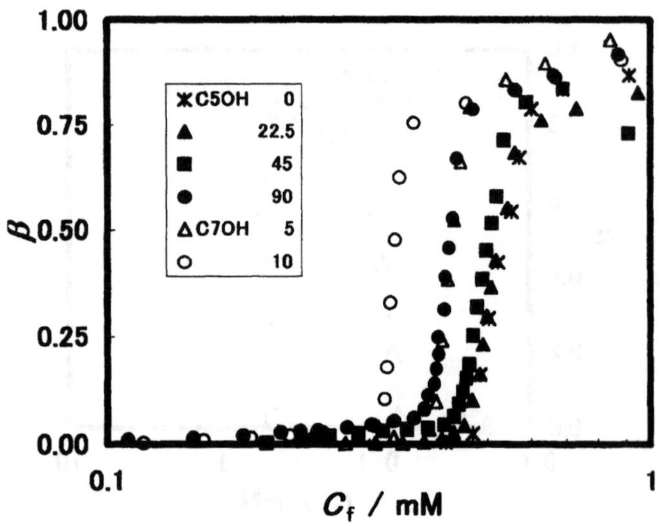

Fig. 4 Binding isotherms in mixed solvents

It is to be noted that all the experimental points fall on a universal line if the abscissa is plotted in $a_A = c_A/c_A^{\circ}$, where c_A° is the saturation concentration of the alkanol in water, as seen in Fig. 5b. The quantity a_A is a sort of thermodynamic activity of the alkanol with the standard state taken as pure alkanol. This means that the binding enhancement by alkanol is uniquely correlated with the activity of the alkanol.

Effects of lower alkanols

In contrast with the medium-chain alkanols just discussed, lower alkanols reduce the binding affinity as in Fig. 6, where the binding isotherms are seen to shift to the right on addition of ethanol. Methanol and propanol showed a similar effect. These lower alkanols should also

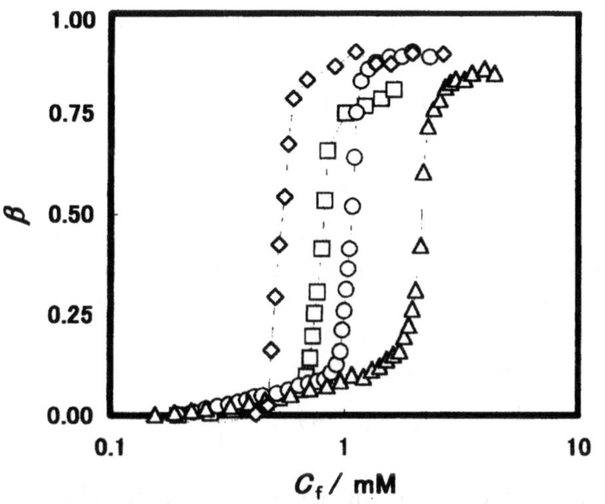

Fig. 6 Binding isotherms in ethanol–water mixed solvents. Ethanol content (% v/v): 0 (*diamonds*); 10 (*squares*); 20 (*circles*); and 30 (*triangles*)

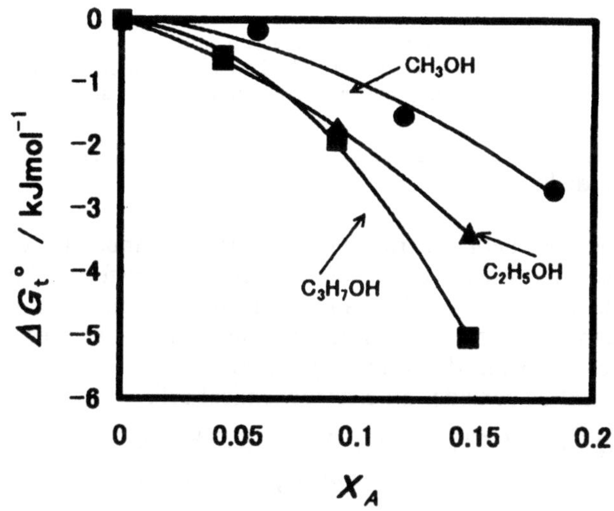

Fig. 7 Free energy of transfer of dibucaine cation versus ethanol mole fraction

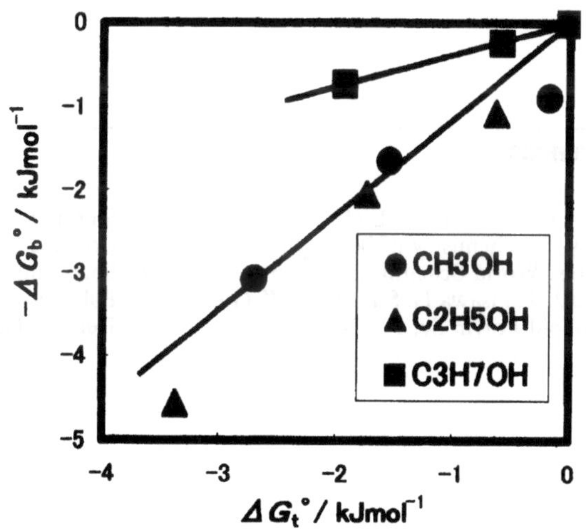

Fig. 8 Correlations between relative free energy of binding and free energy of transfer

have the enhancement effect as reported earlier, albeit very small, compared with medium-chain alkanols. However, this is overwhelmed by a solvent effect which is explained later.

As lower alkanol molecules are added to aqueous dibucaine solution, there is some possibility for dibucaine cations to meet alkanol molecules. This encounter occurs by breaking hydrophobic hydration around a dibucaine cation and releasing the hydrated water molecules, which results in thermodynamic stabilization bringing about a better solvent to dibucaine cation than in water. Dibucaine cations like mixed solvents.

The solvent effect is well represented by the free energy of transfer of dibucaine from water to water–alkanol mixed solvents, and is measured by the same electrode as used in studying the binding of dibucaine to polyion as explained in the Appendix. Figure 7 displays the free energies of transfer of dibucaine to the mixed solvents, which decrease with the alkanol contents with a larger effect being seen for the more hydrophobic alkanol.

The relative free energy of binding, $\Delta G_b^o = -2.303RT\log(uK/u_0K_0)$, in mixed solvents is plotted against the free energy of transfer in Fig. 8, where u_0 and K_0 are binding parameters in pure water solvent. The binding affinities are seen to be well correlated with the free energy of transfer for methanol and ethanol. The solvent effect is so high that dibucaine cations tend to be much more partitioned in the bulk mixed solvent phase than being bound to the polymer. However, the propanol line deviates above the other lower alkanol lines. This may be due to the solvent effect being partly counterbalanced by the binding-enhancement mechanism as explained earlier. The effect of propanol is considered to lie between the effects of medium-chain alkanols and those of methanol and ethanol.

Reference should be made to the relevance to clinical anesthetics. Dibucaine (molecule or ion) experiences various physicochemical environments in body fluid. Binding of dibucaine to an active site may be enhanced or reduced depending on the physiological environments. Especially, the enhancement effects by such materials as fatty acids and many other hydrophobic biomolecules are important since they are abundant in the body fluid. The present study suggests that a simple binding study in vitro has to be carefully interpreted when it is discussed in connection with the actual clinical application where anesthetic molecules experiences much more complicated molecular environments.

Acknowledgements K.S. expresses his sincere congratulations and thanks to Shuji Saito, whose name was ubiquitous when checking references and who also inspired K.S when he started a research project on surfactant–polymer interaction in the 1960s.

Appendix

The free energy of transfer is a thermodynamic idea of a free-energy change of a solute when it is transferred from one solvent to another, and gives information about the change of solvation. For an ionic solute, it is easily measurable in the following manner.

Let us consider an electrochemical cell,

$$\mathrm{Ag/AgCl|KCl\,bridge}|c_\mathrm{w}\;|\mathrm{M}|c_\mathrm{m}\;|\mathrm{KCl\,bridge}\;|\mathrm{AgCl/Ag}\quad (\text{cell B}).$$

This cell is very similar to cell A, with the only difference being that c_m is the solute concentration in a mixed solvent, while c_w is that in an aqueous solution.

Thermodynamically, the electrochemical potential of dibucaine cation is expressed for the two solvents as

$$\tilde{\mu}_\mathrm{w} = \mu^\mathrm{o}_\mathrm{w} + RT\ln a_\mathrm{w} + zF\phi_\mathrm{w} \tag{2}$$

and

$$\tilde{\mu}_\mathrm{m} = \mu^\mathrm{o}_\mathrm{m} + RT\ln a_\mathrm{m} + zF\phi_\mathrm{m}, \tag{3}$$

where the subscripts w and m stand for water and mixed solvents, respectively, and ϕ is the electrostatic potential. When the two solutions are separated by an electrode membrane at equilibrium, $\tilde{\mu}_\mathrm{w} = \tilde{\mu}_\mathrm{m}$. So we have, after rearrangement of equations,

$$\Delta E = -\frac{1}{zF}\left[\left(\mu^\mathrm{o}_\mathrm{m} - \mu^\mathrm{o}_\mathrm{w}\right) + RT\ln\frac{a_\mathrm{m}}{a_\mathrm{w}}\right], \tag{4}$$

where $\Delta E = \varphi_2 - \varphi_1$ is the asymmetric potential of cell B.

The free energy of transfer, $\mu^\mathrm{o}_\mathrm{m} - \mu^\mathrm{o}_\mathrm{w}$, is obtained by measuring the electromotive force of cell B at a condition where the second term on the right-hand side of Eq. (4) vanishes.

References

1. Steiner RF, Garone L (1991) The physical chemistry of biopolymer solutions. World Scientific, London
2. Liu J, Takisawa N, Shirahama K (1999) Colloid Polym Sci 277:247
3. Shirahama K, Liu J, Takisawa N (1999) Colloids Surf A 147:133
4. Shirahama K (1998) In: Kwak JCT (ed) Polymer–surfactant systems, vol 77. Dekker, New York, pp 143–191
5. Satake I, Yang JT (1976) Biopolymers 15:2263

Progr Colloid Polym Sci (2003) 122: 87–94
DOI 10.1007/978-3-540-36114-5

Mihaela Olteanu
Ilie Mandru
Manuela Dudau
Sandu Peretz
Otilia Cinteza

The aqueous liquid/liquid interphases formed by chitosan–anionic surfactant complexes

M. Olteanu · I. Mandru · M. Dudau,
O. Cinteza
University of Bucharest,
Department of Physical Chemistry,
Bucharest, Romania

S. Peretz
Institute of Physical Chemistry
"I.G. Murgulescu",
Department of Colloids,
Bucharest, Romania

M. Olteanu (✉)
4–12 Regina Elisabeta Blvd.,
70346, Bucharest, Romania
e-mail: oltmi@hotmail.com
Tel.: +40-21-3143508

Abstract The solubility behavior of cationic chitosan complexes with two anionic surfactants, sodium dodecyl sulfate (SDS) and sodium bis(2-ethylhexyl)sulfosuccinate (AOT), as well as the formation of gel-like membranes capable of separating two aqueous polymer and surfactant solutions has been investigated. The chitosan–surfactant complexes can exist either as homogenous clear solutions or precipitates or as membranes, depending on the molar ratio between the polymer structural units and the surfactant and the degree of protonation of the polymer. The transitions between the phases were determined and correlated with structural and compositional parameters. The conditions for the formation of new stable chitosan–AOT microcapsules were observed and were compared with those of chitosan–SDS. The two surfactants with different structures behave very differently. SDS forms gel-like membrane with chitosan at a much lower concentrations than AOT, even though the former has a higher critical micellar concentration. It was postulated that in forming a gel-like membrane from chitosan–surfactant complexes a certain role is played by the existence of micelles and their size and shape. The smaller, spherical, more abundant micelles in SDS solutions lead to the formation of mechanically resistant membranes at a lower concentration than AOT. The difference in behavior is more evident in the concentration domain where AOT forms lamellar self-assemblies. The added ethyl alcohol affects both the solubility curves and the membrane formation.

Keywords Chitosan–anionic surfactant complexes · Membranes · Microcapsules · Solubility

Introduction

The binding of surfactants to water-soluble polymers brings about dramatic changes in the physical properties of the polymers, such as solubility, viscosity, gel formation and adsorption to solid surfaces. These changes can be tailored by choosing the surfactant, the ratio between the molar concentration of polymer units to surfactant, the degree of ionization of the polymer and by controlling the pH or salt concentration. The most spectacular changes in polymer properties are met when polyelectrolytes interact with ionic surfactants of opposite charge [1, 2].

The formation of complexes between the components of opposite charge can be characterized by steep binding isotherms of surfactants to polymer. The process is strongly cooperative, suggesting the contribution of the alkyl chain of the surfactants to the binding affinity. In steep binding isotherms, the existence of a critical aggregation concentration [3, 4], often much lower than the critical micellar concentration (cmc) of the surfactant, is a general property of these systems. The complex formation is attributed to ion-pair formation of stoichiometric ratios, but hydrophobic interactions between the alkyl chains seem to play a certain role, with the possibility of charge inversion.

To obtain a polymer–surfactant complex in diluted and semidiluted polymer solutions, the surfactant micellization is the most important, even if perturbed by the presence of the polymer [5]. In the case of hydrophilic polyelectrolytes, the attraction between the polymer and the micelles of opposite charge is purely electrostatic. The ionic micelles are separated from the oppositely charged polyelectrolytes by a layer of water molecules [6]. The precipitation in the mixtures of stoichiometric polyelectrolytes and oppositely charged surfactants and the formation of hexagonal or lamellar structures has also been reported [7].

The polymer–surfactant complexes can be present in the system as soluble entities, particles of colloidal size or gels with good mechanical properties. All these specific behaviors make polymer–surfactant complexes valuable materials for microencapsulation of active components by coacervation, useful in controlled drug delivery, food, beverages, cosmetics and hair care products [8, 9].

Surface properties of polymer–surfactant complexes differ substantially from those of the parent materials. The surface activity of water-soluble polymers is very low, but the complexes with surfactants are more surface-active than the corresponding surfactants in aqueous solutions [10]. The surface tension isotherms of the complexes are located under the corresponding surface tension isotherms of the surfactant alone. The complexes of cationic polymers with sodium dodecyl sulfate (SDS) present a synergetic adsorption at the water/air interface as well as viscoelastic behavior. The viscoelasticity of the solution/air interface confirms the formation of soluble and tensioactive complexes between polymers and oppositely charged surfactants [11].

One of the most important naturally occurring polymers is chitin. It is the second most abundant after cellulose, with the chemical structure 2-amino-2-deoxy-(1→4)-β-D-glucopyranan. Chitosan is a deacetylated chitin, being the only naturally occurring cationic polymer.

The protonation of amino groups makes chitosan water-soluble and the two hydroxyl groups can be chemically modified to give useful products in different fields [12]. Chitosan is legally admitted in many countries as a drug and food additive and therefore it finds numerous practical applications. The majority of them are based on the capability of chitosan for gel formation and to enhance the viscosity. Complexes with anionic surfactants also have good gelation properties [13, 14].

The interaction of SDS with chitosan was studied in terms of the binding isotherm as a function of the degree of deacetylation, which determines the polymer electrical charge density [15]. The capacity to form pervaporation membranes [16] and gel microcapsules and particles was also considered [13, 14]. The present contribution aims to use in aqueous and ethanolic solutions the complexes of chitosan with SDS, a surfactant with a linear alkyl chain, and sodium bis(2-ethylhexyl)sulfosuccinate (AOT), a branched-chain surfactant, to form resistant gelling membranes capable of separating two aqueous solutions.

By contacting drops of chitosan solution protonated with acetic acid with surfactant solution, the system can produce homogeneous solutions or form precipitates or gelling membranes that separate the two aqueous solutions. The mechanical properties of the resulting membrane are strongly dependent on polymer and surfactant concentrations as well as on the degree of chitosan protonation. The membranic gel at the interface is the result of insoluble chitosan–surfactant complex formation by the diffusion of surfactant in the polymeric solution, followed by a cross-linking mechanism, implying electrostatic as well as hydrophobic interactions.

Experimental

Materials

Chitosan with high molecular weight (M_r of about 600,000) was supplied by Fluka and was used without purification. Reagent grade SDS (Merck), AOT (Sigma), acetic acid (Merck) and ethyl alcohol (Merck) were used as received. Bidistilled water was used throughout the study.

Preparation of chitosan and surfactant solutions

Chitosan stock solutions were prepared in 1, 2.5 and 5 wt% acetic acid (pH 2.8, 2.6 and 2.45). The solution with a concentration of 0.1 mol/l monomeric units in 5 wt% acetic acid is very viscous and difficult to handle. The other chitosan solutions were prepared by dilution with acetic acid (the pH varied over the entire chitosan concentration domain between 3.5–5). The solutions of SDS and AOT were prepared from stock solutions of concentrations between 0.1 and 5×10^{-4} mol/l.

Microcapsule preparation and characterization

The chitosan solutions were magnetically stirred at 500 rpm for 48 h on a heated (40 °C) plate (IKA Labortechnik) and then dropped by means of a microsyringe into the surfactant solutions and into 2:1 (v/v) mixtures of surfactant solutions with ethyl alcohol. The syringe needle was 0.3 mm in diameter. The chitosan solution was dropped from approximately 3 cm to the surfactant solution surface so that the chitosan droplet did not remain at the surfactant solution/air interface to form hemispheres. The microcapsules form instantaneously when the chitosan solution droplets reach the surface of the surfactant solution. The microcapsules were directly prepared in spectrophotometry cuvettes for system observation, turbidity determination and visualization of the shape. The microcapsule evolution was followed through video enhanced microscopy with a Philips video camera, for 1 h and then the observations regarding the aspect of the chitosan surfactant complex were used to draw the chitosan surfactant phase diagrams.

Phase diagrams were drawn by using samples whose concentrations in chitosan increased from each other by 0.01 and by 0.001 mol/l in the case of the surfactants.

Pelicule preparation

The chitosan solution was spread on the glass plate. The solution of surfactants was dropped on the chitosan solution and when the pelicule of the insoluble complex was formed, the plate was immersed in the same surfactant solution. After 48 h, they were washed with water.

Turbidity determination

The turbidity determinations were done by means of a Helios UNICAM UV–vis spectrophotometer in order to note the passage from domain A to domain B.

Contact-angle measurements

The contact angle was determined at the pelicule/water/air interface by the gas-bubble method. The pelicule of the insoluble polymer–surfactant complex was put in contact with water. An air bubble was made to adhere on the pelicule surface. The contact angle was calculated from the profile of the air bubble which was captured with the video camera (Philips). The contact angle values are the mean of ten determinations.

Surface tension measurements

The cmc values of surfactants in different conditions were determined from the variation of surface tension with surfactant concentration. The du Noüy method was used with an accuracy of 0.5 mN/m. The inflexion points of the surface tension variation versus the surfactant concentration were determined by titration of the surfactant solution with diluted polymer solution of constant concentration. For the initial samples, the polymer–surfactant ratio was 5 for AOT and 50 for SDS, leading to the formation of an insoluble polymer–surfactant complex. The presence of chitosan in the diluted range does not influence the aggregation of AOT either in 2.5 wt% acetic acid in water or in a 2:1 (v/v) water–ethyl alcohol mixture (Table 1).

All measurements were made at 20 °C.

Results and discussions

During the polymer–surfactant experiments one observes several behavioral domains characterized by specific states depending on the molar ratio between the structural units of the chitosan and the surfactant, namely:

1. Monophase solutions, homogeneous and transparent within domain A.
2. Precipitates, preceded by the appearance of an advanced turbidity within domain B.
3. Formation of gel-like membranes able to separate the two aqueous solutions of polymer and surfactant within domains C (microcapsule that expels its content), D (stable microcapsule) and E (shrunk microcapsule) (Fig. 1). In order to separate the two solutions, the membrane formed at the interface by the polymer–surfactant complex needs to have stability and mechanical resistance.

In the domain where the gel-like membrane forms, the particles present different evolutions owing to the mechanical resistance as well as to the osmotic pressure difference at the interface. At low surfactant concentrations in domain C, the droplets tend to expel the liquid from the interior and form an excrescence (Fig. 1a), while at high surfactant concentrations, domain E (Fig. 1c), the droplets contract. The domain between C and E, or domain D, represents a zone where the particles formed have maximum stability (Fig. 1b). The osmotic pressure is equilibrated in the system without any additive for its regulation. When one analyzes the transition from the homogeneous solutions in domain A to the precipitates in domain B, it can be seen as the solubility limit of a chitosan–surfactant complex. However, some common behavior but also some substantial differences are noticed for the two surfactants. For example, the transition from domain A to domain B (A→B) in the case of SDS for different degrees of

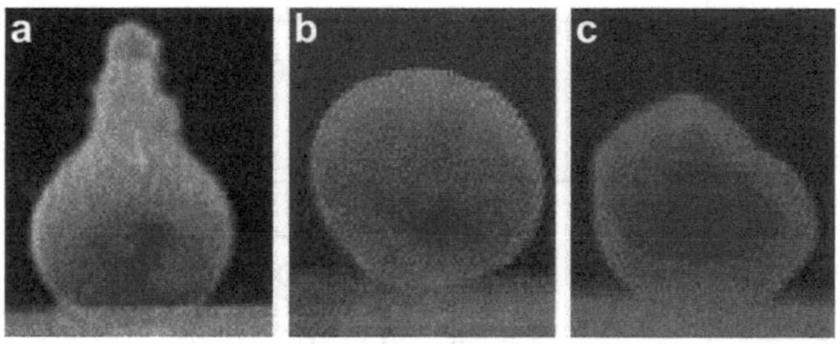

Fig. 1 Aspect of the membranes formed by addition of polymer solution to surfactant solutions: **a** spherical particles with excrescence; **b** spherical particles with maximal resistance; **c** shrinking particles

Table 1 Micellar characteristics of the surfactants sodium dodecyl sulfate (SDS) and sodium bis(2-ethylhexyl) sulfosuccinate (AOT) used for forming complexes with chitosan

Surfactant	Medium			
	Water	Water–ethyl alcohol: 2:1 (v/v)	10^{-3} mol/l chitosan in 2.5 wt% acetic acid	10^{-3} mol/l chitosan in 2.5 wt% acetic acid and water–ethyl alcohol 2:1 (v/v)
	Inflection point of surface tension isotherm 10^{-3} mol/l			
SDS	8.0	15.0	5.0	10.0
AOT	2.0	3.0	3.2	3.0

protonation (2.5 and 1 wt% acetic acid) as well as in 2:1 (v/v) water–ethyl alcohol mixtures presents two characteristic branches (Fig. 2). The solubility is practically unaffected by the surfactant concentration up to a critical value, called the minimal surfactant concentration (MSC), located around 1×10^{-3} mol/l for both degrees of protonation as well as water–ethyl alcohol solutions (Fig. 3).

At a polymer concentration higher than a value defined as the minimal polymer concentration (MPC), the A→B transition or solubility curve became independent of polymer concentration. The MPC increases with the degree of protonation and decreases in water–ethyl alcohol solutions (Table 2).

The behavior of AOT surfactant is different from that of SDS as far as the transition A→B (solubility curve) is concerned (Figs. 4, 5). A certain dependence is observed for the solubility versus surfactant concentration curve on the degree of protonation. Above the MSC values the

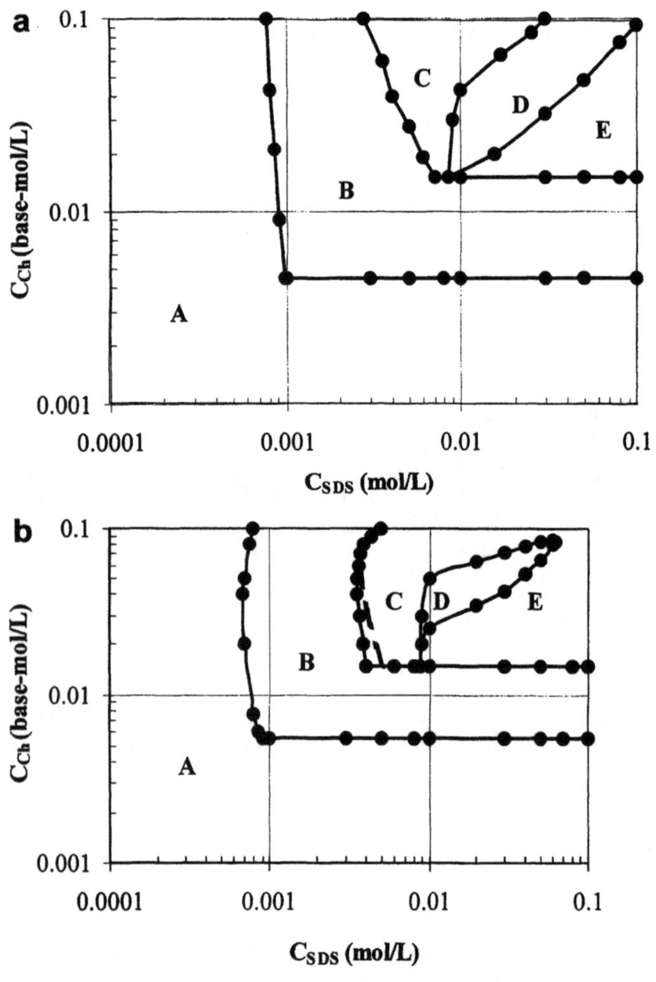

Fig. 2 Solubility and formation domains of gel-like membranes in the protonated chitosan–sodium dodecyl sulfate (SDS) system with **a** 1 wt% acetic acid and **b** 2.5 wt% acetic acid

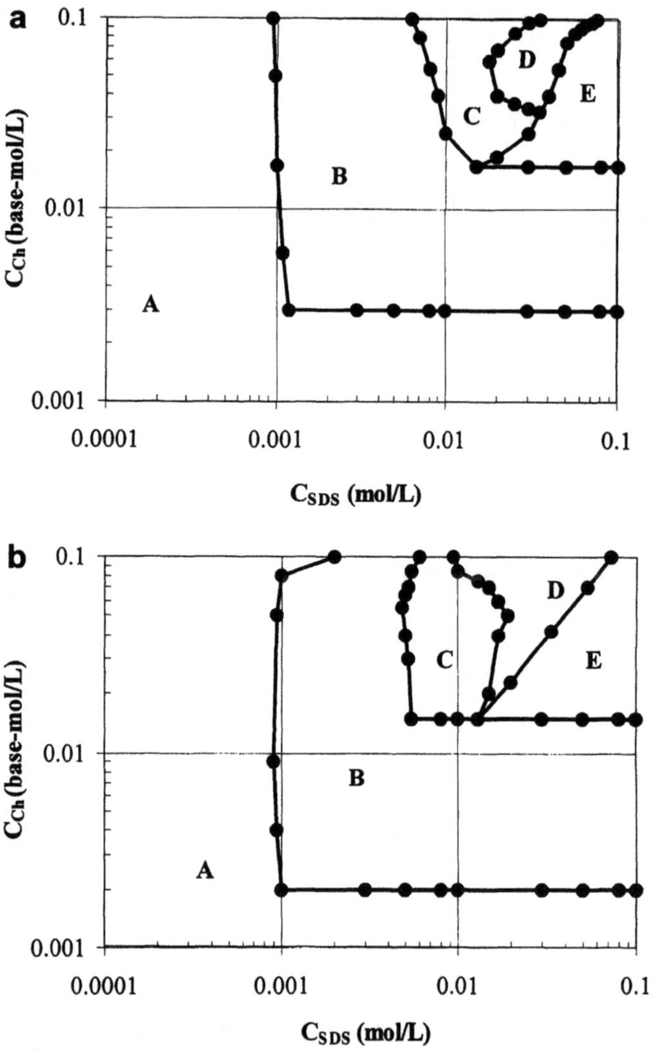

Fig. 3 Solubility and formation domains of the gel-like membranes in SDS–chitosan in 2:1 (v/v) water–ethyl alcohol solutions protonated with **a** 1 wt% acetic acid and **b** 2.5 wt% acetic acid

solubility curve decreases at lower chitosan concentrations. In water–ethyl alcohol solutions, the solubility curve changes so that a domain independent of the concentrations of the two components cannot be sharply defined. The values in Table 2 were obtained as crossing points of extrapolated branches of the A→B curves, except the value from Fig. 4a.

These different solubility behaviors are also found at higher component concentration where gel-like membranes are formed (domains C–E). The transition curves from precipitates to gels able to separate the two liquid phases (transition from domain B to domains C, D and E), for chitosan–SDS systems are quite similar to those from domain A to domain B, especially at higher degrees of protonation (2.5 wt% acetic acid, Figs. 2, 3). However, the transitions are different at lower degrees of protonation (1 wt% acetic acid) where the change from

Table 2 Characteristics of chitosan–SDS and chitosan–AOT complex solubilities and formation of gel-like membranes

Surfactant	Acetic acid concentration (wt%)	Medium	Minimal surfactant concentration (10^{-3} mol/l)	Minimal polymer concentration (10^{-3} base mol/l)	Minimal surfactant concentration for domain D (10^{-3} mol/l)[b]
SDS	1.0	Water	1.00	4.50	8.50
	2.5	Water	0.85	6.00	8.80
	1.0	Water–ethyl alcohol 2:1 (v/v)	1.20	3.00	35.00
	2.5	Water–ethyl alcohol 2:1 (v/v)	1.00	2.00	13.00
AOT	1.0	Water	1.00	6.50	10.00
	2.5	Water	3.50[a]	9.00[a]	15.00
	1.0	Water–ethyl alcohol 2:1 (v/v)	1.00[a]	11.00[a]	11.00
	2.5	Water–ethyl alcohol: 2:1 (v/v)	1.50[a]	8.00[a]	16.00

[a]Values obtained as crossing points of extrapolated branches of A–B curves
[b]At the minimum concentration of chitosan

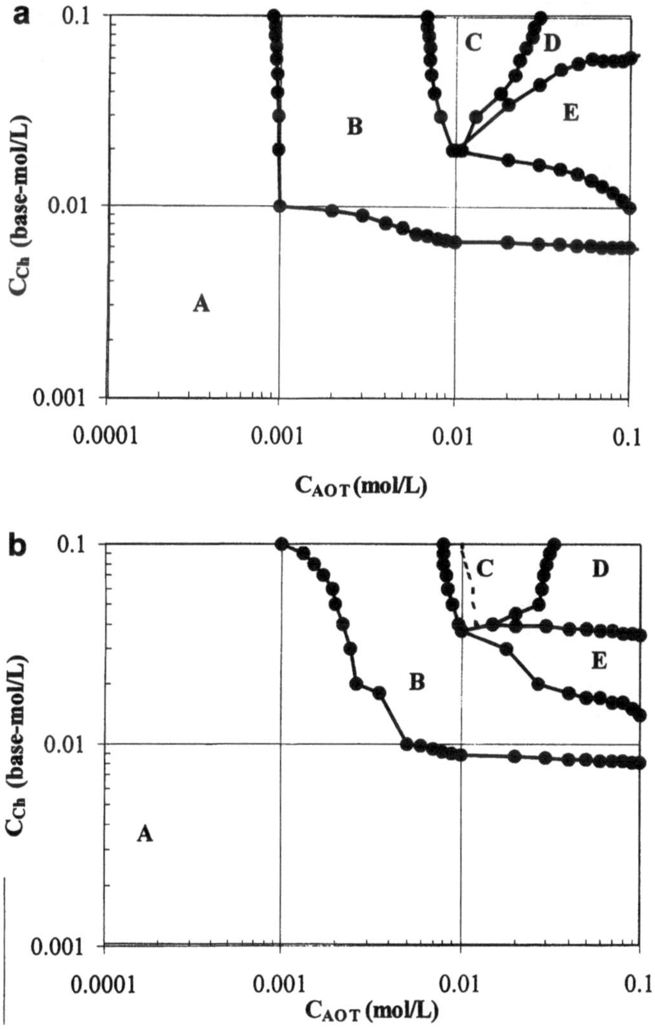

Fig. 4 Solubility and formation domains of the gel-like membranes in the sodium bis(2-ethylhexyl)sulfosuccinate (AOT)–chitosan system protonated with **a** 1 wt% acetic acid and **b** 2.5 wt% acetic acid

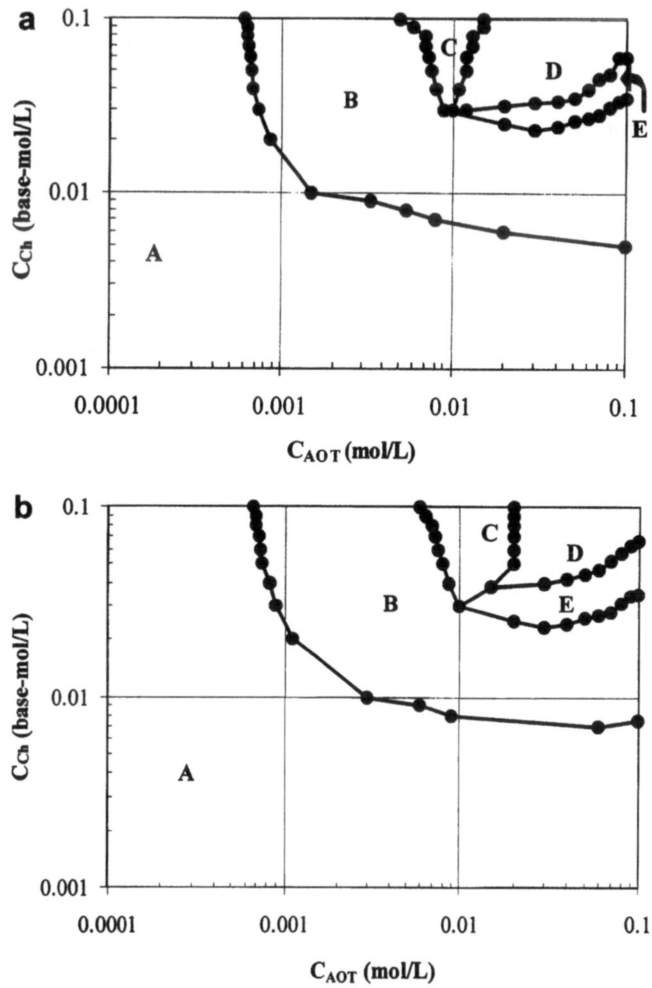

Fig. 5 Solubility and formation domains of the gel-like membranes in AOT–chitosan in 2:1 (v/v) water–ethyl alcohol solutions protonated with **a** 1 wt% acetic acid and **b** 2.5 wt% acetic acid

domain B to domain C is no longer independent of polymer concentration. The behavior of AOT surfactant as far as the transition from domain B to domains C, D and E is concerned (Figs. 4, 5) is clearly different from that of SDS, especially in water–ethyl alcohol solutions.

Among the domains where gel-like membranes are formed, one remarks those where membranes have enough mechanical resistance to ensure the optimal stability for chitosan droplets in surfactant solution (domain D). The extent of these domains depends on the degree of polymer protonation but mainly on the surfactant. The beginning of domain D in aqueous SDS solutions is in the vicinity of the cmc (8×10^{-3} mol/l), but at concentrations higher than the cmc in the case of water–ethyl alcohol solutions. These facts suggest the necessity for the surfactant to be present in the micellar form in order to obtain membranes with good mechanical resistance. In the case of AOT, the minimal concentration where domain D appears (see Table 2) is much higher than 2×10^{-3} mol/l, a concentration that corresponds to the cmc of AOT in water [17].

The size and shape of the D domains are dependent on the degree of chitosan protonation. The increase of this parameter in the systems containing SDS leads to a reduction of domain D (Figs. 2b, 6). This behavior can be explained by the increase in the degree of binding of the surfactant to chitosan along with the increase in the degree of protonation that results in a decrease in the micellar concentration. The micelles ensure a physical cross-linking of chitosan strands.

In the case of AOT, the dependence on the degree of protonation of domain D (Figs. 4a, 5) is different than for SDS. The domain is more extended for high degrees of protonation and the MSC values are very different. Domain D in the case of AOT (Fig. 6) is placed at higher polymer concentrations and even at higher surfactant concentrations.

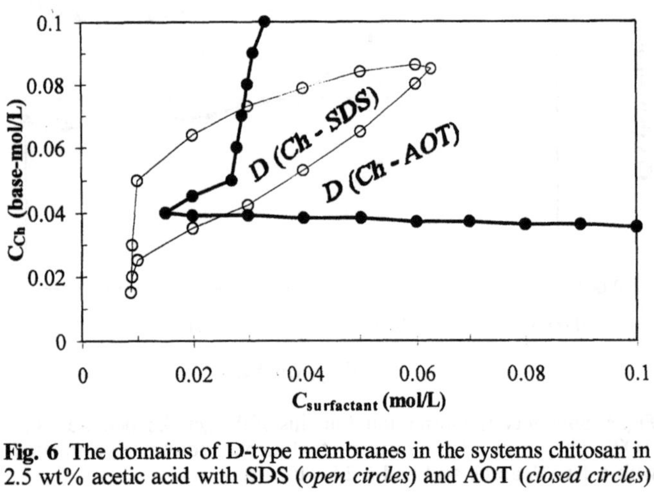

Fig. 6 The domains of D-type membranes in the systems chitosan in 2.5 wt% acetic acid with SDS (*open circles*) and AOT (*closed circles*)

A possible explanation could be given if one admits that the shape of the surfactant micelles plays a certain role in the physical cross-linking of the chitosan–surfactant complex. It is known that SDS forms relatively small spherical micelles (40–60 monomers/micelle), while AOT forms mainly lamellar-type micelles [18] having aggregation numbers much higher than those of SDS. For this reason, the physical cross-linking by a chitosan–surfactant complex interaction takes place at AOT concentrations corresponding to the lamellar micelle formation.

The diagrams in Figs. 2b and 4b were obtained by using a modified working procedure where a small amount of nonmicellar surfactant able to form ionic pairs with the positive centers of the polyelectrolyte was added to the chitosan solution in 2.5 wt% acetic acid (before the dropping starts). The prebinding of chitosan to the surfactant molecules (only up to 10:1 chitosan–surfactant ratio, in order to avoid the precipitation in polymer solution) did not bring about any noticeable modification to domains D and E, while domain B (where precipitates form) became larger. The dotted line in Figs. 2b and 4b indicates the shift of the B→C limit toward higher surfactant concentration.

We presume that the partial screening of the positive charges decreases the polyelectrolyte chain expansion and, therefore, a higher concentration of surfactant micelles is needed in the aqueous solution to diffuse toward the remaining positive centers. This working procedure proves indirectly the role played by micelles in gel formation.

In 2:1 (v/v) water–ethyl alcohol solution the extent of domain D is affected both in size and in the MSC and MPC (Figs. 2, 3). The presence of alcohol, which is a nonsolvent for chitosan, diminishes considerably the A→B transition in the domain above the MSC and changes the domain shape, while the A→B solubility curve for AOT in water–ethyl alcohol is placed at higher polymer concentrations.

The data presented in Figs. 2, 3, 4, 5 and 6 and Table 2 point out the important differences between SDS and AOT. Although AOT has a lower cmc than SDS and one would expect to form complexes and gel-like membranes with chitosan more easily, their behavior is reversed. Thus, SDS forms gel-like resistant membranes more easily, the MSC needed for the apparition of domain D of stable microcapsules being an order of magnitude lower than that for AOT (Table 2). It is possible that this important difference in behavior results from the micellar characteristics of the two surfactants. They may influence the complex formation process of the gel-like membranes by the interaction between the chitosan–surfactant complex and the surfactant in the micellar form.

In the case of the cationic chitosan–SDS system, it was found that the mechanism of physical gelation proceeds by means of spherical micelles with a lower aggregation number than in the aqueous surfactant solution [13, 14]. It was established that polymers interacting strongly with SDS, among which the cationics have a decreasing effect on SDS micelle stability, and the aggregates formed have smaller aggregation numbers than the corresponding micelles [19]. This means that chitosan chains in the gel can be outdistanced as a function of the diameter of the SDS spherical micelles [20], which form over the whole concentration domain studied by us. In the case of the AOT–chitosan system it is possible that a single bilayer could be located between two polycationic chains, leading to a denser and more ordered structure. Several bilayers attached to neighboring polymeric chains could also lead to an ordered, but less dense, structure. The chitosan–AOT gels could be nanostrucutred in both cases.

Studies on membranes of cationic chitosan complexed with anionic surfactants, among which are SDS and a sulfosuccinate-type surfactant [15], support the proposed mechanism. The membranes obtained with disodium cocamido mipa-sulfosuccinate have the same performances as those of SDS as far as the separation factor is concerned, but much lower fluxes, denoting a denser structure and a different mode for membrane aggregation [16].

In the hypothesis of the formation the cationic chitosan–SDS gel via spherical micelles and cationic chitosan–AOT gel via lamellar bilayers, the resulting membranes should differ by their polarities. In order to verify this, pelicules applied on glass were prepared having compositions within domain D. The contact angles at the pelicule/water/air interface were found to be 101° and 108° for SDS and AOT pelicules, respectively.

The aspect of the AOT pelicule was totally different from that of SDS. The first was clear and transparent in comparison with the second, which was opalescent. This fact represents another proof for the different degrees of ordering of the two structures.

The aforementioned interaction mechanism can also explain the important differences between the gel formation limits of AOT (concentrations between 0.08 and 0.1 mol/l, Fig. 4) in comparison with SDS (Fig. 2). The differences are more noticeable with 2.5 wt% acetic acid. The AOT lamellar phase controls both precipitate and gel formation in the presence of chitosan.

Conclusions

The data presented in this study prove the possibility of obtaining gel-like membranes able to separate two aqueous solutions and opens the possibility of developing microencapsulation processes for controlled drug and other active-principle delivery.

The chitosan–SDS and chitosan–AOT systems have some common features, but also some differences. Both systems can exist as homogeneous solutions, as precipitates or can form gel-like membranes, depending on the chitosan–surfactant molar ratio. The transition from homogeneous solution to precipitates is considered as the solubility curve of the chitosan–surfactant complex and is characterized by the existence of a minimal surfactant concentration under which the solubility curve is unaffected by surfactant concentration. The solubility of the chitosan–surfactant mixture is described by a minimal polymer concentration that is dependent on the degree of chitosan protonation and is affected by the presence of ethyl alcohol.

The size of the domain where gel-like membranes are formed, especially domain D, is strongly dependent on surfactant and the degree of protonation of chitosan. Chitosan and SDS produce gel-like membranes at a much lower surfactant concentration than chitosan and AOT. This is very likely due to differences in the micellar structure of the surfactants, even if AOT has a lower cmc than SDS in water. The shape of the surfactant micelle obviously plays the crucial role in the process of physical cross-linking and gel-like membrane formation.

References

1. Goddard ED (1993) In: Goddard ED, Ananthapadmanabhan KP (eds) Interactions of surfactants with polymers and proteins. CRC, Boca Raton, pp 171–201
2. Shirahama K (1998) In: Kwak JCT (ed) Polymer–surfactant systems. Dekker, New York, pp 143–192
3. Shirahama K (1974) Colloid Polym Sci 252:978
4. Shirahama K, Ide N (1976) J Colloid Interface Sci 54:450
5. Hansson P, Lidman B (1996) Curr Opin Colloid Interface Sci 1:604
6. Thalberg K, van Stam J, Lindblad C, Almgren M, Lidman B (1991) J Phys Chem 95:8975
7. Antonietti M, Burger C, Effing JJ (1995) Adv Mater 7:751
8. Illum L (1998) Pharm Res 15:1326
9. Shahidi F, Arachchi JKV, You-Jin Jeon (1999) Trends Food Technol 10:37
10 Asnaeios A, Langevin D, Argillier JF (1996) Macromolecules 29:7412
11. Regismond STA, Winnik FM, Goddard ED (1996) Colloid Surf A 119:221
12. Muzzarelli RAA, Jeuniax C, Gooday CW (1986) Chitin in nature and technology. Plenum, New York

13. Babak VG, Merkovich EA, Galraikh LS, Shtykova EV, Rinando M (2000) Mendeleev Commun Electron Vers 2:83
14. Babak VG, Merkovich EA (2000) 5th World Surfactant Congress, CESIO 2000, pp 657–663
15. Wei YG, Hudson SM (1993) Macromolecules 26:4151
16. Nam SY, Lee YM (1999) J Membr Sci 157:63
17. Li ZX, Lu JR, Thomas RK, Penfold J (1995) Prog Colloid Polym Sci 98:243
18. Rogers J, Winsor PA (1969) J Colloid Interface Sci 30:247
19. Dhara D, Shah DO (2001) J Phys Chem 105:7133
20. Fontell K (1981) Mol Cryst Liq Cryst 63:59

Progr Colloid Polym Sci (2003) 122:95–102
DOI 10.1007/978-3-540-36114-5

Rafael Barreiro-Iglesias
Carmen Alvarez-Lorenzo
Angel Concheiro

Microcalorimetric evidence and rheological consequences of the salt effect on carbopol–surfactant interactions

Abstract The aggregation process of Carbopol 934NF, a Food and Drug Administration approved cross-linked poly(acrylic acid) commonly used in pharmaceutical technology, with nonionic and ionic surfactants in 0.9% NaCl dispersions and artificial lacrimal fluid was analyzed using titration microcalorimetry and rheometry. Titration microcalorimetry experiments indicated that the interaction processes in 0.9% NaCl are thermodynamically similar to those in water, although slightly less intense. In contrast, in artificial lacrimal fluid the carbopol–surfactant interactions were hindered. Carbopol association with Tween 80 (exothermic) or Pluronic F-127 (endothermic) occurs through hydrogen-bonding interactions, which are scarcely debilitated in 0.9% NaCl but strongly impeded in artificial lacrimal fluid. In this medium, the higher pH causes more carboxylic acid groups to become ionized, and the presence of calcium ions blocks them partially. In 0.9% NaCl, there is a shielding effect of the ionic groups of the polymer that promotes the hydrophobic association with sodium dodecyl sulfate and makes the complex formation with benzalkonium chloride more difficult. In both cases, the processes were exothermic. In artificial lacrimal fluid, the interaction with sodium dodecyl sulfate disappears while an endothermic association with benzalkonium chloride takes place. The ion-induced changes in the intensity of the carbopol–surfactant association are reflected in the rheological behavior of the 0.25% acrylic polymer dispersions. The results show that the nature and concentration of ions may have a strong influence on the design and on the physiological behavior of drug delivery systems based on carbopol gels.

Keywords Carbopol · Tween 80 · Pluronic F-127 · Sodium dodecyl sulfate · Benzalkonium chloride

R. Barreiro-Iglesias · C. Alvarez-Lorenzo
A. Concheiro (✉)
Departamento de Farmacia y Tecnología Farmacéutica, Facultad de Farmacia, Universidad de Santiago de Compostela, 15782-Santiago de Compostela, Spain
e-mail: ffancon@usc.es
Fax: +34-981-547148

Introduction

Slightly cross-linked polymer networks that can absorb a large amount of solvent and can undergo sharp conformational transitions in response to changes in temperature, pH or chemical composition of the medium find a wide range of applications [1]. Smart microgel systems are particularly useful as platforms for gelling in situ drug dosage forms, which are easily administered (ocularly, orally, parenterally) in the liquid state [2, 3, 4, 5]. In contact with the physiological environment, the viscosity and elasticity of the system rise strongly, increasing the residence time on the application area and allowing optimum control of the drug release process. Carbopol microgels, constituted by poly(acrylic acid) cross-linked with allylsucrose, are the most commonly used base for gelling in situ devices because of their high biocompatibility [6], mucoadhesion [7, 8], and thickening ability both in aqueous solutions and in polar organic solvents [9].

Recently, it has been shown that the presence of surfactants in the medium may induce or strongly alter the volume-phase transitions of hydrogels, when hydrophobic or electrostatic interactions are established between both components [10, 11, 12, 13]. Polymers and surfactants are commonly formulated together to improve the solubility and bioavailability of the drug, or the stability and rheological properties of the dosage form itself [3, 14]. Although the information regarding the interactions of linear poly(acrylic acid)s with nonionic [15, 16, 17] and ionic [18, 19, 20, 21] surfactants is extensive, much less attention has been paid to cross-linked networks [10, 12, 22, 23]. In addition, most studies were carried out in pure water and may not reflect the microgel–surfactant interactions in the presence of ions under physiological conditions. Also, for some treatments, the drug formulation is prepared to be isoosmotic with the application area fluids, by adding ions before administration. Therefore, the presence of ions, after or before administration, is a common situation that must be considered for drug formulation, especially when swellable hydrogels are used [24]. The information available regarding the salt effect on linear poly(acrylic acid)–surfactant interactions indicates that, in general, ions favor hydrophobic interactions, while they hinder electrostatic bonding, although the effect depends strongly on the nature and concentration of the ions [11, 15, 25].

Recent studies have shown that microcalorimetric techniques are reliable and sensitive tools to study the mechanisms of interaction between a drug or DNA and a polymer [26, 27], or the micellization and aggregation processes of surfactant molecules in polymer dispersions [25, 26, 27, 28, 29]. In contrast to other techniques, calorimetric titration provides direct and continuous information about the interaction process and allows simultaneous determination of all binding parameters (binding constant K, ΔH and ΔG).

The aim of this work was to characterize the interaction process between carbopol microgels and nonionic or ionic surfactants commonly used for drug delivery systems, in two simulated biological media, such as 0.9% NaCl and artificial lacrimal fluid, using isoperibol titration microcalorimetry, by comparing the results with those previously reported for water [30, 31]. The repercussions on the rheological behavior of the carbopol–surfactant gels are also evaluated. The information obtained may be useful to gain insight into the possibilities of the carbopol–surfactant gels as platforms for drug delivery. Additionally, as an important number of drugs, such as some antidepressant or nonsteroidal anti-inflammatory drugs, have amphiphilic character [32], the results may also help to understand the behavior of the carbopol gels loaded with one of these drugs when they enter into contact with biological media.

Experimental

Materials

Carbopol 934NF (62.6% carboxylic groups, molecular weight 3×10^6 Da) was provided by BFGoodrich Europe, UK. Polyoxyethylene(20) sorbitan monoleate (Tween 80), polyoxyethylene–polyoxypropylene–polyoxyethylene triblock copolymer (Pluronic F-127), benzalkonium chloride (BkCl) and sodium dodecyl sulfate (SDS) were from Sigma. Ultrapure water obtained by reverse osmosis (resistivity more than 18.2 MΩ cm; MilliQ, Millipore, Spain) was used. The other chemicals were of analytical grade.

Surface tension measurements

Surface tension measurements of surfactant solutions in water or artificial lacrimal fluid (6.78 g/l NaCl, 2.18 g/l $NaHCO_3$, 1.38 g/l KCl, 0.084 g/l $CaCl_2 \cdot 2H_2O$) were taken using the platinum ring method and a Lauda TD1tensiometer (Lauda, Königshofen, Germany) by applying the needed density corrections.

Titration microcalorimetry

Calorimetric experiments were performed in duplicate using a Tronac-450 isoperibol microcalorimeter and Tronac FS101 calorimetry software (Tronac, Orem, UT, USA). In each experiment, 47.5 ml carbopol solution (0.25% w/w) was placed in a Dewar reaction vessel, and a relatively concentrated surfactant solution (15%) was loaded into a 2-ml calibrated burette. The entire assembly was then immersed into a constant-temperature water bath (298.0 K). After thermal equilibration, the surfactant solution was delivered at a constant rate of 0.3332 ml/min into the reaction vessel, in which a stirrer mixed the two solutions rapidly. The rise or decrease in the temperature of the system was monitored using a thermistor, and was later reproduced using a heating coil in the reaction vessel. The apparent enthalpy was calculated from the applied current and voltage and the heating time. Calibration of the system was assured by titration of tris(hydroxymethyl)aminomethane with HCl. To obtain the heat associated with carbopol ionization, the carbopol dispersion was titrated with ammonium acetate (0.046 M) to produce the same increase in pH as the addition of the surfactant solutions. The integral binding heat for the polymer–surfactant aggregation, process Q_{agg}, was estimated by subtracting from the measured heat produced by addition of the surfactant to the carbopol dispersion, Q_p, the heat effects due to the dilution/demicellization of the surfactant, Q_d, estimated by addition of the surfactant solution into acidic water medium, 0.9% NaCl, or artificial lacrimal fluid and due to the change in the degree of ionization of carbopol carboxylic groups, Q_{ion} [29]:

$$Q_{agg} = Q_p - Q_d - Q_{ion}. \quad (1)$$

The enthalpy of the carbopol dilution (final concentration was 3% lower) was negligible. From the integral heats of the reaction at different times (at least ten points were used), ΔH_{agg} and the apparent binding equilibrium constant, K_{agg}, were estimated, using FS101 software, by iterative convergence applying the least-squares technique [33]. Using K_{agg}, the Gibbs energy change, ΔG_{agg}, per mole of surfactant adsorbed was estimated:

$$\Delta G_{agg} = -RT\ln K_{agg}. \quad (2)$$

Rheological characterization

The rheological behavior of 0.25% carbopol–0.01% surfactant dispersions was evaluated in triplicate at 25 °C using a Rheolyst AR-1000 N rheometer (TA Instruments, Newcastle, UK) equipped with an AR2500 data analyzer, fitted with a Peltier temperature control, a thermostatic concentric-cylinder adapter and a medium coaxial cylinder-recessed end. Oscillatory shear responses (G' or elastic modulus, and G'' or viscous modulus) were determined at 0.1 Pa over the frequency range 0.05–50 rad/s. The linearity of the viscoelastic properties has been verified for all the dispersions investigated.

Results and discussion

Surfactant micellization

The effect of salts on the micellization behavior of surfactants in water has been widely reported [34, 35, 36, 37]. In 0.9% NaCl, the critical micelle concentration (cmc) of Pluronic F-127 and SDS, at 25 °C, is around 0.1 and 0.015%, respectively. No data were found regarding the micellization process of the surfactants studied in lacrimal fluid. The surface tension values obtained for different surfactant solutions prepared in water or in lacrimal fluid are shown in Fig. 1. The cmcs of Tween 80 and Pluronic F-127 were slightly lower in lacrimal fluid than in pure water, and were similar to the values obtained in 0.9% NaCl. This is a consequence of a small salting-out effect that decreases the hydrophilicity of the poly(ethylene oxide) chains and promotes hydrogen-bonding interactions [35, 37].

The decrease in the cmc was particularly important in the case of ionic surfactants, since the presence of ions causes a shielding effect among the ionic groups and promotes the accumulation of the surfactant at the interface [34]. This explains the lower surface tension values and cmc obtained in lacrimal fluid compared to water. The cmc of SDS obtained in lacrimal fluid was similar to the value reported for 0.9% NaCl [36].

The apparent enthalpies associated with the dilution and demicellization processes of Pluronic F-127, Tween 80, SDS and BkCl in water, 0.9% NaCl and artificial lacrimal fluid are shown in Fig. 2. The concentration of the surfactants (15% w/w) in the burette, before being added to the Dewar vessel, was above their cmc. Therefore, when the surfactant solution was slowly added to an aqueous medium without polymer, the micelles broke up until the concentration in the Dewar vessel reached the cmc. Then, micelles were only diluted in a solution of micelles.

For the nonionic surfactants, the demicellization process was exothermic (enthalpy change negative), while for the ionic surfactants an endothermic process was observed. In spite of the differences observed in the cmc of the surfactants in each media, the calorimetric profiles were similar.

The exothermic dilution process of Pluronic solution has been attributed to the hydrogen bonding between the

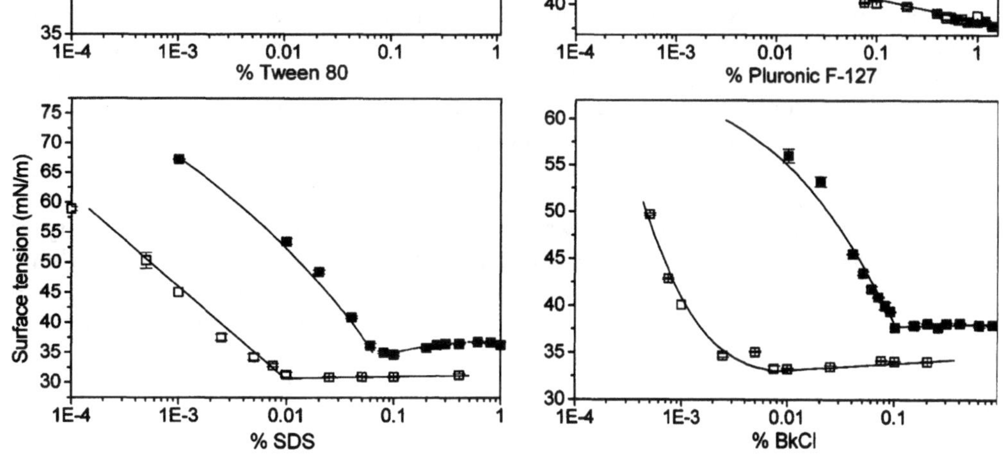

Fig. 1 Surface tension values of the surfactant solutions of polyoxyethylene(20) sorbitan monoleate (*Tween 80*), polyoxyethylene–polyoxypropylene–polyoxyethylene triblock copolymer (*Pluronic F-127*), sodium dodecyl sulfate (*SDS*) and benzalkonium chloride (*BkCl*) in water and in artificial lacrimal fluid

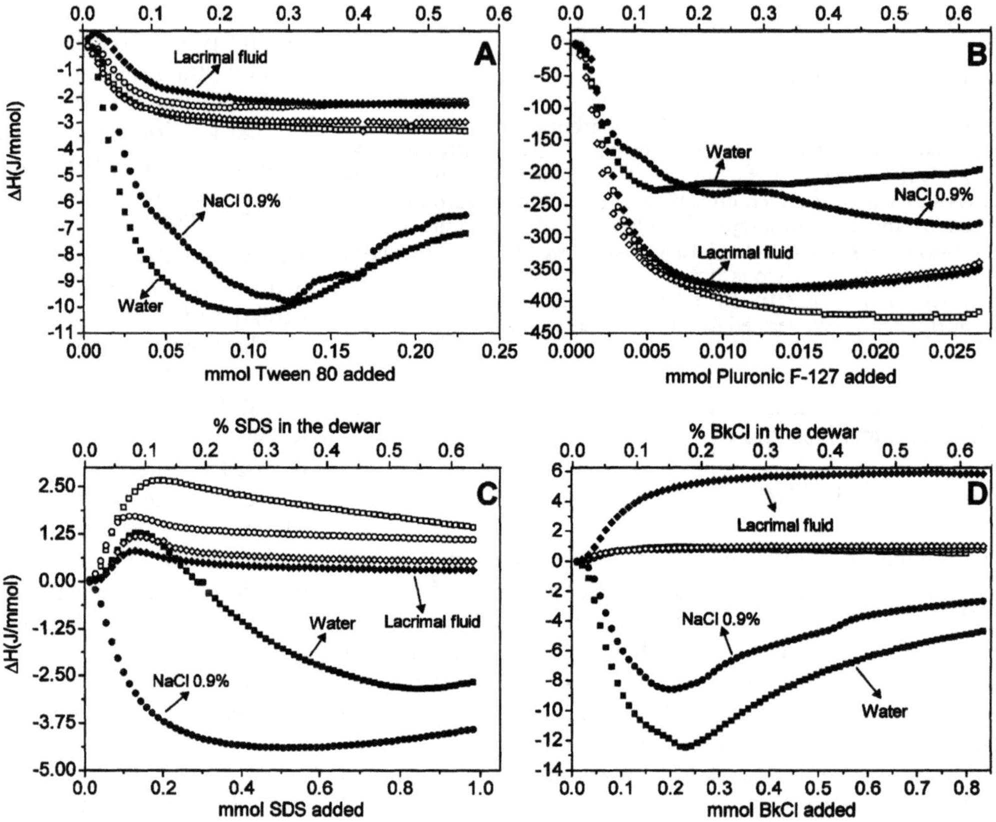

Fig. 2 Calorimetric titration curves observed during the addition of **a** 15% Tween 80, **b** Pluronic F-127, **c** SDS or **d** BkCl solution into a Dewar vessel containing 0.25% carbopol dispersion (*closed symbols*). The corresponding dilution plots of each surfactant observed in water (same results were obtained in diluted HCl solution of the same pH, 3.0, as for the carbopol dispersion), 0.9% NaCl or artificial lacrimal fluid are also given (*open symbols*)

triblock copolymer and water molecules after breakage of water–water and surfactant–surfactant hydrogen bonds [38]. The apparent enthalpy change observed for the demicellization of Pluronic F-127 ($\Delta H_d = -410$ kJ/mol) is in agreement with previously reported micellization enthalpies in water obtained using differential scanning calorimetry [39]. No data were found for Tween 80. The structure of Tween 80 is clearly different from that of Pluronic F-127, and therefore its lower demicellization enthalpy may be attributed to the lack of hydrophobic segments of polyoxypropylene.

The enthalpy of demicellization of SDS was lower in artificial lacrimal fluid than in 0.9% NaCl, which was lower than in water [28, 31]. This may be related to the higher associative capacity of this surfactant in the presence of salts, which implies that the dilution process, as a whole, is accompanied by the breakage of a lower number of micelles than in water. In consequence, the endothermic process reaches the maximum at lower SDS concentrations in the presence of salts. For BkCl, the apparent enthalpy of demicellization was smaller than in the case of SDS, and the salt effect was not detected.

Carbopol–nonionic surfactant aggregation

The association of poly(acrylic acid)s with ethoxylated surfactants is extraordinarily intense [3, 15, 16]. In previous work [40], we observed using IR spectroscopy and differential scanning calorimetry of films, that carbopol interaction with Pluronic F-127 and Tween 80 in water depends markedly on their weight ratio and that, in fact, hydrogen bonding occurs between the carbopol carboxylic groups and the oxygen of the ethylene oxide chains of the surfactants.

The apparent enthalpies of aggregation of each surfactant with carbopol in water, 0.9% NaCl and artificial lacrimal fluid are shown in Fig. 2. In the case of Pluronic F-127, the heat evolved in the Dewar vessel containing carbopol in water or in 0.9% NaCl was lower than in the absence of polymer, suggesting that the polymer–surfactant binding process is endothermic (Fig. 3).The Pluronic F-127–carbopol aggregation occurs thanks to a gain in entropy, resembling the micellization process [30, 38]. In contrast, the apparent enthalpy for Tween 80–carbopol interaction in water or in 0.9% NaCl was more exothermic than the dilution process of the surfactant. This fact may be attributed to the increase in stability of Tween 80 monomers when they adsorb onto carbopol microgels [25]. In artificial lacrimal fluid, the calorimetric profile observed with both surfactants was similar to that obtained in the absence of carbopol, suggesting that the interactions with the polymer in this medium are weak.

The heat exchanges measured after addition of a concentrated surfactant solution to a polymer solution

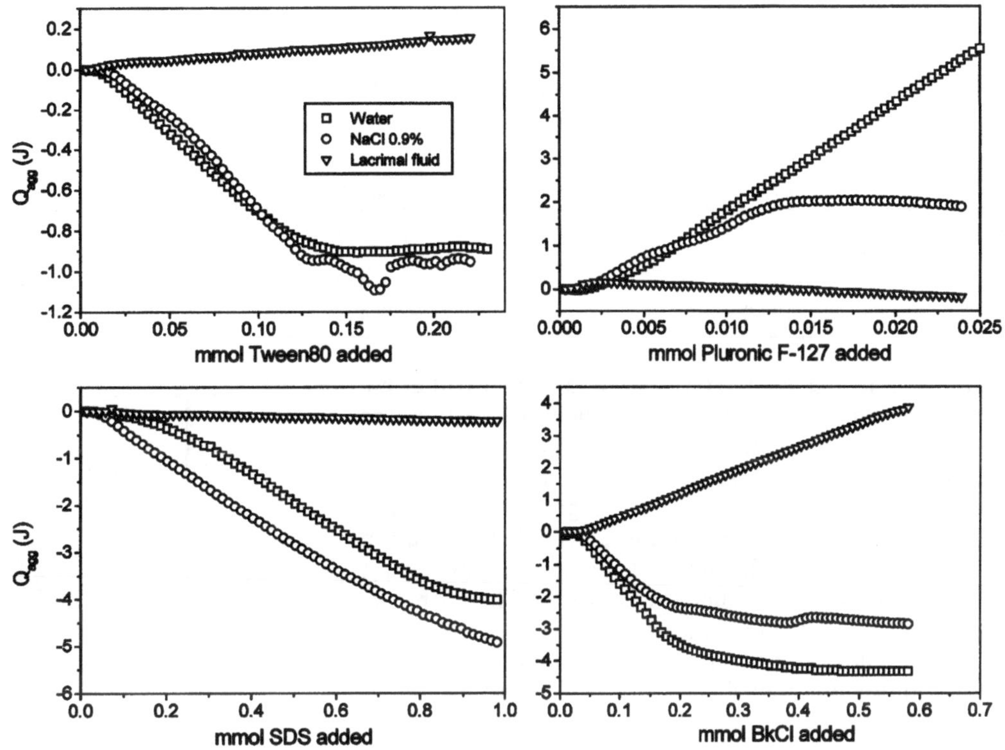

Fig. 3 Carbopol–surfactant aggregation heat, Q_{agg}, estimated as the difference between the heat evolved in the presence and in the absence of carbopol

can be regarded as the sum of the following contributions [29]:

1. Dilution of surfactant and demicellization process.
2. Dilution of polymer.
3. Binding between surfactant and polymer, which may be accompanied by changes in solvation, conformation, and degree of ionization of both components.
4. Interactions between aggregates through adjacent surfactant or polymer molecules.

All those processes, except the dilution effect (negligible in the case of carbopol), are characteristic of the interaction process under given conditions and should be taken into account. The integral heats of reaction, Q_{agg}, estimated as the difference between the heat evolved in the presence of carbopol and the sum of the heats associated with the dilution/demicellization of the surfactant and the ionization process of carbopol carboxylic acid groups are represented in Fig. 3. The initial flat region in Q_{agg} corresponds to a surfactant concentration below the critical associative concentration (cac). In 0.9% NaCl, once the cac is reached (0.06% Pluronic F-127 and 0.01% Tween 80) an almost constant increase (endothermic process) or decrease (exothermic process) in Q_{agg} was observed for Pluronic F-127 and Tween 80, respectively. Above a certain surfactant concentration, the binding reaches saturation point and no changes in Q_{agg} are observed. In lacrimal fluid, the values of Q_{agg} remained close to zero, although the process was slightly exothermic in the case of Pluronic F-127, and endothermic in the case of Tween 80.

The thermodynamic parameters associated with the carbopol–surfactant interaction in each medium are summarized in Table 1. The different orders of magnitude of ΔG_{agg} and ΔH_{agg} impeded the precise estimation of ΔS_{agg}. Nevertheless, the results shown in Fig. 3 and Table 1 indicate that carbopol–nonionic surfactant interaction in 0.9% NaCl is similar to that in water, while in the presence of lacrimal fluid the aggregation process is hindered. Although the main component of lacrimal fluid is NaCl and the ionic strength of the medium is similar to that of 0.9% NaCl, the presence of Ca^{2+} ions and the higher pH of the lacrimal fluid medium should prevent hydrogen bonding between carbopol and nonionic surfactants. Calcium ions may block some carboxylic sites more intensely than sodium ions [15]. In addition, the higher pH causes the other carboxylic acid groups to remain ionized. These phenomena prevent the polymer from being bound by the surfactant. Nevertheless, the concentration of Ca^{2+} ions was below the limit of tolerance of carbopol microgels [41], and no precipitation was observed in the systems studied.

Carbopol–ionic surfactant aggregation

When the SDS solution was added to the carbopol dispersion in water, an initial endothermic demicellization process was observed until the concentration in the

Table 1 Thermodynamic parameters of the carbopol–surfactant aggregation processes, expressed per mole of surfactant and mole of polymer for surfactants polyoxyethylene(20) sorbitan monoleate (*Tween 80*), polyoxyethylene–polyoxypropylene–polyoxyethylene triblock copolymer (*Pluronic F-127*), sodium dodecyl sulfate (*SDS*) and benzalkonium chloride (*BkCl*). Standard deviations were in all cases lower than 2%

Surfactant	0.9% NaCl			Artificial lacrimal fluid		
	K_{agg}	ΔG_{agg} (kJ/mol)	ΔH_{agg} (kJ/mol)	K_{agg}	ΔG_{agg} (kJ/mol)	ΔH_{agg} (kJ/mol)
Tween 80	22.39	−7.676	−258,575	15.85	−6.823	79,773
Pluronic F-127	52.48	−9.780	2,996,933	17.78	−7.10	−66,871
SDS	19.53	−7.392	−307,095	19.95	−7.392	−23,270
BkCl	10.00	−5.686	−729,422	1.01	−1.678	2,313,038

Dewar vessel reached 0.08%, and a new exothermic process became apparent (Fig. 2). When the medium was 0.9% NaCl, the endothermic step was not observed and the process was markedly exothermic overall. The disappearance of the endothermic region indicates that the affinity between carbopol and SDS is higher than in water, and that as the micelles break, the surfactant molecules are adsorbed onto the carbopol. Therefore, the cac in 0.9% NaCl is much lower than in water. The salt produces a shielding effect between the anionic groups of both components that facilitates the electrostatic repulsions being easily overcome by hydrophobic interactions [25]. The lack of interaction observed in lacrimal fluid may be attributed to a higher degree of ionization of carbopol microgels in this medium compared to in 0.9% NaCl or water. The higher pH of lacrimal fluid almost completely prevents the association process.

In the case of BkCl, the calorimetric profile recorded in 0.9% NaCl was similar to that observed in water [31]. The initial endothermic step is not very intense, which indicates that as the micelles break up, the surfactant units mainly interact with carbopol through ionic bonding, although hydrophobic interactions may also occur [20], leading to a strong exothermic effect. The maximum binding enthalpy was reached around 0.2% BkCl in water and in 0.9% NaCl. This finding indicates that the microgel structure of carbopol prevents additional BkCl molecules from interacting with the remaining carboxylic groups inside the microgel (maximum interaction involved around 25–40 mol% carboxylic groups). The slighter exothermic binding process in 0.9% NaCl may be explained by a salt shielding effect among the carboxylic and benzalkonium groups. In contrast, in lacrimal fluid the interaction process was less intense and was endothermic. This suggests that the ionic interactions between carbopol and BkCl are strongly hindered by the blocking effect of the Ca^{2+} ions and the higher degree of ionization of the remaining free carbopol carboxylic groups. Under these conditions, only hydrophobic interactions can appear.

▶

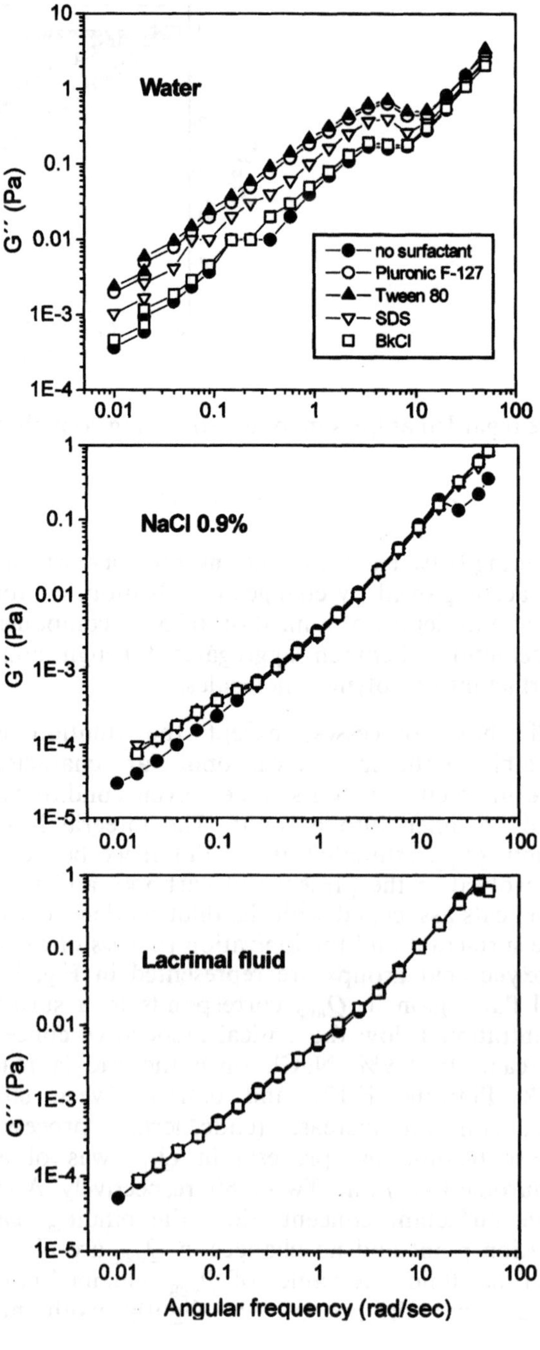

Fig. 4 Surfactant (0.01% w/w) effect on the viscous modulus of carbopol (0.25% w/w) dispersions in water (pH 4), 0.9% NaCl (pH 4) and artificial lacrimal fluid (pH 7.2)

Rheological consequences

The effect of each surfactant above the cac on the viscous modulus (G'') of 0.25% carbopol dispersions is shown in Fig. 4. In water at pH 4, the presence of just 0.01% surfactant promotes the maximum rheological synergistic effect [23], since the bound surfactant molecules act as bridges between several microgels, increasing the consistency of the system.

When the medium was 0.9% NaCl (pH 4), the carbopol-only dispersions had a lower viscous modulus than in water at pH 4 owing to a salting-out effect [24]. In this medium, the increase in G'' caused by the presence of surfactant was less relevant than in water. Nevertheless, the bound surfactant duplicates the values of G'' at low angular frequencies (0.01–0.10 rad/s), which are in the range of those characteristic of the physiological stress caused by skin flexion, blinking, and mucociliary movements [42].

In artificial lacrimal fluid (pH 7.2), the viscous modulus of carbopol-only dispersions is markedly lower than in water, in spite of the increase in the degree of ionization due to the change in pH. This behavior contrasts with the strong increases in viscosity that are observed when the carbopol aqueous dispersions are neutralized with NaOH [23]. The addition of the surfactants does not modify the rheological behavior of carbopol dispersions in artificial lacrimal fluid, owing to the absence of polymer–surfactant interactions, as indicated by titration microcalorimetry experiments.

Conclusions

Both the nature and the concentration of the ions in the medium have a critical effect on the carbopol–surfactant interactions. The results obtained in pure water cannot be directly extrapolated to predict the behavior in biological media in which different ions are present. In 0.9% NaCl medium, the aggregation process was slightly less intense than in water, except for the anionic surfactant. The occurrence of interactions may be used to modulate the rheological behavior of carbopol microgels dispersion in this medium. In contrast, in artificial lacrimal fluid, the carbopol–surfactant interactions are weaker and they do not modify the viscoelasticity of the dispersions.

Acknowledgements This work was financed by the Xunta de Galicia (PGIDT 00PX120303PR, DOG 04/06/97 and PGIDT 01PX12030141F) and the Ministerio de Ciencia y Tecnología, Spain (RYC2001-8). The authors express their gratitude to BFGoodrich Europe for providing free samples of carbopol.

References

1. Osada Y, Kajiwara K (2001) Gels handbook, vol 3. Academic, San Diego
2. Kumar S, Haglund BO, Himmelstein KJ (1994) J Ocul Pharmacol 10:47
3. Lin HR, Sung KC (2000) J Controlled Release 69:379
4. Miyazaki S, Suzuki S, Kawasaki N, Endo K, Takahashi A, Attwood D (2001) Int J Pharm 229:29
5. Srividya B, Cardoza RM, Amin PD (2001) J Controlled Release 73:205
6. Callens C, Adriaens E, Dierckens K, Remon JP (2001) J Controlled Release 76:81
7. Hägerström H, Edsman K (2001) J Pharm Pharmacol 53:1589
8. Valenta C, Kast CE, Harich I, Bernkop-Schnürch A (2001) J Controlled Release 77:323
9. Dittgen M, Durrani M, Lehmann K (1997) STP Pharma Sci 7:403
10. Philippova OE, Hourdet D, Audebert R, Khokhlov AR (1996) Macromolecules 29:2822
11. Mironov AV, Starodoubtsev, Khokhlov AR, Dembo AT, Yakunin AN (1999) Colloids Surf A 147:213
12. Hansson P, Schneider S, Lindman B (2000) Prog Colloid Polym Sci 115:342
13. Sjöström J, Piculell L (2001) Langmuir 17:3836
14. Paulsson M, Edsman K (2002) J Colloid Interface Sci 248:194
15. Saito S (1989) J Am Oil Chem Soc 66:987
16. Anghel DF, Saito S, Iovescu A, Baran A (1994) Colloids Surf A 90:89
17. Anghel DF, Winnik FM, Galatanu N (1999) Colloids Surf A 149:339
18. Shimizu T (1994) Colloids Surf A 84:239
19. Oliveira VA, Tiera MJ, Neumann MG (1996) Langmuir 12:607
20. Fundin J, Hansson P, Brown W, Lidegran I (1997) Macromolecules 30:1118
21. Kogej K, Skerjanc J (1999) Acta Chim Slov 46:269
22. Ashbaugh HS, Piculell L, Lindman B (2000) Langmuir 16:2529
23. Barreiro-Iglesias R, Alvarez-Lorenzo C, Concheiro A (2001) J Controlled Release 77:59
24. Horkay F, Tasaki I, Basser PJ (2001) Biomacromolecules 2:195
25. Kevelam J, van Breemen JFL, Blokzijl W, Engberts JBFN (1996) Langmuir 12:4709
26. Mumper RJ, Duguid JG, Khursheed A, Barron MK, Nitta H, Rolland AP (1996) Pharm Res 13:701
27. Govender T, Ehtezazi T, Stolnik S, Illum L, Davis SS (1999) Pharm Res 16:1125
28. Wang Y, Han B, Yan H, Kwak JCT (1997) Langmuir 13:3119
29. Singh SK, Nilsson S (1999) J Colloids Interface Sci 213:133
30. Barreiro-Iglesias R, Alvarez-Lorenzo C, Concheiro A (2002) Int J Pharm (submitted)
31. Barreiro-Iglesias R, Alvarez-Lorenzo C, Concheiro A (2002) Int J Pharm (submitted)
32. Malmsten M (2001) In: Holmberg K (ed) Handbook of applied surface and colloid chemistry, vol 1. Wiley, Chichester, pp 3–8
33. Eatough DJ, Christensen JJ, Izatt RM (1974) Experiments in thermometric titration and titration calorimetry. Brigham Young University Press, Provo, pp 8–36

34. Woolfrey SG, Banzon GM, Groves MJ (1986) J Colloid Interface Sci 112:583
35. Pandit N, Trygstad T, Criy S, Bohorquez M, Koch C (2000) J Colloid Interface Sci 222:213
36. Magid LJ, Li Z, Butler PD (2000) Langmuir 16:10028
37. Andersson BC, Cox SM, Ambardekar AV, Mallapragada SK (2002) J Pharm Sci 91:180
38. Irwin JJ, Beezer AE, Mitchell JC, Buckton MG, Chowdhry BZ, Eagland D, Crowther NJ (1993) J Phys Chem 97:2034
39. Alexandridis P, Holzwarth JE, Hatton TA (1994) Macromolecules 27:2414
40. Barreiro-Iglesias R, Alvarez-Lorenzo C, Concheiro A (2002) J Therm Anal Calorim 68:479
41. Pochard I, Boisvert JP, Malgat A, Daneault C (2001) Colloid Polym Sci 279:850
42. Edsman K, Carlfors J, Harju K (1996) Int J Pharm 137:233

Progr Colloid Polym Sci (2003) 122: 103–112
DOI 10.1007/978-3-540-36114-5

Lennart Piculell
Jesper Sjöström
Iseult Lynch

Swelling isotherms of surfactant-responsive polymer gels

L. Piculell (✉) · J. Sjöström · I. Lynch
Physical Chemistry 1, Center for Chemistry and Chemical Engineering,
Lund University, Box 124,
22100 Lund, Sweden
e-mail: lennart.piculell@fkem1.lu.se

Abstract Recent progress in the study of surfactant-responsive polymer gels is reviewed. Polymer gels can be designed that drastically swell or shrink in response to small changes in the concentration of an ionic surfactant in an external bath containing the gel. The gel swelling isotherm, where the equilibrium gel volume is measured as a function of the external surfactant concentration, yields useful information on polymer–surfactant interactions. Associating and nonassociating polymer–surfactant pairs may be distinguished, and critical concentrations for surfactant binding can be determined. Generic swelling isotherms for different classes of systems have been established, which apply to chemically widely different gels (cellulose derivatives, vinyl polymers). Hydrophobicity and charge are the essential polymer parameters that determine the features of these isotherms. In order for a surfactant to associate to a nonionic polymer, the polymer hydrophobicity has to exceed a certain threshold. The required threshold hydrophobicity varies with the alkyl chain length, the headgroup, and the counterion of the surfactant. Added salt has large effects on all swelling isotherms.

Keywords Hydrogel · Surfactant-responsive · Amphiphilic · Hydrophobically modified · Polyelectrolyte

Introduction

Polymer–surfactant interactions are manifested and utilized in widely differing contexts. Many experimental techniques have been used to study these interactions over the last 50 years, and the field is the subject of a number of reviews [1, 2, 3, 4]. Responsive polymer gels are gels that drastically change their equilibrium volume in response to external stimuli. Since the breakthrough discoveries by Tanaka and coworkers in the late 1970s, work on responsive gels has expanded into a currently very intense field of research [5, 6, 7]. The subject of the present paper, surfactant-responsive polymer gels, represents the bridge between these two research fields. In addition to the volume response per se, a potentially useful feature of a surfactant-responsive gel is that hydrophobic components may be solubilized in the surfactant aggregates contained in the gel.

In our laboratory, we have over the last decade systematically studied surfactant swelling isotherms of surfactant-responsive gels [8, 9, 10, 11, 12, 13, 14, 15, 16, 17, 18]. The experiments are simple; small gel pieces are equilibrated in quasi-infinite baths of a surfactant at increasing concentration, and the gel dimensions are measured by a video microscope (Fig. 1). We have concentrated on ionic surfactants, since these are more prone to binding to both nonionic and ionic polymers [1, 2, 3]. When an ionic surfactant binds to the gel, this affects the equilibrium volume of the gel through two mechanisms. The net charge of the gel is changed, and the bound surfactant aggregate may bind more than one polymer chain, thus forming an additional cross-link.

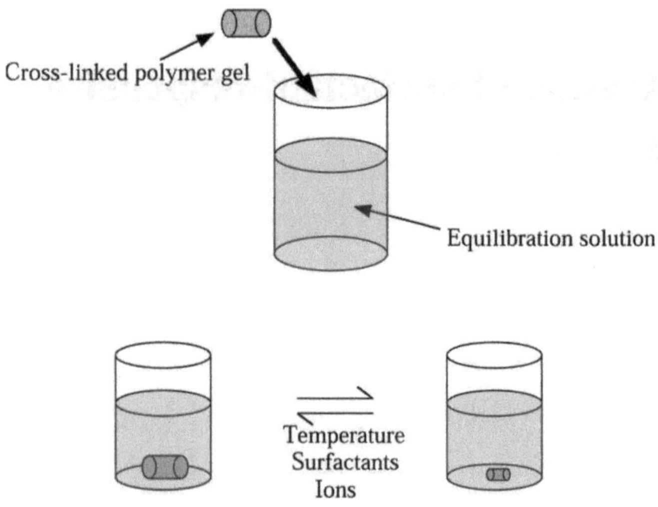

Fig. 1 A surfactant-responsive gel immersed in an equilibration solution changes its volume when the concentration of surfactant in the surrounding bath is changed. Added salt and temperature may influence the surfactant response

Free surfactant, existing as monomeric surfactant salt or free micelles, also affects the swelling by partitioning between the gel and the bath. The surfactant swelling isotherms are generally affected by added salt [11, 15, 16]. The swelling isotherm may also be affected by the temperature at which it is conducted, especially if the gel polymer contains thermally responsive units [10]. The latter feature is, however, not discussed further in the present paper, which only discusses experiments at room temperature.

Our interest in gel swelling isotherms with surfactants is motivated by two main questions. Can polymer–surfactant interactions be used to make surfactant-responsive gels? What can we learn about polymer–surfactant interactions from the study of such gels? To answer these questions, we have studied a large number of gel and surfactant combinations (Table 1). The gel polymers were either cellulose derivatives or synthetic vinyl polymers, representing chemically quite different backbones. The type and the number of hydrophobic and charged functional groups have been varied on both types of polymers. As for the ionic surfactants, cationic and anionic surfactants with different headgroup architectures, a variety of counterions, and varying surfactant chain lengths have been used. We have also studied the effects of added simple salt. The purpose of this paper is to review these studies, highlighting the conditions that need to be satisfied in order for an association to occur, the generic swelling isotherms that apply to different classes of systems, and the trends that occur when varying the system parameters.

Experimental procedures

We give a general description of our experimental procedures. For detailed experimental information we refer to the original publications (Table 1).

Cross-linked hydrogels were made in glass capillaries with an inner diameter, d_0, of 1.4 mm. Gels based on hydroxyethylcellulose (HEC) derivatives were made by cross-linking the linear polysaccharides by divinyl sulfone in alkaline solution [7, 19]. The concentration of polysaccharide at synthesis, c_0, was typically

Table 1 References to investigated combinations of gel and surfactant. hydroxyethylcellulose (*HEC*), ethyl(hydroxyethyl)cellulose (*EHEC*) hydrophobically modified HEC (*HMHEC*), HEC modified by *N*,*N*-dimethyl-*N*-dodecylammonium side chains (*cat-HEC*), poly(ethylene oxide) (*PEO*), polyacrylamide (*PAm*), *N*,*N*-dimethylacrylamide (*DAM*), *N*-isopropylacrylamide (*NIPA*), *N*-*tert*-butylacrylamide (*BAM*), sodium acrylate (*NaA*), sodium octyl sulfate (*SOS*), sodium decyl sulfate (*SDeS*), sodium dodecyl sulfate (*SDS*), sodium tetradecyl sulfate (*STS*), sodium hexadecyl sulfate (*SHS*), sodium octylbenzenesulfonate (*SOBS*), sodium dodecyldi(ethylene oxide) sulfate [*SD(EO)₂S*], dodecylammonium chloride (*C₁₂ACl*), dodecyltrimethylammonium bromide (*C₁₂TABr*), tetradecyltrimethylammonium bromide (*C₁₄TABr*), hexadecyltrimethylammonium bromide (*C₁₆TABr*), hexadecyltrimethylammonium chloride (*C₁₆TACl*), hexadecyltrimethylammonium acetate (*C₁₆TAAc*)

Surfactant	Gel							
	HEC	EHEC	HM HEC	cat-HEC	cat-HM HEC	PEO–PAm copolymer	DAM/NIPA/BAM copolymers	NaA/DAM/NIPA/BAM copolymers
SOS	[15]							
SDeS	[15]				[13]			
SDS	[14, 15]	[9, 10, 12]	[14, 15]	[16]	[12, 13, 15, 16]	[8, 11]	[17]	
STS	[15]			[16]	[13, 15]			
SHS	[15]							
SOBS	[15]					[11]		
$SD(EO)_2S$	[15]		[15]	[16]	[15]		[17]	
$C_{12}ACl$	[15]							
$C_{12}TABr$	[15]	[12]	[15]		[12]			
$C_{14}TABr$	[15]	[12]	[15]		[12]			
$C_{16}TABr$	[15, 17]	[12]	[15]		[12]		[17]	[18]
$C_{16}TACl$	[17]						[17]	[18]
$C_{16}TAAc$	[17]						[17]	[18]

20 g/l. Synthetic vinyl polymer gels, containing approximately 70 g/l monomer (c_0) and 8.6 mg/l cross-linker *N,N'*-methylene bisacrylamide, were prepared by standard methods [20, 21]. After synthesis, the gel rods were cut into approximately 1.4-mm-long cylindrical gel pieces. Residual chemicals were then leached out by immersing the gel pieces in a large excess of deionized water, followed by Millipore water, for at least 1 week.

A washed gel piece was immersed in a flat-bottom vial containing 5 ml aqueous surfactant solution of the desired composition. The diameter of the gel piece was measured by a video camera, calibrated with a 0.1-mm scale, with the help of an image computer program. The evolution of the swelling with time may be followed in an experiment. Interesting nonmonotonic volume changes are sometimes found [13, 15, 16], but these are outside the scope of the present overview, which deals exclusively with equilibrium swelling isotherms. The equilibrium gel size is typically reached in less than 3 days. The degree of swelling is here reported as V/m, where V is the gel volume and m the mass of linear polymer or monomer that the gel piece contained at synthesis. V/m is calculated as $(d/d_0)^3/c_0$, where d_0 is the inner diameter of the glass capillary and d the diameter of the gel immersed in the solution.

For the study of polymer–surfactant interactions, we note some important advantages of the gel swelling experiment compared to experiments on linear polymers.

1. The purification of the polymer from low-molecular-weight (salt) impurities is easily performed for a gel as a simple washing step. Removal of salt impurities is particularly important for polyelectrolytes, since quite small amounts of salt can have a substantial effect on polyion–surfactant interactions.
2. Because the polymer is contained in a separate compartment, the gel piece, it can be studied in an effectively infinite bath of surfactant solution (except, under some circumstances, for oppositely charged systems). Thus, the gel response is immediately obtained as a function of the free surfactant concentration, which is a fundamental quantity. In other words, the gel swelling in some important respects reflects the limiting behaviour of the corresponding linear polymer at infinite dilution.

Slightly hydrophobic nonionic gels

Micelles of ionic surfactants may bind to slightly hydrophobic polymers. This interaction results in a lowering of the critical micelle concentration (cmc) from its bulk value to a new value, which is commonly called the critical association concentration (cac) [1, 3]. In a gel swelling experiment, the association results in a sharp increase in the equilibrium volume of the gel as the surfactant concentration is increased above the cac. The origin of this effect, which is illustrated in Figs. 2, 3, and 4, is that the binding of surfactant results in an excess concentration of surfactant inside the gel. For reasons of electroneutrality, there also has to be an equivalent excess of mobile counterions, providing an additional contribution to the osmotic pressure inside the gel. After the initial increase, maximum swelling is eventually reached on further increase of the surfactant concentration, after which deswelling occurs. The maximum will be discussed further later. Similar swelling isotherms have been obtained for a large number of gel–surfactant pairs, and the cac values obtained from these experiments were found to be in excellent agreement with the values obtained by other techniques [8, 9, 10, 11, 14, 15, 16].

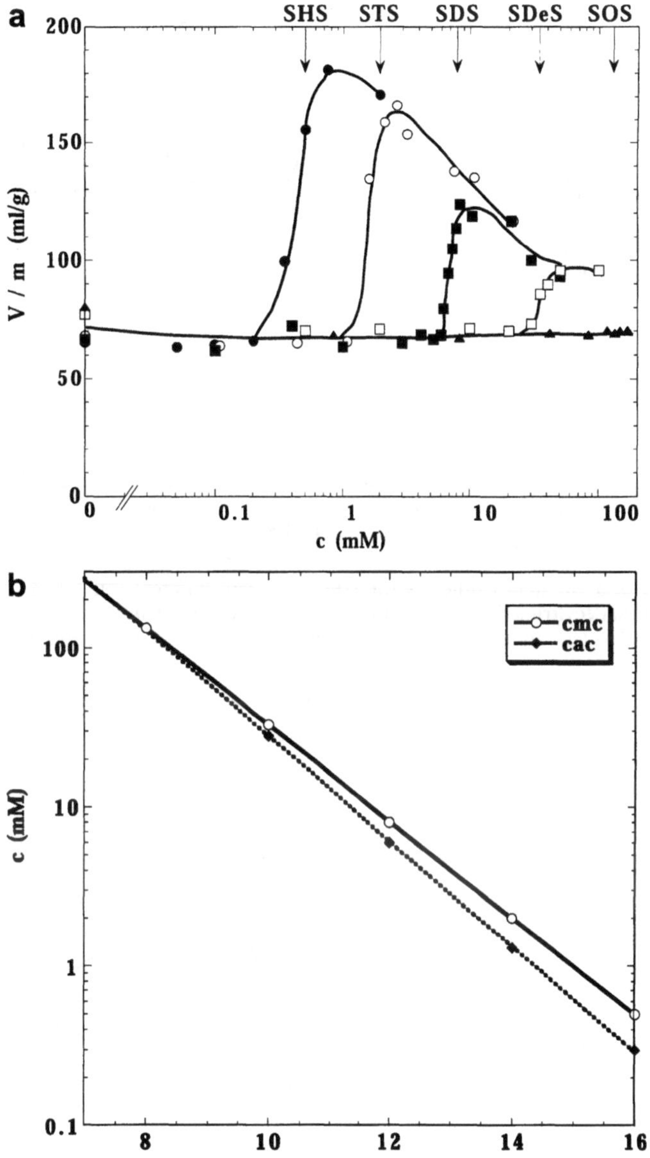

Fig. 2a, b Interactions between hydroxyethylcellulose (*HEC*) gels and sodium alkyl sulfates. **a** Swelling isotherms for sodium hexadecyl sulfate (*filled circles*), sodium tetradecyl sulfate (*STS*) (*open circles*), sodium dodecyl sulfate (*SDS*) (*filled squares*), sodium decyl sulfate (*SDeS*) (*open squares*) and sodium octyl sulfate (*SOS*) (*triangles*). *Arrows* denote critical micelle concentrations (*cmcs*). **b** Variation with hydrocarbon chain lengths of the cmcs and the critical association concentration (*cacs*) extraced from **a**. From Ref. [15]

As is well known, the cac increases (as does the cmc) when the surfactant chain length decreases. Figure 2a shows that the maximum gel swelling also decreases. This effect is general for all systems studied [12, 15]. It could be due to a decrease in surfactant binding, but differences in the concentration of free surfactant at the swelling

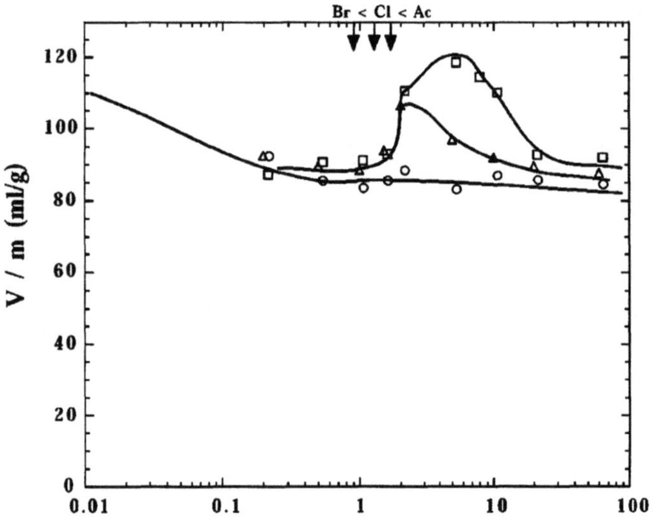

Fig. 3 Swelling isotherms for HEC gels with different ion forms of hexadecyltrimethylammonium ($C_{16}TA^+$): bromide (*circles*), chloride (*triangles*), and acetate (*squares*). *Arrows* denote cmcs. From Ref. [17]

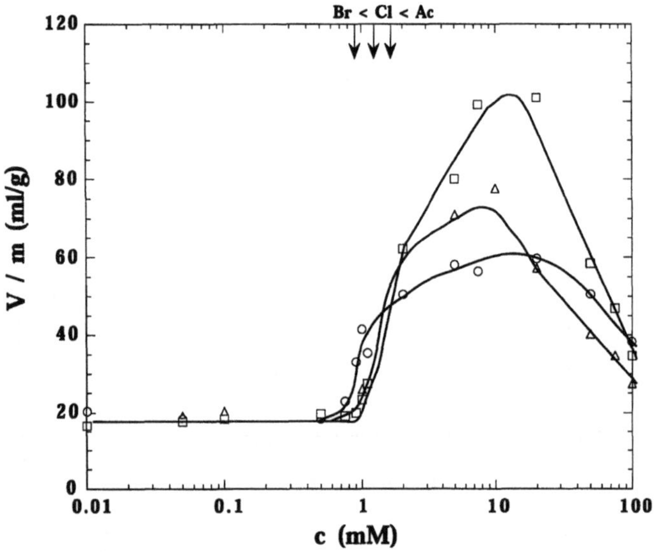

Fig. 4 Swelling isotherms for *N*-isopropylacrylamide (*NIPA*) gels with different ion forms of $C_{16}TA^+$: bromide (*circles*), chloride (*triangles*), and acetate (*squares*). *Arrows* denote cmcs. From Ref. [17]

maximum should also contribute. Free surfactant acts as an external salt, which decreases the swelling of an ionic gel (see later), and the maximum occurs at higher free surfactant concentration for a surfactant with a higher cmc. For sufficiently short surfactant chains, no association occurs; the swelling of an HEC gel is unaffected by added sodium octyl sulfate in the bath, even at concentrations exceeding the cmc [15]. A clue to this effect is given in Fig. 2b. As the surfactant chain length is decreased, the cac approaches the cmc, and the difference seems to vanish (for HEC and sodium alkyl sulfates) at eight carbon atoms. This means that the decrease in free energy on forming micelles at the polymer also vanishes. Evidently, the gel swelling experiment is a good technique to distinguish between associating and nonassociating polymer–surfactant pairs. This is particularly useful in cases – such as for alkyl sulfates and HEC – where the difference between the cac and the cmc is small. A common method to detect polymer–surfactant interactions is to look for a lowering of the cmc in the presence of the polymer, and a small decrease may easily escape notice.

HEC gels do not swell in solutions of cationic alkyltrimethylammonium surfactants with bromide counterions [15], but Fig. 3 shows that small swelling effects occur for the hexadecyltrimethylammonium ($C_{16}TA^+$) surfactant with chloride or acetate counterions [17]. By contrast, *N*-isopropylacrylamide (NIPA) gels swell in $C_{16}TA^+$ solutions regardless of the counterion (Fig. 4) [17], but the swelling varies with the identity of the counterion according to the Hofmeister series [22]. These and similar experiments indicate that a hydrophobicity threshold is required in order for a polymer to bind a given surfactant. The threshold hydrophobicity depends on the alkyl chain length, the headgroup, and the counterion of the surfactant. A higher degree of gel hydrophobicity is required for shorter surfactant chains. The threshold is also higher for cationic trimethylalkylammonium surfactants as compared to anionic alkyl sulfate surfactants. Furthermore, the threshold increases, for cationic surfactants, with the surfactant counterion in the order acetate < chloride < bromide, i.e., in accordance with the Hofmeister series. Evidently, the free-energy gain for forming micelles at the polymer becomes smaller when the counterion binds more strongly to the micelle.

The data in Figs. 2, 3, and 4 indicate that the surfactant concentration at the maximum gel swelling, c_{max}, is correlated with the cmc. Often (Fig. 2), the swelling maximum occurs quite close to the cmc, but sometimes, as in Figs. 3 and 4, c_{max} is much larger than the cmc. Close inspection of the latter data reveals, however, that there is a distinct change in the slope of the swelling isotherm around the cmc. This is expected, since the chemical potential of the surfactant increases much more slowly on increasing surfactant concentration after the cmc; therefore, the binding isotherm must level off.

We interpret the variation in the relative positions of c_{max} and cmc in terms of a saturation binding of surfactant. Ultimately, repulsion between polymer-bound micelles must prevent further binding. Depending on the magnitude of the free-energy difference between polymer-bound and free micelles, this saturation may occur after or, indeed, before the cmc. When saturation occurs after the cmc, the position of c_{max} should depend on two opposing trends: a continued binding, which gives rise to an additional swelling, and an exclusion of free micelles from the gel, owing to repulsion between free and polymer-bound micelles. The latter is presumably the mechanism that ultimately gives rise to a decrease in the gel swelling in all systems. Since no additional contracting force is expected at high surfactant concentrations, the deswelling strongly indicates a decreasing difference in the concentrations of mobile ions inside and outside the gel.

Ionic gels

Ionic gels display a very high degree of swelling in pure water, owing to the large entropy of mixing of the counterions to the gel charges. When oppositely charged surfactant molecules are added to the swelling bath, these start to micellize in the gel at a cac orders of magnitude lower than the cmc. The original simple counterions of the polyion are thus replaced by polyvalent ionic micelles, and the gel deswells. The bound micelles not only eliminate the ionic swelling mechanism, but they also give rise to an attraction between polyion chains, resulting in a collapse of the gel. Stoichiometric polyion–surfactant salts are generally insoluble in pure water, and they only take up limited amounts of water [23]. These features are illustrated in Fig. 5 by swelling isotherms with hexadecyltrimethylammonium bromide ($C_{16}TABr$) for copolymer gels of sodium acrylate (NaA) and *N*,*N*-dimethylacrylamide (DAM) at varying compositions, including the homopolymer NaA and DAM gels [18]. The homopolymer DAM gel does not bind $C_{16}TABr$; the hydrophobicity of the DAM gel is below the threshold for the $C_{16}TABr$ surfactant [17]. As expected, the swelling of a charged surfactant-free copolymer gel increases with increasing content of NaA residues. For a fully collapsed gel, the degree of gel collapse (as compared to the uncharged gel) also increases with increasing NaA content. Thus, both for the swollen and for the collapsed states, the magnitude of the departure from the "inert" gel behaviour increases monotonically with increasing charge density. Interestingly, the gels remain collapsed even at surfactant concentrations orders of magnitude above the cmc of $C_{16}TABr$.

The same type of experiments as in Fig. 5 have also been performed on copolymer gels of NaA and NIPA [18], and the results are given in Fig. 6. NIPA units, as opposed to DAM units, are hydrophobic enough to bind $C_{16}TABr$ (Fig. 4), and this binding immediately shows up in the swelling isotherms for the NaA–NIPA copolymer gels. A cooperative reswelling occurs at a second cac, called cac(2), which is related to the cac of $C_{16}TABr$ for homopolymer NIPA gels. The extent of reswelling increases with increasing NIPA content. At high NIPA contents, both the extent of reswelling and the cac(2) approach the corresponding values for the homopolymer NIPA gel.

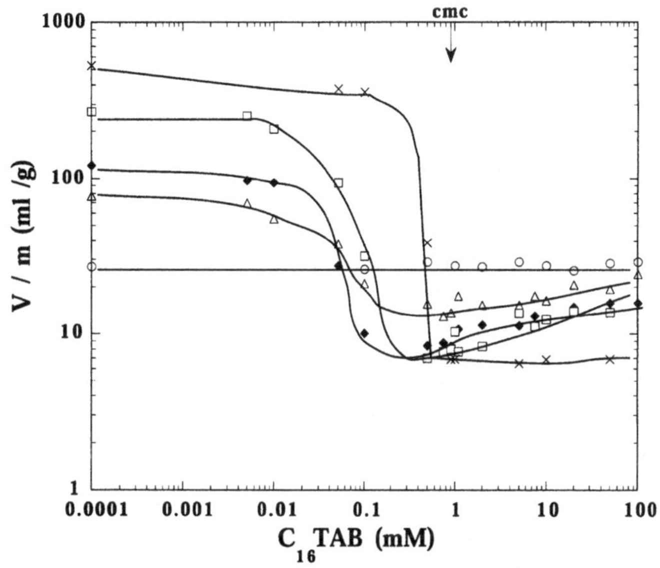

Fig. 5 Swelling isotherms with $C_{16}TABr$ for sodium acrylate (*NaA*)/*N*,*N*-dimethylacrylamide copolymer gels with 0 (*circles*), 5 (*triangles*), 10 (*diamonds*), 20 (*squares*), and 100 (*crosses*) mol% NaA. From Ref. [18]

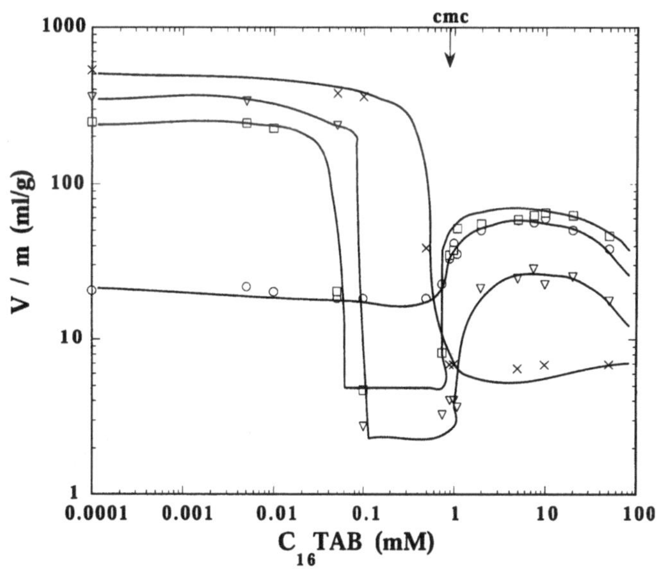

Fig. 6 Swelling isotherms with $C_{16}TABr$ for NaA/NIPA copolymer gels with 0 (*circles*), 20 (*squares*), 50 (*triangles*), and 100 (*crosses*) mol% NaA. From Ref. [18]

The two-step collapse/reswelling behaviour seen in Fig. 6 has also been found for a chemically quite different and more complex system, i.e., a gel of HEC modified by *N*,*N*-dimethyl-*N*-dodecylammonium side chains (cat-HEC) immersed in solutions of anionic surfactants [16]. The swelling isotherms, displaying very sharp changes in swelling at the cac and the cac(2) are shown in Fig. 7. This system represents the most dramatic surfactant-response that we have encountered so far. The initial surfactant binding step gives rise to a decrease in the gel volume by a factor of 400. The similarly sharp reswelling at the second binding step causes the gel volume to increase by a factor of 80 for sodium tetradecyl sulfate. Figure 7 also illustrates that both the cac and the cac(2) decrease – as expected – when the cmc of the surfactant decreases. By comparing with the results in Figs. 5 and 6, we conclude that the hydrophobicity of the cat-HEC polymer is sufficient (above the threshold) to give rise to an additional cooperative surfactant binding at the cac(2). For the alkyl sulfate surfactants, the interaction with the HEC backbone should contribute (Fig. 2), but the similarity to the swelling isotherm with sodium dodecyldi(ethylene oxide) sulfate [SD(EO)$_2$S] – which does not bind to HEC [15] – indicates that the main source of the hydrophobic interaction is the presence of the hydrophobic moieties on the cationic substituents. Note also that with STS and SD(EO)$_2$S, the swelling maximum on reswelling occurs far below the cmc. This indicates that the hydrophobic interaction in this case is strong enough to result in a saturation binding before free micelles are formed.

Effects of hydrophobic side chains

Self-associating properties may be conferred to a water-soluble polymer if the polymer chains are modified by a small fraction of strongly hydrophobic moieties, typically short alkyl side chains, along the polymer backbone [24, 25]. If the degree of modification is kept low (typically a few mole per cent of the repeating units), such a hydrophobically modified polymer (HMP) remains water-soluble, but its thickening or viscosifying effect is significantly enhanced in the semidilute regime, compared to that of the nonmodified polymer. HMPs also interact strongly with added surfactants [25]. It is by now well established that HMP–surfactant mixtures contain mixed micellar aggregates involving the surfactant and the hydrophobic side chains, or "hydrophobes", of the HMP. The mixed-micelle formation has profound consequences for the surfactant binding isotherm. Since the hydrophobes of the HMP are typically associated in micellelike aggregates even before surfactant is added, no cac is required in order for the surfactant to bind. Rather, the binding isotherm may be understood in terms of a partitioning of the surfactant between the bulk solution, where it exists as free monomers, and the mixed micelles. The extent of surfactant incorporation in the micelles at a given total concentration is essentially determined by the cmc of the surfactant [14, 26]. The mixed micelle becomes dominated by the surfactant only when the concentration of monomeric surfactant approaches the cmc. The resulting binding isotherm features a noncooperative binding region at concentrations far below the cmc, followed by a steep cooperative binding as the free surfactant concentration approaches the cmc.

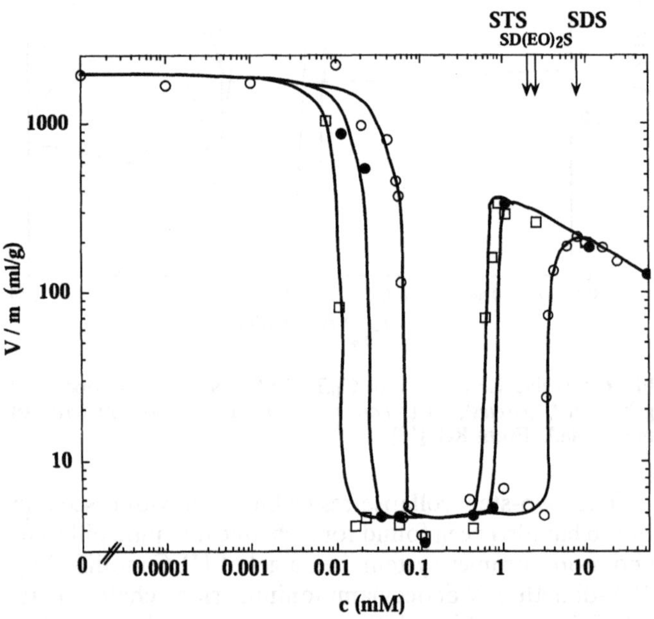

Fig. 7 Swelling isotherms for gels of HEC modified by *N*,*N*-dimethyl-*N*-dodecylammonium side chains (*cat-HEC*) with the anionic surfactants SDS (*open circles*), sodium dodecyldi(ethylene oxide) sulfate [*SD(EO)$_2$S*] (*filled circles*), or STS (*squares*). *Arrows* denote cmcs. From Ref. [16]

Naturally, hydrophobic modification affects the surfactant swelling isotherm of a polymer gel. Even if the surfactant does not bind to the polymer backbone, it will generally bind to the micelles formed by the HMP hydrophobes. The swelling isotherms for uncharged hydrophobically modified HEC (HMHEC) gels with two different anionic surfactants, sodium dodecyl sulfate (SDS) and SD(EO)$_2$S, are compared in Fig. 8 [15]. Swelling isotherms for nonmodified HEC with the same surfactants have shown that HEC binds SDS, but not SD(EO)$_2$S [15]. The two isotherms in Fig. 8 show the same features. At low surfactant concentrations, the added surfactant gives rise to a small, but significant, shrinking of the gel. Evidently, the mixed micellization initially gives rise to an additional attraction between the polymer chains. As the surfactant concentration approaches the cmc, however, the gel starts to swell. Two effects should contribute to this swelling. The first is the ionic effect discussed earlier, which becomes important at large degrees of surfactant binding. The other effect is due to the vanishing of the mixed micellar cross-linking. The number of mixed micelles becomes larger at high degrees of surfactant incorporation, and the number of hydrophobes per mixed micelle thus becomes smaller [14]. Eventually, each micelle is attached to only one polymer chain.

A swelling maximum appears in Fig. 8 for the same reasons as discussed earlier (saturation of the polymer

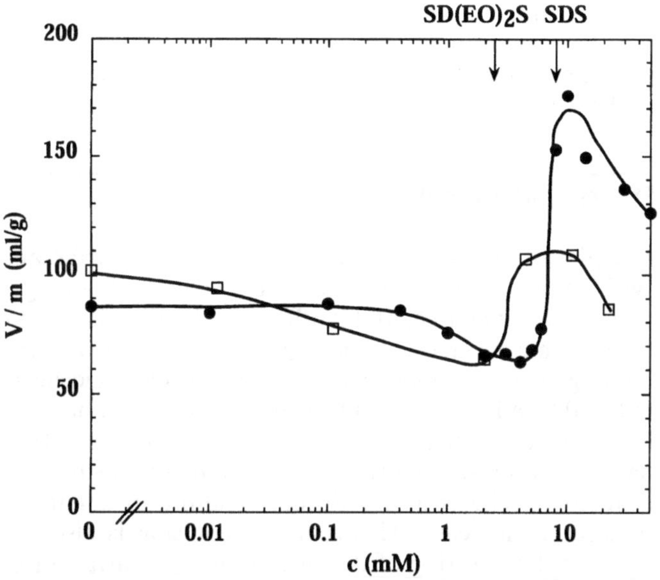

Fig. 8 Swelling isotherms for hydrophobically modified HEC (*HMHEC*) gels with the anionic surfactants SDS (*circles*) or SD(EO)$_2$S (*squares*). *Arrows* denote cmcs. From Ref. [15]

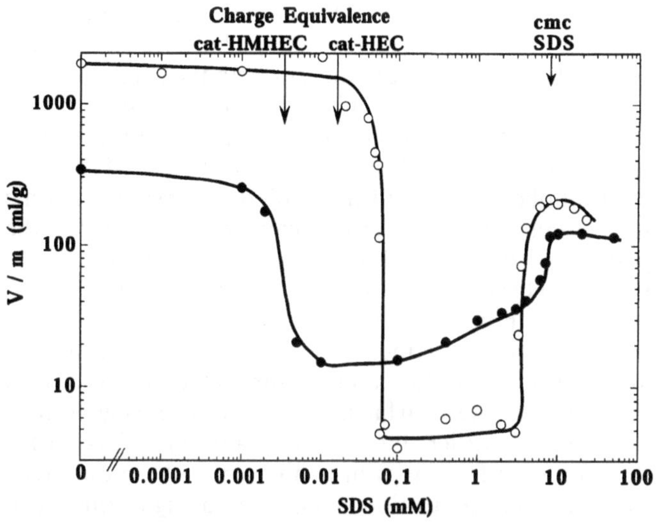

Fig. 9 Swelling isotherms for cat-HEC (*open circles*) or cat-HMHEC (*filled circles*) gels with SDS. Charge equivalence points indicate where the number of SDS molecules in the bath is equal to the number of charges on the gel. From Ref. [16]

and formation of free micelles). There is a quantitative difference in the maximum degree of swelling for the two surfactants. Interestingly, the trend is opposite to that shown in Figs. 2a and 7, where the surfactant with the lowest cmc gave the largest swelling. This difference can be attributed to the fact that SDS binds also to the backbone HEC chain, which gives rise to a larger saturation binding.

The effects of hydrophobic modification on the swelling isotherm for a polyelectrolyte gel with oppositely charged surfactant are illustrated in Fig. 9. The isotherm for the cat-HEC gel with SDS, shown Fig. 7, is compared with that for a cat-HMHEC gel. The chemical differences between the two cellulose derivatives is that in cat-HMHEC the cationic substituents contain hydrophobic dodecyl groups; moreover, the degree of cationic substitution is lower, by roughly a factor of 4. Both gels show shrinking and redissolution, but there are qualitative differences that may be attributed to the presence or the absence of hydrophobes. In cat-HEC, a finite cac (about 0.02 mM SDS) has to be reached before the binding and deswelling begins. By contrast, the deswelling of the cat-HMHEC gel begins immediately on surfactant addition. Although it is not evident from the logarithmic plot in Fig. 9, the volume decrease is, in fact, almost linear in the surfactant concentration from zero surfactant down to the fully collapsed state at about 0.05 mM SDS.

At this stage, it is appropriate to comment on the possibility to conduct gel experiments under conditions of an infinite reservoir of surfactant. Under the standard conditions of our experiments, we cannot assume that the bath concentration of surfactant remains unaffected by binding to the gel if there is strong binding at low surfactant concentrations. This is, in fact, often the case for oppositely charged pairs during the collapse transition. The slope of the collapse thus typically reflects the amount of surfactant consumed by the gel. This is, however, not a principal difficulty for an oppositely charged system, if there is a finite cac for the binding. The condition of an infinite bath can then always be satisfied, with a sufficiently large bath. It becomes problematic, however, for charged HMP gels such as cat-HMHEC. There, the surfactant binding begins from zero free surfactant concentration, and the free surfactant concentration remains extremely small throughout the entire collapse of the gel [27].

The redissolution is also different for the two types of gel. The cat-HEC gel stays collapsed over a large surfactant concentration range (0.06–3 mM), whereas the cat-HMHEC starts to reswell immediately after the swelling minimum has been reached. Both the immediate onset of a deswelling and the subsequent immediate onset of a reswelling reflect the continuous binding of the anionic surfactant [the absence of a cac or a cac(2)] to the HMP. Close to the cmc, an additional increase in swelling of the cat-HMHEC gel is seen. This last step is attributed to the binding of SDS to the HEC backbone, compare with the results for HMHEC with SDS in Fig. 8.

The cat-HMHEC polymer is so hydrophobic that it precipitates out of solution when a small amount (about 15 mM) of a simple salt, such as NaCl, is added [27]. A corresponding collapse of a cat-HMHEC gel occurs if a similar concentration of a simple salt is added to the external solution [12, 16]. This salt sensitivity contributes to a striking collapse/reswelling behaviour of cat-HMHEC gels with cationic surfactants (Fig. 10) [12]. In order to understand this behaviour, we should first note that cationic surfactants, just like anionic surfactants, join in mixed micelles with the cat-HMHEC hydrophobes. We recall that different types of cationic surfactants generally form mixed micelles. (From the point of view of mixed micellization, it is instructive to view cat-HMHEC as a polymer-modified cationic surfactant, or "polysoap", where the surfactant headgroups are joined by a polymer chain.) However, the extent of incorporation of the cationic surfactant in the mixed micelles at a given surfactant concentration is governed, as discussed earlier, by the surfactant cmc. This means that for a surfactant with a high cmc, the monomeric surfactant concentration builds up to high concentrations before there is strong incorporation in the mixed micelles. Therefore, the swelling isotherms for cationic surfactants should initially follow the collapse isotherm for a simple salt, until, close to the surfactant cmc, strong surfactant binding sets in. The latter results, as discussed earlier, in an increased charge of the mixed micelles and a dissolution of mixed micellar cross-links, both contributing to a reswelling of the gel. These expectations are clearly borne out in the swelling isotherms in Fig. 10, where the isotherm with NaCl is compared with those of cationic surfactants with different alkyl chain lengths. There is a clear tendency that the surfactants, at their swelling minima, give a stronger collapse than a simple salt. This additional surfactant effect is in accordance with the minimum preceding the reswelling for nonionic HMP gels with ionic surfactants (Fig. 8).

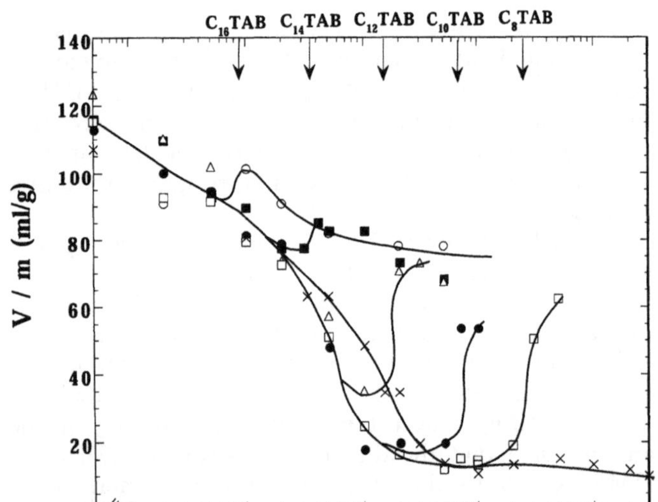

Fig. 10 Swelling isotherms for cat-HMHEC gels with hexadecyltrimethylammonium bromide (*$C_{16}TAB$*) (*open circles*), tetradecyltrimethylammonium bromide (*$C_{14}TAB$*) (*filled squares*), dodecyltrimethylammonium bromide (*$C_{12}TAB$*) (*open triangles*), decyltrimethylammonium bromide (*$C_{10}TAB$*) (*filled circles*), octyltrimethylammonium bromide (*C_8TAB*) (*open squares*), or NaCl (*crosses*). *Arrows* denote cmcs. From Ref. [12]

Effects of added salt

Added salt has two general effects on the surfactant swelling isotherm. It reduces the swelling of any gel carrying a net charge, and it affects the binding isotherm for an ionic surfactant. The driving force for surfactant binding to a net oppositely charged (mixed) micelle is reduced by added salt, resulting in less surfactant binding at a given concentration of free surfactant. On the other hand, the repulsion between a surfactant and a similarly charged micelle is also reduced, resulting in enhanced binding in this case. This reduced repulsion is also the cause of the lowering of the cmc of an ionic surfacant in the presence of salt.

Swelling isotherms for nonionic HEC gels with SDS in the presence of varying amounts of NaCl are shown in Fig. 11 [15]. Quite similar swelling isotherms have been found for a gel of polyacrylamide with pendant poly(ethylene oxide) side chains (PEO–PAm) with anionic surfactants [11]. These results demonstrate that the effect of added salt can be quite pronounced. At low levels of added salt (up to about 30 mM, in the present case), the qualitative features of the swelling isotherm remain unchanged, but the onset of swelling is moved towards lower surfactant concentrations and the swelling maxi-

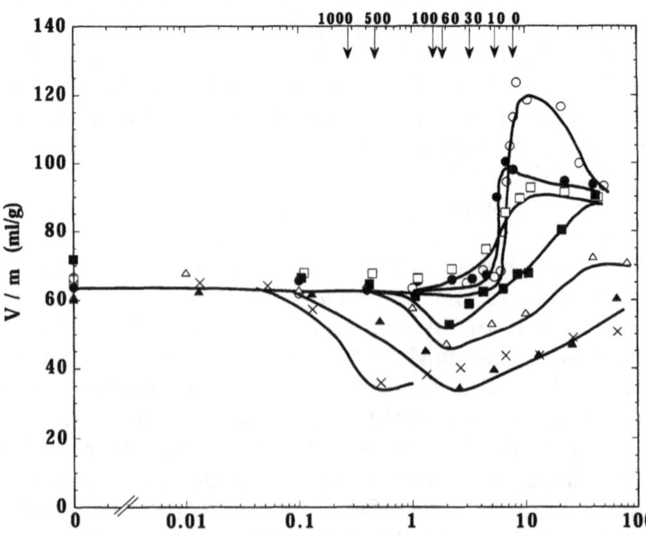

Fig. 11 Swelling isotherms for HEC gels with SDS at different levels of added NaCl (mM): 0 (*open circles*), 10 (*filled circles*), 30 (*open squares*), 60 (*filled squares*), 100 (*open triangles*), 500 (*filled triangles*), or 1,000 (*crosses*). *Arrows* denote cmcs for SDS at the various salt concentrations. From Ref. [15]

mum becomes less pronounced. These effects are readily understood as consequences of a lowering of both the cmc and the cac, and a less pronounced ionic swelling at a given degree of surfactant binding.

A different type of swelling isotherm is seen at high levels of added salt (above about 30 mM). Here the gel first shrinks and then reswells with increasing concentration of surfactant. In our experiments on the PEO–PAm copolymer gel, which showed the same features, we also measured surfactant binding isotherms in addition to the swelling isotherms [11]. We found that the surfactant binding started at the onset of gel shrinking, and then increased monotonically throughout the binding isotherm. The onset of shrinking at high levels of salt thus indicates the cac, and an increasing level of surfactant binding first gives rise to a deswelling and then a reswelling of the gel. The conclusion is that at high levels of added salt, a micellar cross-linking of the polymer chains, with each micelle binding (on average) more than one active network chain, dominates the swelling behaviour. At high levels of surfactant binding, however, this cross-linking is lost, since each micelle now interacts with only one active chain. (This nonmonotonic variation in micellar cross-linking is quite analogous to that found for HMP gels.) This interpretation implies that the amount of bound surfactant must continue to increase even far above the surfactant cmc; see the positions of the cmcs relative to the swelling isotherms in Fig. 11. The binding isotherms in Ref. [11] showed that this was indeed the case.

Swelling isotherms for the ionic cat-HEC gel with SDS, at various levels of added salt, are shown in Fig. 12 [16]. We recall that this is a system where the gel interacts both

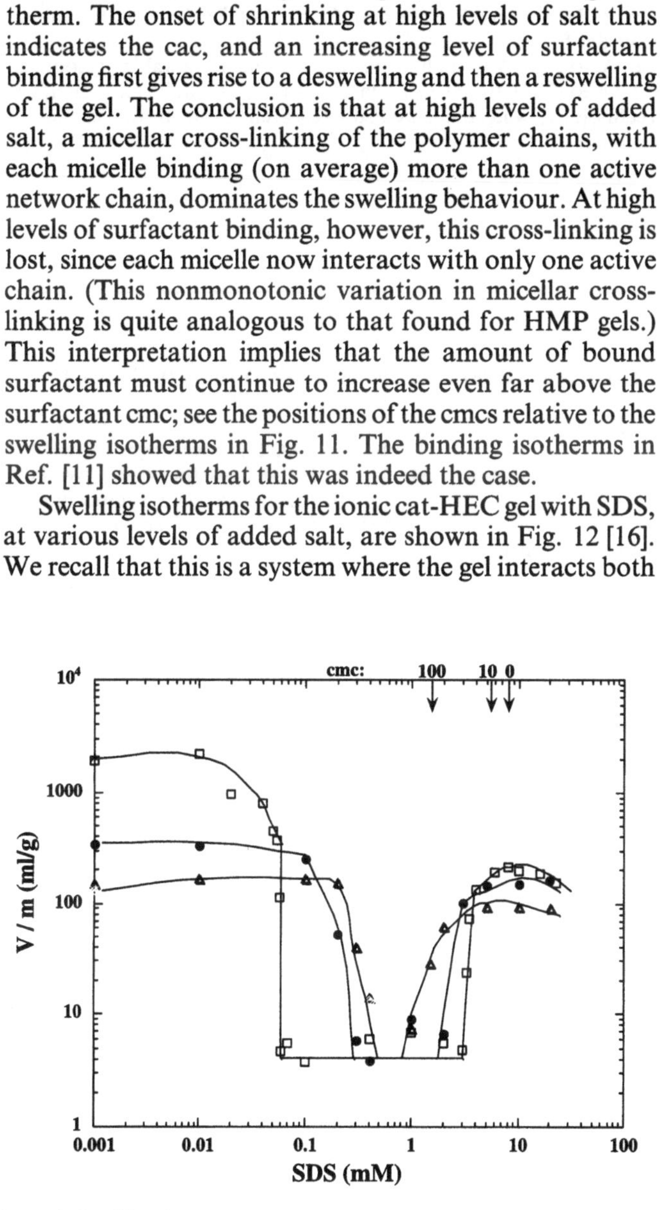

Fig. 12 Swelling isotherms for cat-HEC gels with SDS at different levels of added NaCl (mM): 0 (*squares*), 10 (*circles*), or 100 (*triangles*). *Arrows* denote cmcs for SDS at the various salt concentrations. From Ref. [16]

electrostatically and hydrophobically with the surfactant; hence the appearance of the reswelling at high surfactant concentrations. Both in the swollen state, below the cac, and in the reswollen state at high surfactant concentration, the gel swelling decreases with added salt, as expected for an ionic gel. By contrast, the volume of the fully collapsed gel is unaffected by added salt. The latter indicates that the fully collapsed state represents a stoichiometric polyion–surfactant complex, additionally stabilized by the hydrophobic interactions, which therefore neither swells nor shrinks with added salt.

Both the cac and the cac(2) are affected by added salt, which provides a possibility to tune the stability ranges of the swollen, collapsed, and reswollen states [16]. The cac increases, since the driving force for micellization at the oppositely charged polymer decreases in the presence of salt. By contrast, the cac(2) decreases, since the cooperative binding of ionic surfactant to a net neutral, stoichiometric polyion–surfactant complex is facilitated by added salt. As a consequence of the two opposing salt trends of the cac and the cac(2), the surfactant concentration interval, where the collapsed state is stable, shrinks. However, this concentration interval never vanishes completely because of the hydrophobic polymer–surfactant attraction [15]. For a purely electrostatic polyion–surfactant interaction (e.g., NaA gel–cationic surfactant system), it is possible to dissolve the complex by adding enough salt [28, 29].

At very high salt concentrations (not shown) the effects of the ionic substituents vanish [16]. The swelling isotherm is then dominated by the hydrophobic interacions between SDS and cat-HEC, and it approaches the swelling isotherm for SDS with HEC, shown in Fig. 11.

The salt sensitivity of the surfactant swelling isotherms for the charged and cat-HMHEC gels has also been investigated [12, 16]. Here even more complex trends can be found which, nevertheless, are readily understood in terms of the surfactant binding isotherm and the two general salt effects mentioned at the beginning of this section.

Conclusions

The gel swelling method is a simple and sensitive technique to study polymer–surfactant interactions. Even small changes in polymer dimensions, resulting from these interactions, are readily detected as changes in macroscopic gel dimensions. Associating and nonassociating polymer–surfactant pairs may be distinguished, and cac values may be measured. Phase separation and redissolution phenomena in mixtures of linear polymers and surfactants are mirrored by collapse and reswelling phenomena in the corresponding gel–surfactant systems. The molecular interpretation of all these phenomena is greatly facilitated by the fact that the chemical potential of the surfactant is controlled by the external reservoir in the gel swelling experiment.

Investigations on a variety of gels, based on chemically different polymer backbones, have established generic interaction patterns and swelling isotherms with ionic surfactants. The features of the swelling isotherms are essentially determined by the hydrophobicity of the polymer backbone and on its content of covalently attached charges or strongly hydrophobic side chains. A threshold hydrophobicity of the polymer backbone is required for surfactant binding, and this threshold seems to vary, according to a universal pattern, with the alkyl chain length, the headgroup, and the counterion of the ionic surfactant.

From a responsive-gels perspective, our results show that different patterns of swelling or shrinking on increasing the surfactant concentration may be obtained by proper choice of gel polymer, surfactant, and added salt. Owing to the existence of an interaction threshold, slightly hydrophobic gels are surfactant-selective, and the selectivity may be tuned by changing the hydrophobicity of the gel polymer. Strongly hydrophobic moieties on the gel polymer give rise to continuous surfactant binding and more continuous swelling isotherms. Dramatically responsive gel–surfactant systems may be designed, especially for oppositely charged systems, where very large changes in the gel volume can result from small changes in the surfactant concentration. Charged gels can be made that do, or do not, reswell on increasing the surfactant concentration or adding simple salt.

Acknowledgements We are grateful to Olof Rosén and Dominique Hourdet for their contributions to work cited in this article, and to the Centre for Amphiphilic Polymers from Renewable Resources, the Swedish National Graduate School of Colloid and Interface Technology, and the Swedish Research Council for economic support.

References

1. Goddard ED (1986) Colloids Surf 19:255
2. Goddard ED (1986) Colloids Surf 19:301
3. Lindman B, Thalberg K (1993) In: Goddard ED, Ananthapadmanabhan KP (eds) Interactions of surfactants with polymers and proteins. CRC, Boca Raton, pp 203–276
4. Kwak JCT (1998) Polymer–surfactant systems. Dekker, New York
5. Dusek K (1993) Responsive gels: volume transitions I. Advances in polymer science, vol 109. Springer, Berlin Heidelberg New York
6. Dusek K (1993) Responsive gels: volume transitions II. Advances in polymer science, vol 110. Springer, Berlin Heidelberg New York
7. Sjöström J (2002) Doctoral thesis, Lund University, Lund, Sweden
8. Piculell L, Hourdet D, Iliopoulos I (1993) Langmuir 9:3324
9. Piculell L, Thuresson K, Ericsson O (1995) Faraday Discuss 101:307
10. Rosén O, Piculell L (1997) Polym Gels Networks 5:185
11. Rosén O, Piculell L, Hourdet D (1998) Langmuir 14:777
12. Rosén O, Sjöström J, Piculell L (1998) Langmuir 14:5795
13. Sjöström J, Piculell L (2000) Langmuir 16:4770
14. Piculell L, Nilsson S, Sjöström J, Thuresson K (2000) In: Glass JE (ed) Associative polymers in aqueous media. ACS symposium series 765. American Chemical Society, Washington, DC, pp 317–335
15. Sjöström J, Piculell L (2001) Langmuir 17:3836
16. Sjöström J, Piculell L (2001) Colloids Surf A 183–185:429
17. Lynch I, Sjöström J, Piculell L (in preparation)
18. Lynch I, Sjöström J, Piculell L (in preparation)
19. Balazs EA, Leshchiner A (1986) US Patent 4,582,865
20. Matsuo ES, Tanaka T (1988) J Chem Phys 89:1695
21. Lynch I (2000) Doctoral thesis, University Collage Dublin, Ireland
22. Ninham BY, Yaminsky V (1997) Langmuir 13:2097
23. Svensson A, Piculell L, Cabane B, Ilekti P (2002) J Phys Chem B 106:1013
24. Glass JE (2000) Associative polymers in aqueous media. ACS symposium series 765. American Chemical Society, Washington, DC
25. Piculell L, Thuresson K, Lindman B (2001) Polym Adv Technol 11:1
26. Piculell L, Guillemet F, Thuresson K, Shubin V, Ericsson O (1996) Adv Colloid Interface Sci 63:1
27. Guillemet F, Piculell L (1995) J Phys Chem 99:9201
28. Sasaki S, Koga S, Imabayashi R, Maeda H (2001) J Phys Chem B 105:5852
29. Ilekti P, Piculell L, Tournilhac F, Cabane B (1998) J Phys Chem B 102:344

Progr Colloid Polym Sci (2003) 122: 113–121
DOI 10.1007/978-3-540-36114-5

Paola Roscigno
Gerardino D'Errico
Ornella Ortona
Roberto Sartorio
Luigi Paduano

A comparison study between sodium decyl sulfonate and sodium decyl sulfate with respect to the interaction with poly(vinylpyrrolidone)

P. Roscigno · G. D'Errico · O. Ortona
R. Sartorio · L. Paduano (✉)
Dipartimento di Chimica,
Universita' degli Studi di Napoli
Federico II, Via Cinthia Monte S.Angelo,
80126 Naples, Italy
e-mail: padu@chemistry.unina.it
Fax: +39-081-674090

Abstract Surface tension, fluorescence quenching, and intradiffusion measurements have been performed on sodium decyl sulfonate and sodium decyl sulfate aqueous solutions in the presence and in the absence of poly(vinylpyrrolidone). Despite their chemical similarity, these two surfactants show differences in their behavior even in binary aqueous solution. In fact, the cmc values are different as are their aggregation numbers. Furthermore, the overall picture of the properties we have studied suggests the presence of a preferential interaction of poly (vinylpyrrolidone) toward sodium decyl sulfate with respect to sodium decyl sulfonate, for which no complex formation has been detected. The aqueous poly(vinylpyrrolidone)–sodium decyl sulfate system has been extensively studied in order to obtain details on the micellar aggregates bound on the polymer. This includes an estimation of the number of micellelike clusters per polymer molecule and the number of surfactant molecules present in these micelles.

Keywords Sodium decyl sulfonate · Sodium decyl sulfate · Poly(vinylpyrrolidone) · Polymer–surfactant complex

Introduction

The great interest in polymer-surfactant systems has been driven by many current or possible applications in pharmaceutical formulations, personal care products, food products, industrial detergents, paints and coatings, but it has also been inspired by fundamental interest in intermolecular interactions [1, 2]. Since the first reports by Saito [3, 4], Jones [5, 6], Arai and coworkers [7, 8] and Shirahama et al. [9], a great number of publications and several reviews on this field have been published [10, 11]. In general, the subtle balance of hydrophilic, hydrophobic and ionic interactions governs the aggregation between the polymer and surfactant [10, 11]. However, it has been customary to classify the polymer–surfactant interactions according to polymer or surfactant charge and with respect to surfactant concentration.

The early works between the 1950s and the 1970s on this topic were focused on a variety of nonionic polymers, such as poly(ethylene oxide) (PEO), poly(vinylpyrrolidone) (PVP), methylcellullose (MeC), poly(vinyl alcohol) (PVOH), poly(vinyl acetate) (PVAc) and poly(propylene oxide) (PPO) [12, 13]. In their review of 1973, Breur and Robb [14] noted that "reactivity index" could be assigned to nonionic polymers on the basis of their affinity to anionic surfactants. The "reactivity" followed the sequence PVOH < PEO < MeC < PVAc < PPO≈PVP. Although the authors generally refer to "anionic surfactants", in most of the cases the experimental works on these polymers have been performed using sodium dodecyl sulfate (SDS) as anionic surfactant [11].

The high reactivity of PVP towards anionic surfactants was explained in terms of the slight effective positive charge present on the pyrrolidonic group. There are several papers that claim complex formation between PVP and alkyl sulfate surfactants [8, 15]. To our knowledge, experimental studies on systems containing

PVP and alkyl sulfonate, excluding two recent papers of the authors, are absent in the literature [16, 17]. However, despite this lack and owing to the chemical similitude between the alkyl sulfonate and the alkyl sulfate surfactants, PVP is assumed to be the most efficient nonionic polymer in interaction with all anionic surfactants, no matter what the kind of polar head. The authors in previous studies have shown that this general conviction could be incorrect. In order to show the differences between surfactants belonging to the alkyl sulfonate and alkyl sulfate series, we present here a detailed study on the aqueous solutions of both sodium decyl sulfonate (C10S) and sodium decyl sulfate (C10OS), in the presence and in absence of PVP. Surface tension, static fluorescence and intradiffusion measurements were performed on C10S–water and C10OS–water binary systems and on PVP (1 wt%)–C10S–water and PVP (1 wt%)–C10OS–water ternary systems for a wide range of surfactant concentrations.

Despite their chemical similarity, the two surfactants show some differences even in the binary systems. For instance, in aqueous solution C10S and C10OS exhibit different value of the critical micellar concentration (cmc) and of the micelle aggregation number.

However, the most remarkable difference between C10S and C10OS is the different behavior that they exhibit with respect to the interaction with PVP. The experimental evidence reported in this paper suggests a preferential interaction of PVP towards C10OS with respect to C10S. In contrast with the physico-chemical properties observed for the system containing C10OS, the behavior of the PVP–C10S–water system is similar to that of C10S in water. However, the PVP–C10OS–water system clearly shows the presence of a polymer–surfactant complex.

Experimental

Materials

PVP (average molecular mass 24,000 g mol^{-1}) was purchased from Fluka. C10S and C10OS (stated purity above 99%) were from Sigma. All solutes were reagent grade and were used without further purification. The molar masses of the surfactants C10S and C10OS and the water used were 244.33, 260.32 and 18.016 g mol^{-1}, respectively. All solutions were prepared by weight with doubly distilled water. The samples to be measured by Fourier transform pulse-field-gradient spin-echo NMR (FT-PGSE-NMR) were prepared using heavy water (Aldrich, purity above 99.8%). Solubilized tetramethylsilane (TMS, Sigma, purity 99.9%) was used in the micellar composition range to measure the micelle intradiffusion coefficient.

Surface tension

The surface tension, γ, was measured with an accuracy of 0.1 dyn cm^{-1}, by the De Noüy ring method, using a KSV Sigma 70 digital tensiometer, equipped with an automatic device to select the rising velocity of the platinum ring and to set the time lag between two measurements. The temperature was kept constant at 25.00 ± 0.01 °C. The tensiometer was calibrated with water and acetone [18]. The surface tension was measured for the PVP–C10OS–water and PVP–C10S–water systems along with their corresponding surfactant–water systems. In the ternary systems, the PVP concentration was kept constant at 1 wt%. The concentration of the surfactants ranged from infinite dilution to 0.258 mol kg^{-1} (Table 1).

Table 1 Surface tension of sodium decyl sulfonate (C10OS)–water and sodium decyl sulfate (C10S)–water systems in the absence and in the presence of 1 wt% poly(vinylpyrrolidone) (PVP)

C10S–water		PVP–C10S–water		C10OS–water		PVP–C10OS–water	
m_{C10S} (mol kg^{-1})	γ (dyn cm^{-1})	m_{C10S} (mol kg^{-1})	γ (dyn cm^{-1})	m_{C10OS} (mol kg^{-1})	γ (dyn cm^{-1})	m_{C10OS} z(mol kg^{-1})	γ (dyn cm^{-1})
0.0000	70.5	0.0000	57.9	0.0000	69.1	0.0000	65.9
0.0006	67.4	0.0009	54.3	0.0013	64.2	0.0008	62.5
0.0014	65.8	0.0025	53.6	0.0022	61.8	0.0023	59.5
0.0020	62.7	0.0044	51.9	0.0046	54.9	0.0036	58.0
0.0042	58.9	0.0060	51.1	0.0058	53.0	0.0048	56.0
0.0066	55.3	0.0080	49.9	0.0096	47.4	0.0067	54.3
0.0105	54.0	0.0109	45.5	0.0118	44.9	0.0097	51.2
0.0187	52.1	0.0203	41.8	0.0168	39.2	0.0146	46.7
0.0269	48.5	0.0286	38.7	0.0271	33.7	0.0201	45.4
0.0379	44.0	0.0387	36.7	0.0476	37.6	0.0515	40.9
0.0501	40.0	0.0463	36.5	0.0809	37.8	0.0826	38.4
0.0693	36.7	0.0581	37.7	0.1080	37.7	0.1076	37.9
0.0898	36.8	0.0775	38.2	0.1497	37.2	0.1292	37.6
0.1042	36.8	0.0975	37.5	0.1813	37.9	0.1478	37.3
0.1173	36.2	0.1147	37.0	0.2242	37.6	0.1895	36.9
0.1867	36.9	0.1260	37.0			0.2260	36.6
		0.1810	37.0			0.2583	36.2

Intradiffusion measurements

The intradiffusion coefficients were measured by the FT-PGSE-NMR technique at 25 °C [19]. A spectrometer operating in the ^{1}H mode at 80 MHz, and equipped with a pulsed magnetic field gradient unit, made by Stelar (Mede, Italy), was employed. The temperature was maintained within 0.1 °C with a Stelar variable-temperature controller, model VTC87. The expression for the individual spin-echo peak amplitude for a given line is

$$A = A_o \exp\left[-\gamma_{mg}^2 g^2 D \delta^2 (\Delta - \delta/3)\right] , \quad (1)$$

where A_o is a constant for a specific set of experimental conditions, γ_{mg} is the gyromagnetic ratio of the proton, D is the intradiffusion coefficient of the species responsible for the NMR signal, g is the strength of the applied gradient, and Δ and δ are the time parameters in the pulse sequence. The time between the 90° and 180° pulses was kept constant so that the relaxation effect must not be accounted for. The duration of the gradient pulses, δ, was varied over a suitable range to observe the decay of the spin-echo signal, A. Equation (1) was fitted by a nonlinear least-squares routine to the decay of A as a function of δ.

To evaluate the values of the intradiffusion coefficients, g must be known. Measurements to establish its value were performed on heavy water with trace amounts of light water ($D_{HDO} = 1.872 \times 10^{-9}$ m^2 s^{-1}). The experimental errors for the intradiffusion coefficients were generally less than 3%. In order to correct the intradiffusion constants obtained in deuterated solutions back to those in normal water, it is necessary to multiply by the factor 1.23, the ratio between the self-diffusion coefficients of normal and deuterated water [20].

The intradiffusion coefficients were measured for the ternary PVP–C10OS–water and PVP–C10S–water systems at constant PVP concentration and for the corresponding surfactant–water binary systems. The surfactant concentration ranged from 0.0098 to 0.14 mol kg^{-1} (Table 2). To evaluate the diffusion coefficients of the micelles, a trace amount of TMS was added to the system. Because of its strong hydrophobicity, TMS shows negligible solubility in water and is entirely confined in the hydrophobic micellar core. Thus, the observed TMS intradiffusion coefficients correspond to the micelle intradiffusion coefficients [21]. The TMS intradiffusion coefficients are also collected in Table 2.

Fluorescence quenching

In order to evaluate the mean aggregation number of micelles and that of micelle-like clusters bound to PVP, fluorescence quenching measurements were performed using a JASCO FP-750 spectrofluorimeter at 25 °C. The fluorescence probe pyrene and the quencher dodecylpyridinium chloride (DpyCl) were used in all the systems. The excitation wavelength was set at 290 nm and the fluorescence intensity was detected at 383 nm. This latter wavelength corresponds to the third vibronic peak appearing in the emission spectrum of pyrene. The aggregation numbers were measured under the assumption that the intramicellar "static" quenching between the probe and the quencher, both following a Poisson distribution, is dominant. The pyrene concentration was 2×10^{-6} mol dm^{-3}, and the fluorescence intensity at 383 nm was monitored for decreasing amounts of the DpyCl (acceptor), starting from a molar concentration ratio R = [quencher]/[micelle] = [Q]/[M]~1 [22, 23, 24].

The fluorescence intensity in the presence of the inhibitor is given by

$$I = I_o \exp(-[Q]/[M]) , \quad (2)$$

where I_o is the fluorescence intensity in the absence of inhibitor. In the approximation of the phase separation model the following equation holds:

$$[M] = (m - \text{cmc})/n , \quad (3)$$

where m is the total molal concentration of the surfactant. The aggregation number, n, is obtained from the slope of $\ln(I/I_o)$ versus $[Q]/(m\text{–cmc})$ [25, 26]. The values of the aggregation numbers measured for all the systems studied are reported in Table 3.

Table 2 Intradiffusion coefficients of C10S and C10OS in the systems C10S–D2O and C10OS–D2O in the absence and in the presence of 1 wt% PVP

C10S–D_2O		PVP–C10S–D_2O		C10OS–D_2O			PVP–C10OS–D_2O		
m_{C10S} (mol kg^{-1})	$D_{C10S} \times 10^9$ (m^2 s^{-1})	m_{C10S} (mol kg^{-1})	$D_{C10S} \times 10^9$ (m^2 s^{-1})	m_{C10S} (mol kg^{-1})	$D_{C10OS} \times 10^9$ (m^2 s^{-1})	$D_m \times 10^9$ (m^2 s^{-1})	m_{C10OS} (mol kg^{-1})	$D_{C10OS} \times 10^9$ (m^2 s^{-1})	$D_m \times 10^9$ (m^2 s^{-1})
0.0000	0.634			0.0000	0.605				
0.0171	0.487	0.0186	0.446	0.0147	0.461		0.0123	0.347	
0.0213	0.445	0.0209	0.488	0.0177	0.445		0.0182	0.337	
0.0240	0.452	0.0257	0.420	0.0205	0.436		0.0269	0.312	
0.0268	0.442	0.0301	0.400	0.0235	0.415		0.0310	0.295	
0.0358	0.428	0.0355	0.397	0.0311	0.397		0.0311	0.290	
0.0409	0.410	0.0510	0.365	0.0348	0.402		0.0371	0.279	
0.0507	0.385	0.0585	0.345	0.0383	0.358		0.0408	0.267	
0.0596	0.325	0.0780	0.265	0.0489	0.274		0.0461	0.279	
0.0783	0.275	0.1000	0.202	0.0556	0.243		0.0549	0.244	
0.0932	0.241	0.1206	0.138	0.0746	0.208	0.0912	0.0573	0.241	
0.1014	0.225	0.1272	0.139	0.0915	0.162	0.0903	0.0618	0.228	
0.1074	0.198			0.1117	0.141	0.0893	0.0689	0.209	
0.1216	0.170			0.1212	0.139	0.0883	0.0760	0.168	
0.1456	0.173			0.1399	0.136	0.0876	0.0772	0.183	
							0.0915	0.161	0.0951
							0.0981	0.163	
							0.1075	0.132	0.0884
							0.1177	0.134	0.0731

Table 3 Distinctive results of the experimental data for the C10OS–water and C10S–water systems in the absence and in the presence of 1 wt% PVP

	Critical micelle concentration (mol kg^{-1})	Aggregation number	T_1 (mol kg^{-1})	T_2 (mol kg^{-1})	Aggregation number of micellelike clusters on PVP	Number of aggregates for each PVP molecule
C10OS–water	0.024	57 ± 5				
PVP–C10OS–water			0.003	0.09	15 ± 2	10 ± 2
C10S–water	0.04	35 ± 3				
PVP–C10S–water			Absent	Absent		

Results

The experimental data will be analyzed according to the current hypothesis on the surfactant–polymer interaction. When characterizing the interaction between a surfactant and a polymer, three characteristic surfactant concentrations, at a given polymer concentration, have been defined: T_1 (critical aggregation concentration), T_2^* and T_2 [5, 10]. T_1 corresponds to the onset of cooperative binding of surfactant to the polymer (a concentration lower than the cmc of the binary surfactant–water system). T_2^* corresponds to the polymer saturation by surfactant and T_2 to the concentration at which the pure micelles (unbound to the polymer) form in the bulk.

Surface tension

The surface tension, γ, is plotted versus m in Figs. 1 and 2 for the C10S–water and C10OS–water systems and the corresponding ternaries containing 1 wt% PVP are also shown. With regards to the binaries, the surfactant added at concentrations below the cmc has the effect to decrease γ; then as the cmc is reached, the surface tension curve changes its slope and approaches a flat line. The cmc values detected for C10S and C10OS in water are about 0.04 and 0.024 mol kg^{-1}, respectively, in good agreement with literature data [27, 28].

PVP–C10OS–water

Figure 1 shows that the trend of γ versus m of the PVP–C10OS–water system is quite different with respect to that of the C10OS–water system. The γ curve of the ternary system crosses over the surface tension curve of the corresponding surfactant–water system at $C_{C10OS} = 0.003$ mol kg^{-1} and at $C_{C10OS} = 0.09$ mol kg^{-1}, namely at T_1 and T_2.

Similar behavior has been found by Arai and Murata [18] ($T_1 = 0.008$ mol dm^{-3} and $T_2 = 0.020$ mol dm^{-3}), although in that case the system contained PVP ($M_w = 90{,}000$ mol g^{-1}) at 0.10 wt%. It is to be noted that in the present work we evaluated the T_1 value as the first intersection between the γ trend for the binary C10OS–water and the ternary PVP–C10OS–water systems, as previously proposed by Arai and Murata [18]. The absence of a marked change of slope in the γ versus m curve in our system is probably an indication of the moderated cooperativity of the complex formation between C10OS and PVP.

PVP–C10S–water

Figure 2 shows that the presence of PVP in the system does not change substantially the behavior of the surfactant concentration dependence of the surface tension with respect to that of the C10S–water system. In fact as the surfactant concentration is raised the γ values of the ternary systems overlap those of the binary mixtures, so the only breakpoint detectable corresponds to the cmc of the C10S–water system. The lack of other transition points for the PVP–C10S–water system suggests the absence of polymer–C10S aggregate in solution.

The lower γ values (Figs. 1, 2) observed in dilute ternary solutions, with respect to the corresponding

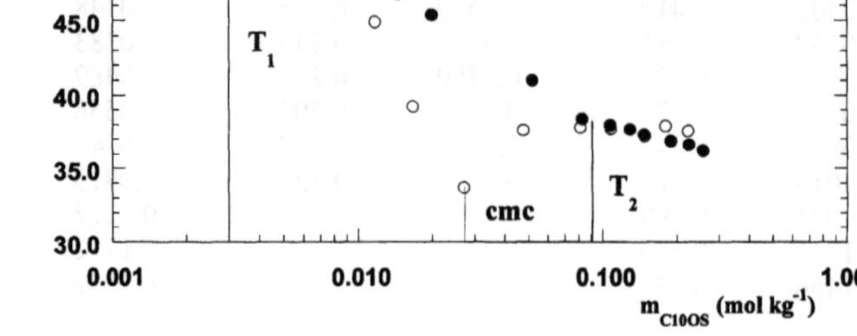

Fig. 1 Surface tension of sodium decyl sulfonate (C10OS)–water in the absence (*open circles*) and in the presence (*solid circles*) of 1 wt% poly(vinylpyrrolidone) (PVP)

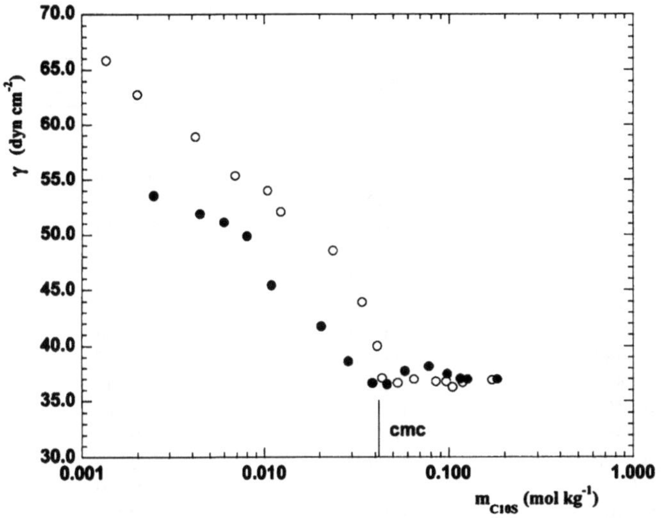

Fig. 2 Surface tension of sodium decyl sulfate (C10S)–water in theabsence (*open circles*) and in the presence (*solid circles*) of 1 wt% PVP

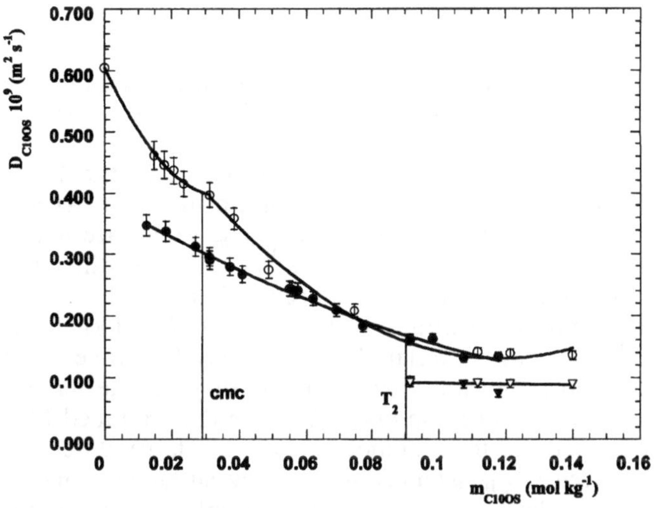

Fig. 3 Surfactant intradiffusion coefficients in C10OS–D_2O (*open circles*) and in C10OS–PVP(1 wt%)–D_2O (*solid circles*) systems, and tetramethylsilane intradiffusion coefficients in C10OS–D_2O (*open triangles*) and in C10OS–PVP (1 wt%)–D_2O systems (*solid triangles*)

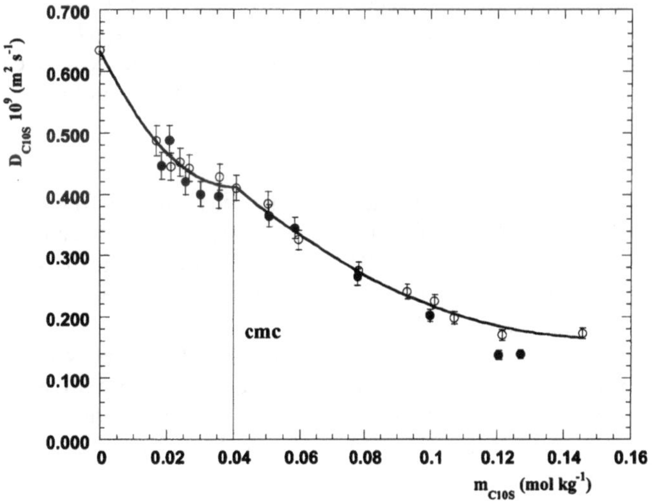

Fig. 4 Surfactant intradiffusion coefficients in C10S–D_2O (*open circles*) and in C10S–PVP–D_2O (*solid circles*) systems

binary systems, can be reasonably assigned to the hydrophobic contribution of the polymer to the air–solution interface composition. In fact, PVP is a moderate surface tension agent [29]. A solution of 1 wt% PVP in water has $\gamma = 62.1$ dyn cm^{-1}, whereas the surface tension of pure water is 72.0 dyn cm^{-1}.

Intradiffusion

The intradiffusion coefficients of the surfactant measured for the ternary PVP–C10OS–water, D^{T}_{C10OS}, and PVP–C10S–water, D^{T}_{C10S}, systems at constant PVP concentration (1 wt%) are shown in Figs. 3 and 4. In the same figures the corresponding binary intradiffusion coefficients, D^{B}_{C10OS} and D^{B}_{C10S}, are also reported. It is to be noted that in the ternary PVP–surfactant–water systems it was not possible to monitor the NMR PVP signal because of the quite low polymer relaxation times. Then only the surfactant intradiffusion coefficients were measured.

C10OS–D_2O and *PVP–C10OS–D_2O* systems

As shown in Fig. 3, the concentration dependence of the surfactant intradiffusion coefficient in the binary system shows a marked change of slope at the cmc. The large concentration dependence of the intradiffusion coefficients above the cmc is generally attributed to the increasing number of micelles in the system as the surfactant concentration is raised. Owing to their size, the micelles have lower diffusion coefficients than the monomeric surfactant and produce an obstruction effect on the motion of the other micelles and of the free monomers.

In the premicellar composition range, the intradiffusion coefficients may be fitted as a function of the square root of the ionic strength, I, as is usual for electrolyte solutions [30]:

$$D^{B} = D^{B^{\infty}}(1 - \alpha I^{1/2}) \ , \qquad (4)$$

where $D^{B^{\infty}}$ is the binary infinite dilution intradiffusion coefficient, which can be calculated through the Nernst relationship, and α is a fitting parameter [31].

As is well known, the intradiffusion coefficients are the concentration-weighted average of the diffusion coefficients of the free monomers, D^{B}_{f}, and the micelles, D^{B}_{M}, according to [30]

$$D^{\mathrm{B}} = \frac{m_{\mathrm{f}}}{m} D_{\mathrm{f}}^{\mathrm{B}} + \frac{n m_{\mathrm{M}}}{m} D_{M}^{B} , \quad (5)$$

where m, m_f and m_M are the total surfactant, the free monomer and the micelle molalities, and n is the aggregation number.

The C10OS intradiffusion coefficients in the PVP–C10OS–water system are also shown in Fig. 3. At low surfactant concentration the D^{T}_{C10OS} values are markedly smaller than those of D^{B}_{C10OS}. This difference reduces as the surfactant concentration is raised. For a surfactant concentration higher than 0.07 mol kg^{-1}, the D^{T}_{C10OS} values overlap the binary ones. It is to be noted that no significant changes in the slope of the ternary system have been observed. Because the presumed PVP obstruction effect cannot explain the large difference between D^{T}_{C10OS} and D^{B}_{C10OS} at low surfactant concentration, the observed trend has to be interpreted in terms of C10OS–PVP interaction.

In the surfactant concentration region where only interaction between the surfactant and the polymer occurs, between T_1 and T_2, the diffusion data can be analyzed, as suggested by Jonsson et al. [32] and Carlsson et al. [33], according to a two-site model. The observed surfactant intradiffusion coefficients, D^{T}, is a population average of the intradiffusion of the free surfactant and the surfactant bound to the polymer:

$$D^{\mathrm{T}} = \frac{m_{\mathrm{f}}}{m} D_{\mathrm{f}}^{\mathrm{T}} + \frac{m_{\mathrm{b}}}{m} D_{b}^{T} . \quad (6)$$

Here m_f and m_b are the concentration of the free and polymer-bound surfactant, and m is the total concentration of the surfactant. D^{T}_{f} and D^{T}_{b} are the intradiffusion coefficients of the free surfactant and polymer–surfactant complex, respectively.

Since D^{T}_{b} is significantly smaller than D^{T}_{f} (usually D^{T}_{b} is approximated by the polymer intradiffusion coefficient, approximately 0.072×10^{-9} m^2 s^{-1} [16]), the second term on the right side of Eq. (6) can be neglected and Eq. (7) is obtained:

$$D^{\mathrm{T}} \cong \frac{m_{\mathrm{f}}}{m} D_{f}^{T} . \quad (7)$$

Assuming that D^{T}_{f} can be approximated with the D^{B}_{C10OS} values, Eq. (7) can be used to evaluate the amount of C10OS bound to the polymer. Hence Eq. (7) was applied to the PVP–C10OS–water system in the concentration range between T_1 ($m_{C10OS} \cong$ 0.003 mol kg^{-1}) and T_2 ($m_{C10OS} \cong 0.09$ mol kg^{-1}) as measured through surface tension. The results are reported in Table 4 and are presented in Fig. 5, where m_b and m_f are plotted against m.

As already mentioned, for both C10OS–water and PVP–C10OS–water systems TMS was added, in order to measure the intradiffusion coefficient of the micelles bound on the polymer [19]. However, at a C10OS concentration below 0.09 mol kg^{-1} (T_2) the TMS dissolution was not observed as suggested by the absence of its characteristic peak in the NMR spectrum [34]. Furthermore it is to be noted that the value of the micelle intradiffusion coefficients in the presence and in the absence of the polymer are the same, as reported in Table 2 and shown in Fig. 3. This suggests that the micelles bound to the polymer are smaller than the free micelles and could not allow the TMS dissolution in their core.

Table 4 Binding concentrations from intradiffusion coefficients of C10OS in D_2O–PVP at 1 wt%

m_{C10OS} (mol kg^{-1})	m_b (mol kg^{-1})	m_f (mol kg^{-1})
0.0000	0.0000	0.0000
0.0040	0.0014	0.0026
0.0080	0.0029	0.0051
0.0100	0.0038	0.0062
0.0120	0.0046	0.0074
0.0180	0.0075	0.0105
0.0220	0.0096	0.0124
0.0280	0.0129	0.0151
0.0320	0.0154	0.0166
0.0380	0.0193	0.0187
0.0420	0.0220	0.0200
0.0480	0.0263	0.0217
0.0580	0.0341	0.0239
0.0700	0.0441	0.0259
0.0800	0.0530	0.0270
0.0900	0.0623	0.0277

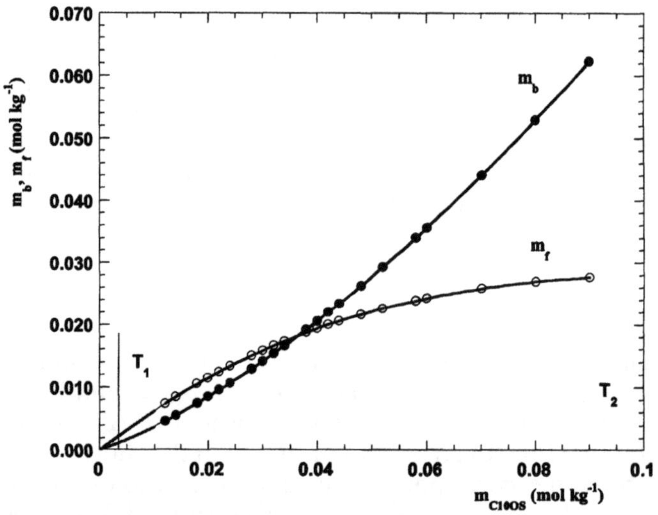

Fig. 5. Calculated concentrations of bound, m_b, and free, m_f, C10OS in 1 wt% PVP–D_2O solution versus the total C10OS concentration

C10S–D_2O and *PVP–C10S–D_2O* systems

The intradiffusion coefficients of the surfactant in the presence and in absence of PVP are reported in Fig. 4.

The concentration dependence of D^{B}_{C10S} in the C10S–D_2O system is nearly the same as that observed in the binary system containing C10OS, so the arguments briefly discussed in the previous paragraph for the C10OS–D_2O system still hold for the C10S–D_2O system.

In the ternary PVP–C10S–D_2O system the values of the surfactant diffusion coefficient are not affected by the presence of the polymer. In contrast with respect to that observed in the system containing C10OS and PVP, the intradiffusion coefficients of C10S are substantially equal to those measured in the absence of the polymer (Fig. 4). This is a strong indication of the absence of PVP–C10S direct interactions.

Fluorescence quenching

Measurements of fluorescence quenching on the C10OS–water and C10S–water systems were performed at $m_{C10OS}=0.1$ mol kg^{-1} and $m_{C10S}=0.2$ mol kg^{-1}, respectively, namely at 5 times their cmc. The aggregation numbers of pure micelles were evaluated through Eq. (2) (Table 3).

In the PVP–C10OS–water ternary system the measurements were carried out at a C10OS concentration of $m_{10OS}=0.01$ mol kg^{-1}, i.e. within the T_1 and T_2 range, so that the free micelles are absent and do not interfere with the bound clusters onto the polymer. In this case Eq. (3) was modified by replacing the cmc with T_1.

Fluorescence quenching measurements were also performed on the PVP–C10S–water system at $m_{C10S}=0.01$ mol kg^{-1}, although as discussed later no polymer–surfactant interaction was detected.

C10OS–water and PVP–C10OS–water systems

The aggregation number obtained for the pure C10OS micelle is 57 ± 5, in agreement with the value reported in Ref. [35]. In the PVP–C10OS–water system, aggregates are also present well below the cmc of C10OS, with an aggregation number of 15 ± 2. This value is much smaller than the measured aggregation number for pure C10OS. A smaller number of surfactants involved in the micelle-like cluster formation with respect to the pure micelle observed in the absence of polymer is a common feature of the polymer–surfactant systems. D'Aprano et al. [36] measured the frequency dependence of ultrasonic relaxation in PVP–SDS. This technique yields relaxation times which can be used, again in combination with a mass-action model, to calculate the surfactant aggregation number and the aggregate size polydispersity. The aggregation number reported for the PVP–SDS complexes was 19 ± 3, which is much smaller than the accepted value of 65 for pure SDS micelles [10].

Recently, experiments carried out on the PVP–lithium perfluorononanoate system have shown a monotonic change of the aggregation number of the clusters bound to the polymer from 5 to 22, but these are, however, smaller than the pure micelles, whose aggregation number is about 35 [17].

C10S–water and PVP–C10S–water systems

The evaluation of the aggregation numbers in these systems led to a significant value only for the C10S–water system ($n=35 \pm 3$) (Table 3). This number is in quite good agreement with that obtained by the diffusion coefficient measurements reported in Ref. [37].

In the presence of polymer, in contrast with that observed in the system containing decyl sulfate, $\ln(I/I_o)$ versus $[Q]/(m - T_1)$ is a leveled-off line, which implies that no aggregates are present in the system. When the C10S concentration was raised to 0.2 mol kg^{-1}, i.e. above the binary cmc, micelles with $n=33 \pm 3$ were detected in the presence of PVP.

Discussion

The analysis of the data collected in this report allows several considerations. The behavior of the two surfactants in aqueous solution is different. In fact, the cmc values are 0.040 mol kg^{-1} for C10S and 0.024 mol kg^{-1} for C10OS, whereas the aggregation numbers of the micelles are 35 and 57, respectively. These values are in good agreement with those reported in the literature. For the alkyl sulfonate surfactant series, the micelle aggregation numbers have been found to follow [37]

$$n = (21 \pm 1) + (12.8 \pm 0.8)(n_c - 8) \quad n_c \geq 8 , \qquad (8)$$

where n_c is the number of carbons in the alkyl chain. For fewer than five carbons in the hydrophobic tail, no micellization was observed.

In contrast, for the alkyl sulfate series, the aggregation number of the micelle increases with n_c according to [35]

$$n = (42 \pm 2) + (7.8 \pm 0.4)(n_c - 8) \quad n_c \geq 8 . \qquad (9)$$

Thus, for the same length of the hydrophobic tail, the surfactants of the alkyl sulfonate series form micelles with smaller values of the aggregation numbers than the alkyl sulfates. The difference in the behavior of the two surfactants under study is also confirmed by the intradiffusion coefficients. In fact, above the cmc the D^{B}_{C10S} values are larger than the D^{B}_{C10OS} ones. This is probably due, according to Eq. (6), to a larger amount of free C10S present in the system with respect to C10OS. Furthermore, the C10OS micelles are larger than those formed by C10S. Both these reasons lead to the difference observed in the value of the diffusion coefficients of the two surfactants as their concentration is raised.

The differences just highlighted could be due to the absence of an oxygen atom in the polar head of decyl sulfonate with respect to decyl sulfate. As a result of this absence, it is presumed that the hydration shell of the polar head extends over a larger part of the alkyl chain in C10S, which "shortens" its hydrophobic tail with respect to that of the equivalent sulfate surfactant. In turn, this leads to a reduction of the hydrophobic driving force of the micellization process of decyl sulfonate.

Recently, Huibers [38] has estimated the charge distribution in common ionic surfactant molecules using quantum chemical methods. He has shown that the headgroup charges distribute to the rest of the molecule, with a significant fraction of the charge on the α-methylene group (the first CH_2 group of the alkyl tail, attached to the headgroup). This charge is as large as −0.40 for the sulfonate, owing to the highly polar C–S bond, whereas it is lower and of the opposite sign (0.18) for the sulfate surfactants. Owing to this difference Huibers suggested that the headgroup charge repulsion influences the self-assembly processes, which reflects on the cmc and probably on the micelle aggregation number for the different surfactant classes we are dealing with.

In view of the systems containing PVP, the overall picture of the experimental evidences collected on the PVP–C10S–water mixture leads to the exclusion of any detectable interaction between the surfactant and PVP. In this system, only a PVP excluded-volume effect could be present [33, 39, 40, 41]. This effect, not detectable in our diffusion data because of the low PVP concentration, has been clearly found in a similar system, PVP (10 wt%)–octyl sulfonate (C8S)–water where a large decrease of the octyl sulfonate intradiffusion coefficients was observed [16].

In contrast, in the PVP–C10OS–water system there is clear evidence of a strong cooperative association between the polymer and the surfactant. The onset of binding of decyl sulfate onto the polymer occurs, according to the surface tension measurements, at quite low surfactant concentration ($T_1 = 0.003$ mol kg^{-1}) with respect to the cmc of the binary water–surfactant system (cmc = 0.024 mol kg^{-1}). The fluorescence measurements indicate that this complex is formed by a micelle-like clusters bound to the polymer whose aggregation number is substantially lower, about 15, than those observed in the absence of polymer, about 57. The presence of a polymer–surfactant complex reduces the actual surfactant concentration in solution and the formation of pure micelles shifts to higher stoichiometric concentrations of C10OS. In the presence of PVP (1 wt%), the system attains the condition of free decyl sulfate concentration, according to its cmc, to form pure micelles at $m_{C10OS} \cong 0.08$ mol kg^{-1}. In fact at this value of the surfactant concentration, m_f = cmc (Table 3, Fig. 5).

Finally, the average number of aggregates bound onto each PVP molecule can be evaluated, once the aggregation number of the micelle-like clusters on the PVP and the amount of C10OS needed to saturate the polymer have been determined. On the basis that the clusters bound on the polymer are formed by about 15 molecules and that the concentration of surfactant to saturate the polymer is about 0.08 mol kg^{-1}, the average number of clusters bound to each PVP ($M_w = 24{,}000$ g mol^{-1}) molecule is about 10. This finding implies that the PVP chain binds several micelles (smaller than a free micelle) distributed within the polymer coil and separated by lengths of the free chain as originally suggested by Shirahama et al. [9] and confirmed experimentally by Cabane and coworkers [42, 43].

Conclusion

This paper has provided a comparison between C10OS and C10S with respect to their interaction with PVP in water. The analysis of the experimental results has highlighted differences in the behavior of these two surfactants even in the binary aqueous solution. Despite their chemical similarity the values of the cmc and the aggregation number of the pure micelles of decyl sulfate and decyl sulfonate are shown to be quite different. Owing to this evidence it is possible to suggest that the driving force of micellization is weaker for decyl sulfonate than for decyl sulfate.

The difference between these two surfactants with respect to the interaction towards PVP is remarkable. Differently from C10OS, C10S does not show any detectable interaction with the polymer. This is in contrast with the wide conviction that all anionic surfactants form complexes with PVP.

Decyl sulfate has been shown to complex with PVP, forming micellelike clusters smaller than the free micelles. About ten decyl sulfate cluster bind onto the polymer and are distributed along the PVP molecule. Since the clusters are small aggregates strung on the polymer coil, separated by lengths of the free chain, it seems reasonable to suggest that PVP and C10OS form a "necklace" complex.

Acknowledgement This research was carried on with the financial support of the Italian MURST.

References

1. Taber JJ (1980) Pure Appl Chem 52:1323
2. Goddard DE, Ananthapadmanabhan KP (1998) In: Kwak JCT (ed) Polymer–surfactant systems, vol 77. Dekker, New York, pp 143–192
3. Saito S (1954) Kolloid Z 137:93
4. Saito S (1958) Kolloid Z 158:120
5. Jones MN (1967) J Colloid Interface Sci 23:36
6. Jones MN (1968) J Colloid Interface Sci 26:532
7. Arai H, Horin S (1969) J Colloid Interface Sci 30:372
8. Arai H, Murata M, Shinoda K (1971) J Colloid Interface Sci 37:223
9. Shirahama K, Tsuji K, Takagi T (1974) J Biochem 75:309
10. Goddard ED, Ananthapadmanabhan KP (1993) Interactions of surfactants with polymers and proteins. CRC, Boca Raton
11. Shirahama K (1998) In: Kwak JCT (ed) Polymer–surfactant systems, vol 77. Dekker, New York, pp 143–192
12. Isemura T, Imanishi A (1958) J Polym Sci 33:337
13. Cabane B, (1977) J Phys Chem 81:1639
14. Breuer MM, Robb ID (1972) Chem Ind 13:530
15. Shirahama K, Ide N (1976) J Colloid Interface Sci 3:54
16. Roscigno P, Paduano L, D'Errico G, Vitagliano V (2001) Langmuir 17:4510
17. Ortona O, D'Errico G, Paduano L, Sartorio R (2002) Phys Chem Chem Phys 4:2604
18. Arai H, Murata MJ (1973) J Colloid Interface Sci 44:475
19. Stilbs P (1987) NMR Spectrosc 19:1
20. Goldarmer EV, Hertz HG (1970) J Phys Chem 74:3734
21. Jansson M, Stilbs P (1985) J Phys Chem 89:4868
22. Infelta PP, Gratzel M, Thomas JK (1975) J Phys Chem 78:190
23. Tachiya M (1975) Chem Phys Lett 33:289
24. Gehlen MH, De Schryver FC (1993) Chem Rev 93:199
25. Turro NJ, Yekta A (1978) J Am Chem Soc 73:1904
26. McNeil R, Thomas JK (1981) J Colloid Interface Sci 83:57
27. Mukerjee P, Mysels KJ (1971) Critical micelle concentrations of aqueous surfactant systems. National standards reference data series no 36. US Government Printing Office, Washington, DC
28. Rassing J, Sams PJ, Wyn-Jones EJ (1974) J Chem Soc Faraday Trans II 70:1247
29. Sesta B, Segre AL, D'Aprano A, Proietti N (1997) J Phys Chem B 101:198
30. Annunziata O, Costantino L, D'Errico G, Paduano L, Vitagliano V (1999) J Colloid Interface Sci 216:16
31. Tyrrell HJV, Harris R (1994) Diffusion in liquids. Butterworth, London
32. Jonsson B, Wennerstrom H, Nilsson PG, Linse P (1986) Colloid Polym Sci 264:77
33. Carlsson A, Karlstrom G, Lindman B (1989) J Phys Chem 93:3673
34. Jansson M, Stilbs P (1985) J Phys Chem 89:4868
35. Ranganathan R, Peric M, Bales BL (1998) J Phys Chem B 102:8436
36. D'Aprano A, La Mesa C, Persi L (1997) Langmuir 13:5876
37. Annunziata O, Paduano L, Costantino L, D'Errico G, Vitagliano V (1999) J Colloid Interface Sci 216:8
38. Huibers PDT (1999) Langmuir 15:7546
39. Lekkerkerker HNW, Poon WCK, Pusey PN, Stroobants A, Warren PB (1992) Europhys Lett 20:559
40. Annunziata O, Paduano L, Pealrstein AJ, Miller DG, Albright JG (2000) J Am Chem Soc 122:5916
41. Vergara A, Paduano L, Vitagliano V, Sartorio R (2000) J Phys Chem B 104:8086
42. CabaneB, Duplessix R (1982) J Phys (Paris) 43:1529
43. Cabane B, Duplessix R, Zemb T (1985) J Phys (Paris) 46:2161

Progr Colloid Polym Sci (2003) 122: 122–129
DOI 10.1007/978-3-540-36114-5

Regine von Klitzing
Branko Kolarić

Influence of the polycation architecture on the oscillatoric forces of aqueous free-standing polyelectrolyte/surfactant films

R. von Klitzing (✉) · B. Kolarić
Stranski-Laboratorium für Physikalische und Theoretische Chemie,
Technische Universität Berlin,
Strasse des 17. Juni 112, 10623 Berlin, Germany
e-mail: klitzing@chem.tu-berlin.de
Tel.: +49-30-31426774
Fax: +49-30-31426602

Abstract In the present study, free-standing films were formed from aqueous solutions of surfactant and polyelectrolyte and they were investigated by a thin film pressure balance. The surfactant was either a nonionic alkylpolyglucoside or cationic hexadecyltrimethylammonium bromide. Linear poly(diallyl dimethylammonium chloride) (PDADMAC) and poly(ethylene imine) (PEI) of different degrees of branching were investigated. The resulting disjoining pressure isotherms show steps in film thickness owing to oscillatoric structural forces. With respect to the polyelectrolyte architecture the step size, Δh, shows different scaling behavior with the polyelectrolyte concentration, *c*. While Δh in the thickness of the films containing linear polyelectrolytes scales as $c^{-1/2}$, for the branched polyelectrolytes Δh scales as $c^{-1/3}$. The effect of electrostatic interactions on the film stratification was studied by changing the ionic strength or the polymer charge density. For this purpose the ratio between cationic and neutral monomers was varied in the case of the strong PDADMAC or in the case of weak PEI the pH was changed. Both increasing the ionic strength and decreasing the polymer charge density leads to a reduction in the oscillation amplitude of the disjoining pressure. At high ionic strengths or for neutral polymers the steps in the film thickness vanish.

Keywords Foam film · Branched polyelectrolytes · Linear polyelectrolytes · Disjoining pressure · Small-angle scattering

Introduction

Interactions between polymers and surfactants in an aqueous medium are of great scientific interest because of their importance for many industrial applications (e.g. paints, cosmetics, detergents). Further they present a model system to study the interactions between biologically relevant polymers (e.g. DNA, proteins) and amphiphiles (e.g. lipids). In basic research, the diverse interactions between both classes of compounds are of interest. Depending on the charge combination of the polyelectrolyte and the surfactant, the mixed solutions show a fascinating variety of phases [1]. In the case where surfactant and polymer are oppositely charged, aggregation is mainly induced by electrostatic attraction; however, if one of the compound is neutral, the interactions leading to the formation of complexes are more complex and hydrophobic interactions become important [2, 3]. Further, in composite systems the interactions between polymers and surfactants become more and more interesting. For instance, polymers are used as stabilizers for foams and emulsions. To use polymers in an efficient way in surfactant systems, it is necessary to control and to manipulate the properties of the colloidal mixture. Therefore, the behavior of the building blocks, i.e. the single surfactant films, must be well known. The present study focuses on single, free-standing aqueous surfactant films (so-called foam films) containing polyelectrolytes. This class of mixed films is interesting in two ways. On one hand, the interaction between surfactant and polyelectrolyte affects the film stability and, on the other hand, the film presents a cavity where the polymers are entrapped. The films are between 5 and 120-nm thick and therefore the physical dimension of the total system is of the same order of magnitude as the characteristic lengths (e.g. the correlation length) of the polymeric system. This raises the question whether the chain structuring in the film is changed or not.

For several decades the conformation and ordering of polyelectrolyte chains in aqueous media has been investigated by many experimental methods [5, 6, 7, 8] and has been described by theoretical calculations [9, 10, 11, 12, 13]. In the semidilute regime, the chains overlap, which leads to a typical correlation peak in the scattering spectrum. Its position in the reciprocal space corresponds to a correlation length, ξ, in the real space. This can be interpreted as an average distance between the chains (typically several tens of nanometers). In general, this correlation length is assumed to be correlated to the mesh size of a transient network of entangled polyelectrolyte chains. The correlation length is predicted to scale with the polyelectrolyte concentration as $c^{-1/2}$, which has been demonstrated in many experimental studies.

To get information about the effect of the polyelectrolytes on the interactions in the film, the disjoining pressure is measured. It is the excess pressure between the pressure in the film and the pressure of the corresponding liquid. This difference is due to interaction between the film interfaces, which is mainly determined by electrostatic repulsion between (charged) interfaces, van der Waals attraction, steric repulsion between adsorbed surfactant molecules and structural forces. Disjoining pressure isotherms (disjoining pressure as a function of film thickness) were measured with a so-called thin film pressure balance (TFPB). Recent TFPB measurements of mixed aqueous polyelectrolyte/surfactant films resulted in a stepwise thinning above a certain polyelectrolyte concentration [14, 15, 16, 17, 18, 19, 20, 21, 22], whereas the corresponding pure surfactant film drains in a continuous way. This behavior has been found for different classes of polyelectrolytes and polymer-like aggregates, like flexible and stiff chains [20] and wormlike surfactant micelles [16]. Qualitatively, this effect is the same for all classes of mixed polyelectrolyte/surfactant films investigated and it is independent of the polyelectrolyte and surfactant charge [17, 18]. For a certain polyelectrolyte, the step size in the isotherms of the mixed films does not depend on the surfactant used. This leads to the conclusion that the steps in the film thickness are related to the ordering of the polyelectrolyte chain in the film.

The scope of the paper is to show the effect of polymer molecular architecture on the interactions within the free-standing films. Therefore TFPB measurements were carried out for free-standing films made from aqueous polyelectrolyte/surfactant solutions. The polyelectrolytes were either linear or branched. Further the effects of polymer molecular weight, degree of polymer charge, ionic strength, and surfactant were checked.

Experimental

Materials

The linear polyelectrolyte used for the present studies was a statistic copolymer consisting of positively charged diallyl dimethylammonium chloride (DADMAC) monomers and neutral *N*-methyl-*N*-vinylacetamide (NMVA) monomers. This was a gift from Werner Jaeger from the Fraunhofer Institute in Golm. Details about the synthesis and the characterization are described in Ref. [23] for poly(DADMAC) (PDADMAC) and in Ref. [24] for poly(NMVA), and the copolymers. Beside the pure PDADMAC) and the neutral poly(NMVA) chains with linear charge densities of f=0.75, 0.5 and 0.24 were available. The molecular weight was between 5,000 and 709,000 g/mol.

The polycation poly(ethylene imine) (PEI) was a gift from BASF and is available at three different degrees of branching with 100% secondary N atoms (linear), and 50 and 38% secondary N atoms. Depending on the number of bonds to the next carbon atoms one distinguishes between primary, secondary and tertiary *N* atoms. The chemical structures are shown in Fig. 1. While PDADMAC is a strong polyelectrolyte, the charge density of PEI can be easily changed by the pH. The concentrations of the polyelectrolytes are given in units of moles of monomer per liter.

The surfactant C_{12}/C_{14}–alkylpolyglucoside (APG) was a gift from Henkel (Dusseldorf, Germany). It is a mixture of nonionic APGs with alkyl chains of 12 and 14 C atoms, respectively, and an average number of glycoside units of 1.7. It was used at a fixed concentration of 1.7×10^{-5} mol/l, below the critical micelle concentration (cmc). The cationic hexadecyltrimethylammonium bromide ($C_{16}TAB$) was purchased from Sigma-Aldrich. It was used

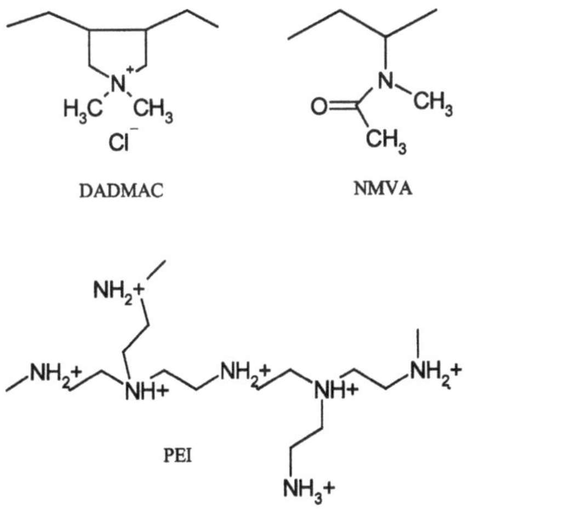

Fig. 1 Chemical structure of the polyelectrolyte poly(ethylene imine) (*PEI*) and the monomers diallyl dimethylammonium chloride (*DADMAC*) and *N*-methyl-*N*-vinylacetamide (*NMVA*)

at a fixed concentration of 10^{-4} mol/l. The free-standing polyelectrolyte/surfactant films were made from aqueous solutions containing PDADMAC or PEI and one of the two surfactants. All the solutions were prepared from Milli Q water.

Thin film pressure balance

The disjoining pressure isotherms (disjoining pressure as a function of the film thickness) were measured with the porous-plate technique using a TFPB developed by Mysels and Jones [25] and Exerowa and coworkers [26, 27]. In this apparatus the film is formed from an aqueous polyelectrolyte/surfactant solution on the hole (diameter of about 1 mm) of a porous glass plate. The film holder is enclosed in a metal cell, which allows the film to be pressurized using a syringe pump. During the film drainage, the capillary pressure causes a sucking of film liquid into the plateau borders until the disjoining pressure begins to affect the dynamics. In the equilibrium condition, the capillary pressure and the disjoining pressure compensate each other. The film is illuminated by white light and observed with video microscopy. Parallel to the video microcopy, the film thickness was determined by the interferometrical method of Scheludko [28]. Under the present experimental conditions, the reflected intensity decreases with decreasing film thickness. It is assumed that the equilibrium film thickness is reached when the reflected intensity stays constant over a period of 20 min. The dynamic measurements were also carried out with this apparatus.

Results and discussions

Effect of concentration and molecular weight

Linear polyelectrolytes

The disjoining pressure isotherms of a pure $C_{12}G_2$ film and the isotherm of a mixed film formed from aqueous solutions of nonionic C_{12}/C_{14}–APG and positively charged PDADMAC are shown in Fig. 2.

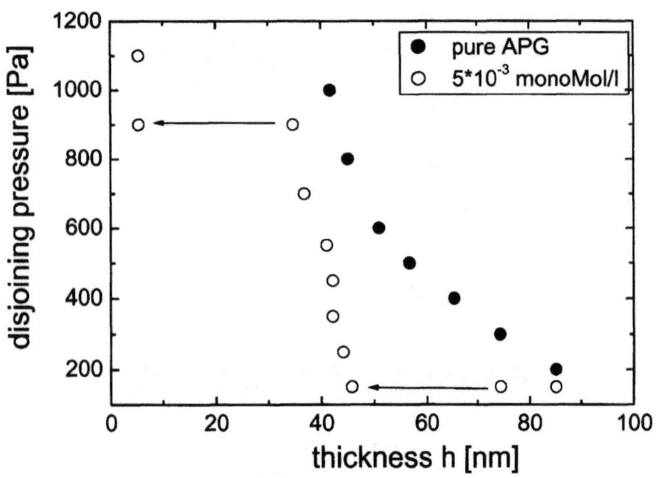

Fig. 2 Disjoining pressure isotherms of pure $C_{12}G_2$ films at a concentration of 10^{-5} mol/l and of mixed PDADMAC/alkylpolyglucoside (*APG*) films at a PDADMAC concentration of 5×10^{-3} monomol/l and at an APG concentration of 1.7×10^{-5} mol/l

The surfactant concentration is well below the cmc in both cases. The pure surfactant film thins out continuously with increasing pressure. The mixed polyelectrolyte/surfactant solution was clear and there was no sign of any aggregation between polyelectrolyte and surfactant. At a certain polyelectrolyte concentration, the film starts to thin stepwise. At a concentration of 5×10^{-3} monomol/l, two steps of around 30 nm are observed. The steps are not reversible. By decreasing the pressure no step backward to the former branch of the isotherm is possible, but the thickness does not change in a significant way. Since the surfactant is nonionic but carries partial negative charges at the sugar head groups, it is assumed that there is no strong complexation between the surfactant and the polyelectrolyte, but a slight attraction. The measurements were also carried out at other polyelectrolyte concentrations. Generally, the number of steps increases with increasing PDADMAC concentration and the steps becomes smaller [18]. In the present case, the film thickness is 5 nm at the highest pressure, independent of the PDADMAC concentration. This is the thickness of the Newton black film of pure surfactant films, which indicates that most of the polyelectrolyte chains are pressed out of the film. The Newton black film is stabilized by steric interactions between the molecules which are adsorbed at the film surfaces. Recent fluorescence spectroscopy experiments with fluorescein-labeled poly(allylamine hydrochloride) showed that even in these thin films a few polyelectrolyte chains remain [29]. The step size is independent of the surfactant used [17], which indicates that the stratification is induced by the polyelectrolytes and not by polymer/surfactant interactions. Therefore, the corresponding aqueous solutions were investigated by small-angle neutron scattering (SANS) [22]. The spectrum shows a scattering peak which is shifted to a higher q value with increasing polymer concentration, while the scattering intensity decreases and the peak becomes broader. The occurrence of a structure peak indicates interactions between the polyelectrolyte chains. Under the assumption that the Bragg equation is applicable, a correlation length, ξ, can be calculated from the peak position q_{max} by $\pi = 2\pi/q_{max}$. The correlation length of the polyelectrolytes in solution is compared with the step size in the disjoining pressure isotherm in Fig. 3. Both lengths are almost identical within experimental error. This leads to the conclusion that the interactions between the polyelectrolytes in solution which induce the structure peak in the small-angle-scattering spectrum are the same as the interactions between the polyelectrolytes in the free-standing film which lead to the stratification. The disjoining pressure isotherms are considered as parts of a damped pressure oscillation. Oscillation forces between two interfaces are also predicted by theoretical models for liquid crystals [30] and for polymer systems [31, 32].

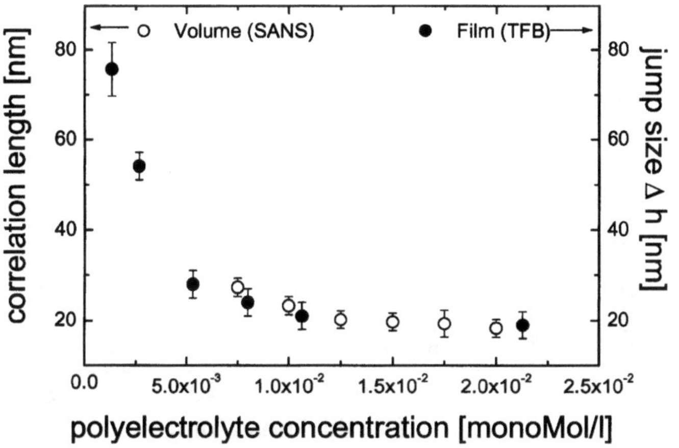

Fig. 3 Step size in the film thickness and correlation length calculated from the position of the correlation peak in the small-angle neutron scatteing spectra of the corresponding aqueous PDADMAC solutions as a function of the PDADMAC concentration

Structural forces which cannot be explained by Derjaguin–Landau–Verwey-Overbeek forces are observed for many other classes of systems entrapped between two planar surfaces. Damped oscillation forces are measured for solvents between two mica interfaces using a surface force apparatus (SFA) [33]. This phenomenon also occurs in films of liquid crystals [34] or in films of latex suspensions [35]. In these cases the molecules or particles can be considered as form-invariant hard spheres. But structural forces occur also in surfactant films at concentrations above the cmc. The micellar systems have been investigated using a SFA between solid interfaces [8, 36] or with a TFPB between liquid interfaces [37]. In all these cases the period of the oscillation corresponds more or less to the diameter of the molecules, particles or aggregates (plus screening length), respectively. The structural forces are explained by layer-by-layer expulsion of the molecules, particles or aggregates. This prompts the question which characteristic length of the polyelectrolyte system corresponds to the period of the oscillatoric curve. The correlation length in solution is related to the mesh size of a transient network predicted by the isotropic model of de Gennes et al. [9]. Above a critical concentration, c^*, the chains start to overlap and form a network which breaks down and is rebuilt within nanoseconds. In free-standing films another mechanism could be responsible for the stratification of the films: layers of polymer coils could be entrapped between the interfaces and layer after layer is pressed out of the film in analogy to the expulsion of layers of micelles. Such a model was suggested by Milling [38], who observed oscillatoric forces in aqueous polyelectrolyte solutions using an atomic force microscope. The decisive parameter to distinguish between the two models is the molecular weight. The diameter of the coils scales with the chain length but the mesh size in the model of overlapping polymer chains does not. The disjoining pressure isotherms for different molecular weights between 5,000 and 709,000 at a fixed monomer concentration of 5×10^{-3} monomol/l is shown in Fig. 4. The PDADMAC/$C_{12}G_2$ films were unstable at high molecular weight, while films containing $C_{16}TAB$ were stable. As already mentioned the step size is not influenced by the choice of the surfactant. The step size is independent of the polymer molecular weight. SANS measurements of the corresponding solutions show a structure peak whose q position is also not influenced by the molecular weight. These results confirm the isotropic model of overlapping chains. The expulsion of one mesh size of the polyelectrolyte network is more complex than for small molecules, particles or aggregates. One idea is that the thickness of the film core fits to a whole-number multiple of ξ. The network is a transient one, i.e. it breaks down and it is rebuilt, and its reorganization rate is much higher than the time resolution of the TFPB experiments. Up to a certain pressure the network reorganizes itself with, foer example, n mesh sizes. If the pressure is further increased the network rebuilds itself with $n-1$ mesh sizes and the polyelectrolyte chains which do not contribute to the network are pressed out. The polyelectrolyte concentration is reduced in comparison with the concentration in the surrounding meniscus which induces (attractive) depletion forces between the film surfaces. The film thins out until the thickness of the film core fits to $n - 1$ mesh sizes.

Further, the similarity of Δh and ξ leads to two conclusions:

1. Though the film is only few times thicker than the correlation length (small n) of the bulk, this length

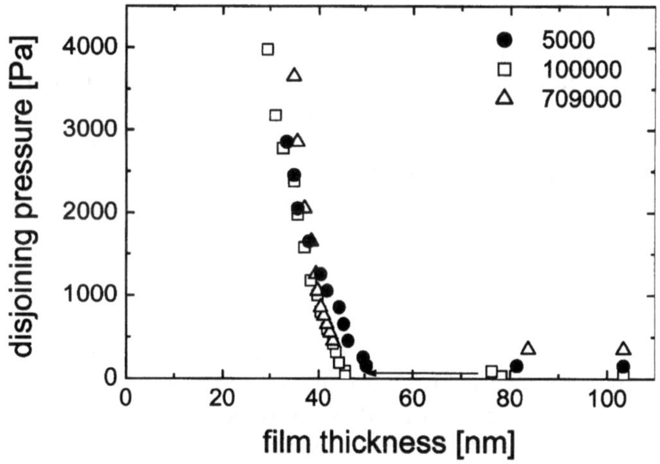

Fig. 4 Disjoining pressure isotherms of free-standing hexadecyltrimethylammonium bromide ($C_{16}TAB$)/PDADMAC films at different molecular weight of PDADMAC. The PDADMAC concentration was 5×10^{-3} monomol/l and the $C_{16}TAB$ concentration was 10^{-4} mol/l

does not change in the free-standing film. This means that the film does not present a geometrical confinement for the polymer system. This phenomenon has not been clarified yet. It could be explained by image charges located in the gas phase extending the network to infinity to both sides of the film to avoid finite-size effects in the film.

2. The polyelectrolyte concentration in the film is the same as in the bulk solution, since the correlation length is quite sensitive to concentration.

It is remarkable that at a molecular weight of 5,000 g/mol the steps occurs. The contour length at this molecular weight is smaller than the step size, which would mean that the monomer concentration is still below the critical overlap concentration. One possible explanation could be that at a molecular weight of 5,000 g/mol, the correlation length does not correspond to the mesh size of a network, but to the distance between different coils. To our knowledge the PDADMAC/C_{16}TAB system is the only system where a stepwise thinning has been observed below the critical overlap concentration, c^*. Usually, the mixed polyelectrolyte/surfactant films show only stratification above c^*.

Branched polyelectrolytes

The branched polyelectrolyte used in the study was the weak polyelectrolyte PEI. To avoid the effect of differences in the backbone the results of branched PEI are compared not only with those of the strong polyelectrolyte PDADMAC but also with the results given by linear PEI. The isotherms of linear (100% secondary N atoms) and of branched PEI (38% secondary N atoms) are shown in Fig. 5. The polymer concentration is 3.4×10^{-1} monomol/l and is much higher than in the case of PDADMAC. In general, at a certain polymer concentration, the films containing PEI are less stable than the films of PDADMAC. The film stability increases monotonically with the PEI concentration. The isotherm of the linear PEI shows two steps which occur at the same low pressure of about 20 Pa. The film was not in equilibrium before the first step from 50 to 18 nm; hence, this step size has no further physical meaning. The isotherm of the film containing the branched PEI also shows two steps, but they occur at different pressures. The pressure which is needed to induce the second jump (18–10 nm) is around 2,500 Pa. This difference in the isotherms is explained by stronger structural forces induced by a higher charge density, which itself leads to higher electrostatic repulsion. This relation is explained in more detail in the next section. In the case of the linear PEI, Δh scales with the polymer concentration again as $\Delta h \propto c^{-1/2}$ and the step size does not depend on the molecular weight, the same as in films containing PDADMAC. In contrast to the linear polyelectrolyte chains, Δh scales with the polyelectrolyte concentration as $\Delta h \propto c^{-1/3}$ as shown in Fig. 6. This is a typical scaling law for an isotropic ordering of colloids (particles, aggregates or polymer coils) in three dimensions. For instance, the position of the structure peak, q_{max}, for a dilute polymer solution scales with the concentration as $q_{max} \propto c^{-1/3}$. The structure peak is related to the distance between the centers of mass of the noninterdigitating polymer coils. In foam films containing micelles the step size scales with $c^{-1/3}$, which is explained by a layering of micelles followed by a layer-by-layer expulsion. In analogy to this picture, the present results are explained by a layering of the spherelike branched polymers within the film. In contrast to linear polyelectrolytes, the molecular weight has a pronounced influence on the step size. PEI of a molecular weight of 25.000 g/mol induces steps about twice as large as PEI with a molecular weight of 5,000 g/mol. On one hand, this is due to a larger diameter in the case of the higher molecular weight but, on the other hand, the concentra-

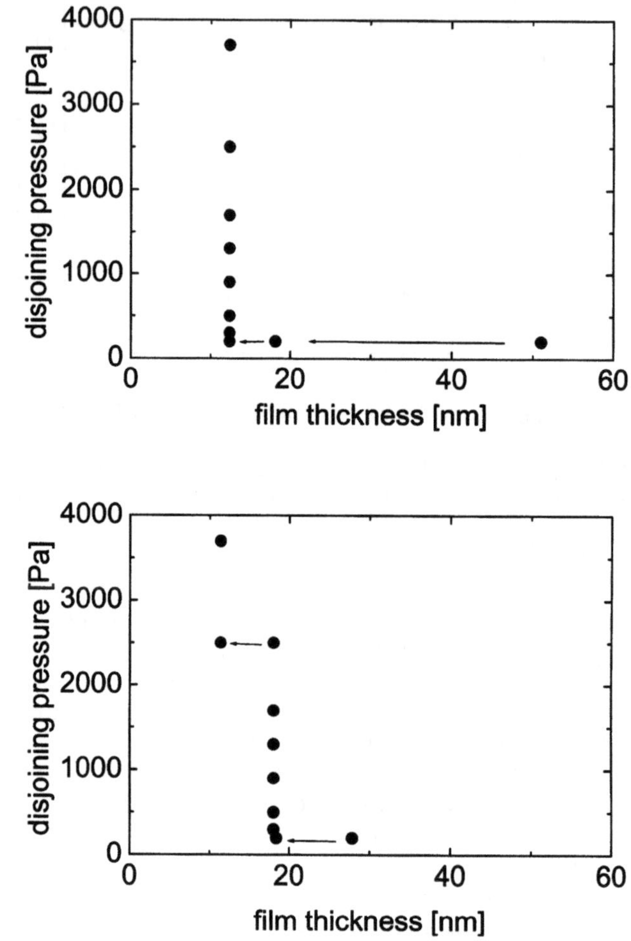

Fig. 5 Disjoining pressure isotherms of C_{16}TAB films containing linear (*top*) and branched (*bottom*) PEI. The PEI concentration was 3.4×10^{-1} monomol/l and the concentration C_{16}TAB 10^{-5} mol/l

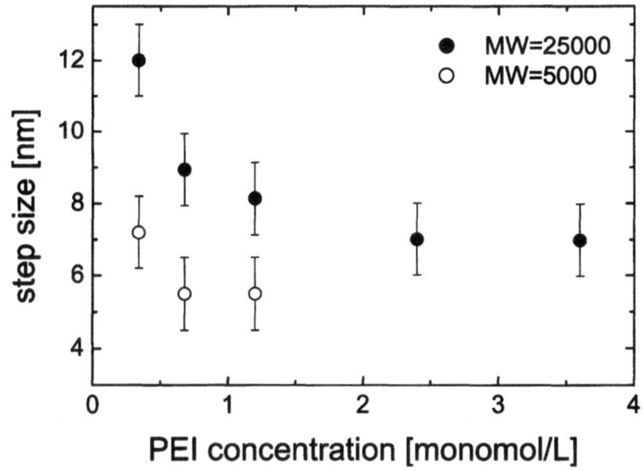

Fig. 6 Step size of film thickness as a function of PEI concentration

tion of polymers is lower, since the same number of monomers was fixed in the experiments. Under the assumption that the branched PEI presents an irregular spherical dendrimer, the diameter of one PEI molecule of t molecular weight 5.000 g/mol is about 3 nm and the PEI(25,000) molecule has a diameter of about 4.5 nm. These values are much smaller than the measured step sizes in the disjoining pressure isotherms. A reason for this difference could be the electrostatic repulsion between the spherical molecules expressed by the Debye length, δ, around the cationic "spheres". At an ionic strength of, for example, 0.34 mol/l, which corresponds to a concentration of 0.34 mol/l of charged monomers the Debye length is of the order of 0.5 nm. If all monomers of a PEI(5,000) molecule were charged, the minimum distance between two PEI centers of mass would be 4 nm (molecular diameter $+2\delta$), which is factor of 2 below the step size. A molecular distance of about 7 nm would mean a Debye length of about 2 nm and therefore a charge density of 12%. This is not in agreement with the pH measurements in solution at pH 4, which shows a charge of about 74%, measured by a streaming potential. Under the assumption that the average number of monomers per chain is 116 for $M_{w=}$5,000 and 580 for $M_w = 25{,}000$ a simple geometrical consideration leads to a mean distance between the centers of mass of 8.6 and 14.7 nm, respectively, at a monomer concentration of 0.34 monomol/l. The experimental data are a bit smaller (7 and 12 nm, respectively). The difference between calculation and experiment could be explained by the fact that the PEI molecules are pressed against each other within the film. Another explanation could be that most of the counterions are entrapped within the branched molecules, which reduce the number of free counterions, leading to an increase in the Debye length. An increase of the Debye length by a factor of 4 from 0.5 to 2 nm would mean that 12% of the counterions are free.

Effect of electrostatics

Ionic strength

The decrease and broadening of the structure peak in polyelectrolyte solution indicates a reduced position correlation between the polyelectrolyte chains with increasing concentration. Together with the polyelectrolyte concentration the counterion concentration is also increased, leading to a stronger electrostatic screening of the polymer charges and therefore to a reduction in repulsion. The structuring becomes "softer" and the peak broadening can be considered as a increasing polydispersity of the correlation length. This observation leads to the conclusion that the electrostatic repulsion between the polymers plays an important role for the structuring of the polymer chains in solution and the question is how the film transition is influenced by this factor. The electrostatic interactions are manipulated by the ionic strength and by the degree of charge along the polymer chains. The disjoining pressure isotherms for two ionic strengths are shown in Fig. 7. At a higher ionic strength, the pressure which is needed to induce the step is lower than at a lower ionic strength. This result confirms the assumption that the structuring becomes softer with increasing ionic strength (counterions or additional salt ions). The correlation length seems not to be affected by the addition of salt, which is in agreement with SANS data on polyelectrolyte solutions of different ionic strengths [22] where the position of the structure peak does not change.

On the other hand, the increase n the number of counterions (by increasing the polyelectrolyte concentration) leads to an increase in the pressure which is necessary to overcome the energy barrier to induce a step

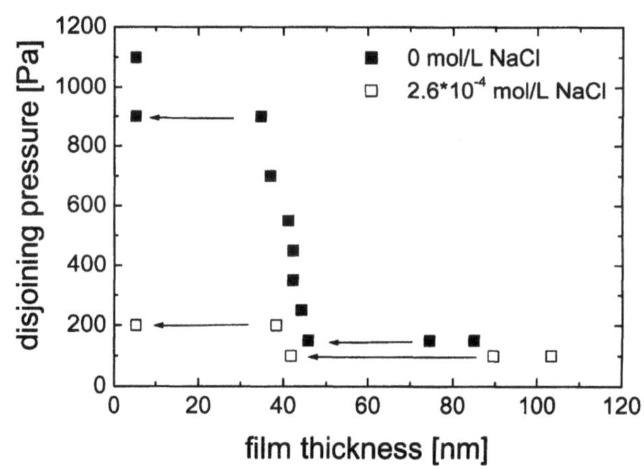

Fig. 7 Disjoining pressure isotherms of APG/PDADMAC films at an NaCl concentration of 2.6×10^{-4} mol/l and without additional salt. The APG concentration was 1.7×10^{-5} mol/l and the PDADMAC concentration was 5×10^{-3} monomol/l

in the film thickness [18]. This is an effect of increasing polyelectrolyte concentration which reduces the correlation length. It means that the film is thinner when the steps occur and the damping of the amplitudes is less pronounced in this thickness regime than in a film of a lower polyelectrolyte concentration where the steps occur at higher film thickness where the damping is already strong. This increase in pressure amplitude with the polyelectrolyte concentration has been predicted by the model of Yethiraj [32].

Polymer charge density

A decrease in polymer charge density has qualitatively the same influence on the pressure amplitude as increasing the salt concentration. The isotherms of a film containing fully charged PDADMAC and 24% charged PDADMAC are shown in Fig. 8. For the weakly charged PDADMAC, the steps are induced at a lower pressure, which leads to the fact that all steps occur at the same low pressure. All thicknesses occur at the same time, while in films containing fully charged PDADMAC the steps occur one after the other at different pressure. In contrast to the experiments at different ionic strengths, the correlation length is sensitive to the polymer charge density. With decreasing charge density the structure peak in the SANS solution spectra is shifted to lower q values corresponding to an increase in correlation length. An increasing step size can also be estimated from the isotherms; however, it is difficult to get proper values for the step size in the case of weakly charged PDADMAC, since the intermediate thicknesses cannot be stabilized. Immediately, the next smaller thickness occurs.

In the case of weak polyelectrolytes the charge density was varied by changing the pH. The isotherms for branched PEI (32% secondary N atoms) at pH 4 and 9 at a fixed PEI concentration of 0.34 monomol/l are shown in Fig. 9. At pH 4, three steps are observed. The first two steps (from 103 to 28 nm and from 28 to 19 nm) occur at a low pressure of about 200 Pa. The third step is induced at a higher pressure of 2,500 Pa. The film at a thickness of 103 nm was not equilibrated; therefore, the first large step cannot be analyzed. The thinnest film, at above 3,700 Pa, is still a common black film which is stabilized by electrostatic repulsion between both interfaces. Since both the surfactant and the polymer are positively charged, the electrostatic repulsion in the film is well pronounced. The charge density per chain is about 74% in solution. At pH 9, only one large step, at a low pressure of 200 Pa, was observed. Again, the film was not in equilibrium before the step occurred. The drainage ends up at a thickness of 5 nm, which corresponds to a Newton black film which is stabilized by steric repulsion. At pH 9, the PEI chains have a degree of charge of 17%. Since the polymer is less charged than at a pH 4, the electrostatic repulsion is reduced and the film overcomes the electrostatic energy barrier. The reduction of the numbers of steps and the occurrence of the step at a low pressure confirms the results described already for PDADMAC. The electrostatic repulsion between the chains is reduced and the structuring of the polymers in the film becomes softer at pH 9. A change in pH means at the same time a change in ionic strength since the pH is changed by HCl and NaOH. PEI in Milli-Q water (pH 5.5) at these high concentrations has a pH between 9 and 10, since PEI is a weak base. To get pH 4 one has to add a certain amount of HCl. The effect of increasing the ionic strength does not seem to be decisive from pH 9 to pH 4, since the isotherm shows more pronounced structural forces owing to charge density effects. But, for instance, at pH 2 the amplitude of the oscillatoric forces is reduced, which means that the effect of increasing the ionic strength becomes more important

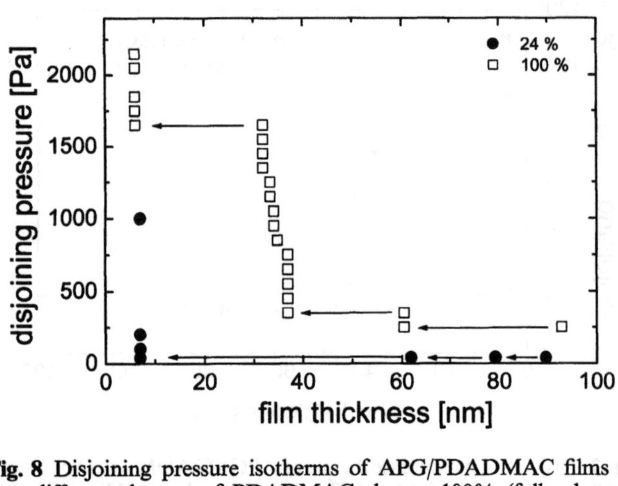

Fig. 8 Disjoining pressure isotherms of APG/PDADMAC films at two different degrees of PDADMAC charge: 100% (fully charged chain) and 24%. The APG concentration was 1.7×10^{-5} mol/l and the PDADMAC concentration was 10^{-2} monomol/l

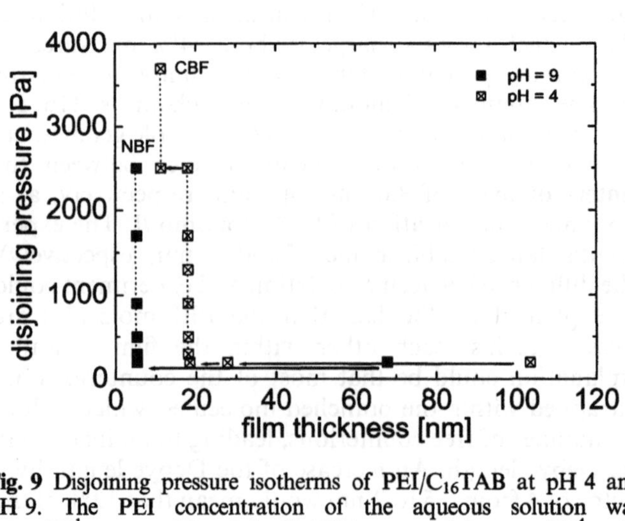

Fig. 9 Disjoining pressure isotherms of PEI/C_{16}TAB at pH 4 and pH 9. The PEI concentration of the aqueous solution was 3.4×10^{-1} monomol/l and the C_{16}TAB concentration was 10^{-4} mol/l

than the influence of increasing charge density. The degree of branching also has an influence on the charge density as mentioned already. Since the charge per volume is higher in a branched polymer than in a linear one, the electrostatic repulsion between the molecules is stronger. This leads to a stronger structuring of the branched polymers and to higher amplitudes in the pressure oscillations.

Conclusions

TFPB measurements of thin foam films containing linear (PDADMAC, PEI) and branched polyelectrolyte (PEI) were made. Both classes of polyelectrolytes lead to a film stratification. While linear polyelectrolytes produce steps in the film thickness which scale with the polymer concentration as $\Delta h \propto c^{-1/2}$, branched polyelectrolytes lead to a scaling behavior of $\Delta h \propto c^{-1/3}$. The step size increases with increasing molecular weight at a fixed polymer concentration, while for linear polyelectrolytes Δh is independent of molecular weight. These results lead to the conclusion that the linear polyelectrolytes overlap within the film and form a transient network, while the branched polyelectrolytes form layers of spherelike molecules between the two interfaces. For both groups of polyelectrolyte the chain structuring is reduced by decreasing the electrostatic repulsion. This was induced by increasing the ionic strength or decreasing the polymer charge density. The latter parameter was changed by using either strong polyelectrolytes with different ratios of charged and neutral monomers, P(DADMAC-*stat*-NMVA), or by changing the pH in the case of weak polyelectrolytes (PEI).

Acknowledgements The authors thank W. Jaeger and T. Schimmel (Fraunhofer Institute for Applied Polymer Research) for the PDADMAC and the streaming potential measurements. G. Grimm and U. Steuerle are acknowledged for the PEI and the the Deutsche Forschungsgemeinschaft (Schwerpunkt "Polyelektrolyte mit definierter Molekülarchitektur") for financial support.

References

1. Kwak JCT (ed) (1998) Polymer–surfactant-systems. Surfactant science series, vol 77. Dekker, New York
2. Vasilescu M, Anghel DF, Almgren M, Hansson P, Saito S (1997) Langmuir 13:6951
3. Anghel DF, Saito S, Baran A, Iovescu A (1998) Langmuir 14:5342
4. Nierlich M, Williams C, Boué F, Cotton JP, Daoud M, Farnoux B, Jannink G, Picot C, Moan M, Wolff C, Rinaudo M, de Gennes PG (1979) J Phys 40:701
5. Förster S, Schmidt M (1995) Adv Polym Sci 120:51
6. Nishida K, Kaji K, Kanaya T (1995) Macromolecules 28:2472
7. Essafi W, Lafuma F, Williams CE (1999) Eur Phys J B 9:261
8. Claesson PM, Bergström M, Dedinaite A, Kjellin M, Legrand J-F, Grillo I (2000) J Phys Chem B 104:11689
9. Gennes P-G de, Pincus P, Velasco RM, Brochard F (1976) J Phys 37:37
10. Odijk T (1977) J Polym Sci Polym Phys Ed 15:688
11. Skolnick J, Fixman M (1977) Macromolecules 10:944
12. Dobrynin AV, Colby RV, Rubinstein M (1995) Macromolecules 28:1859
13. Barrat JL, Joanny J-F (1996) Adv Chem Phys 94:1
14. Bergeron V, Langevin D, Asnacios A (1996) Langmuir 12:1550
15. Asnacios A, Espert A, Colin A, Langevin D (1997) Phys Rev Lett 78:4974
16. Espert A, Klitzing R von, Poulin P, Colin A, Zana R, Langevin D (1998) Langmuir 14:4251
17. Klitzing R von, Espert A, Asnacios A, Hellweg T, Colin A, Langevin D (1999) Colloid Surf A 149:131
18. Kolarić B, Jaeger W, v Klitzing R (2000) J Phys Chem B 104:5096
19. von Klitzing R (2000) Tenside Surfactants Deterg 37:338
20. von Klitzing R, Espert A, Colin A, Langevin D (2001) Colloid Surf A 176:109
21. Theodoly O, Tan JS, Ober R, Williams CE, Bergeron V (2001) Langmuir 17:4910
22. von Klitzing R, KolariÆ B, Jaeger W, Brandt A (2002) Phys Chem Chem Phys 4:1907
23. Dautzenberg H, Görnitz E, Jaeger W (1998) Macromol Chem Phys 199:1561
24. Ruppelt D, Kötz J, Jaeger W, Friberg SE, Mackay RA (1997) Langmuir 13:3316
25. Mysels KJ, Jones MN (1966) Discuss Faraday Soc 42:42
26. Exerova D, Schedulko A (1971) Chim Phys 24:47
27. Exerova D, Kolarov T, Khristov KHR (1987) Colloid Surf A 22:171
28. Schedulko A (1967) Adv Colloid Interface Sci 1:391
29. Toca-Herrera J-L, v Klitzing R (2002) Macromolecules 35:2861
30. Bock H, Schoen M (2000) J Phys Condens Mater 12:1545
31. Chatellier X, Joanny JF (1996) J Phys II 6:1669
32. Yethiraj A (1999) J Chem Phys 111:1797
33. Israelachvili J (1992) Intermolecular and surface forces. Academic, New York, p 266
34. Ivanov IB, Dimitrov DS (1988) In: Ivanov IB (ed) Thin liquid films. Dekker, New York, p 379
35. Perez E, Proust JE, Ter-Minassian-Saraga L (1988) In: Ivanov IB (ed.) Thin liquid films. Dekker, New York, p 891
36. Kekicheff P, Richetti P (1992) Phys Rev Lett 68:1951
37. Bergeron V, Radke CK (1992) Langmuir 8:3020
38. Milling AJ (1996) J Phys Chem 100:8986

Progr Colloid Polym Sci (2003) 122:130–140
DOI 10.1007/978-3-540-36114-5

Hans-Jürgen Tiefenbach
Helmut Durchschlag
Gregor Schneider
Rainer Jaenicke

Thermodynamic analysis of serum albumin denaturation by sodium dodecyl sulfate

K.-J. Tiefenbach · H. Durchschlag (✉)
G. Schneider · R. Jaenicke
Institute of Biophysics and Physical Biochemistry, University of Regensburg, Universitätsstrasse 31, 93040 Regensburg, Germany
e-mail: helmut.durchschlag@biologie.uni-regensburg.de
Tel.: +49-941-9433041
Fax: +49-941-9432813

Abstract Binding of variable amounts of sodium dodecyl sulfate (SDS) to bovine serum albumin (BSA) may be followed by analytical ultracentrifugation. The molar masses of the protein–surfactant complexes obtained from sedimentation equilibria allow binding isotherms to be established. UV absorption spectroscopy, fluorescence and circular dichroism are used for probing the conformational stability of proteins upon addition of surfactants, similar to the approaches used for denaturation by strong chaotropic reagents or heat. Measurements allow the determination of the equilibrium constant and the free energy, ΔG, for the transition from the folded (native) to the unfolded (denatured) state. ΔG at 25 °C in the absence of a surfactant or denaturant may be taken as a measure for the conformational stability of a protein. Though the results depend on the validity of certain assumptions regarding the folding/unfolding mechanism, as well as the evaluation method and the experimental conditions, the thermodynamics of surfactant-induced unfolding may be estimated and compared to the results of similar binding and unfolding experiments with other denaturants. The data presented in this work refer to BSA as a model. The results show diminished ΔG values for SDS-induced denaturation, compared to unfolding experiments with chaotropic denaturants or heat denaturation.

Keywords Surfactant binding · Protein denaturation and unfolding · Analytical ultracentrifugation · Spectroscopy · Stabilization energy

Introduction

In the biosciences several major areas of research make use of the interactions between surfactants and biological systems [1, 2, 3, 4]. For example, ionic and nonionic surfactants have been applied in biochemical work for a variety of purposes, including the solubilization and analysis of simple and conjugated proteins and more complex structures, such as tissues and cellular components, particularly biomembranes. Insight into the stability and stabilization of the native conformation of a given protein, including the mechanisms of its denaturation and the occurrence of multiple interactions, may be obtained by comparing the impact of surfactants and chaotropic agents on the protein. Understanding the action of surfactants and other denaturing agents on biological material is also required for the optimal choice of surfactants and denaturants for specific tasks in the processing of proteins.

Usually surfactants bind to proteins to a different extent depending on the surfactants and proteins under consideration. Most synthetic ionic surfactants (e.g. the alkyl sulfates) cause protein denaturation, whereas natural ionic surfactants (e.g. the cholates) and nonionic surfactants [e.g. the poly(oxyethylene) alkylphenols] normally do not cause significant structural changes. Both ionic and hydrophobic interactions between proteins and surfactants are responsible for the manifold effects reported in the literature.

In the case of ionic surfactants, the behavior of proteins at moderate surfactant concentrations ("initial binding") may differ substantially from that at excess of surfactants. Owing to the electrostatic nature of the interactions, the effects of pH and ionic strength are more significant for ionic than for nonionic surfactants. The size and structure of protein–surfactant complexes vary with the type of protein (e.g. nonconjugated or conjugated), with the number of disulfide bonds stabilizing its tertiary structure, and with the nature of the surfactant used.

A wide range of physicochemical techniques can be used in structural studies of protein–surfactant complexes and their constituents (proteins, monomeric and micellar surfactants); spectroscopic, hydrodynamic, scattering, electrophoretic, chromatographic and calorimetric methods have been applied, providing a wealth of local and global characteristics of the molecules under varying experimental conditions. Evidently, the given techniques can also be used to study the nature of the conformational changes occurring at different levels of surfactant binding and, after special adaptations, also the kinetics of these alterations. Though the precise structure of protein–surfactant complexes is still unknown, in particular the hydrodynamic and spectroscopic approaches can be advantageously used to investigate the thermodynamics of surfactant-induced protein unfolding by quantifying the stabilization energies of proteins [4, 5].

The anionic surfactant sodium dodecyl sulfate (SDS) is routinely used in biochemistry for analyzing the polypeptide composition of protein mixtures, preferably by electrophoretic and hydrodynamic techniques. Therefore, this surfactant has been the primary target of previous physicochemical studies of the structure of complexes formed between this amphipathic ligand and proteins, on the one hand, and the interactions accompanying the binding process, on the other [6, 7, 8, 9, 10, 11, 12, 13, 14, 15, 16]. After treatment with SDS, proteins are commonly denatured and, in the case of oligomeric structures, separated into subunits; protein disulfides can be cleaved by reducing agents such as 2-mercaptoethanol or dithiothreitol.

Studies of protein–surfactant complexes have focused on serum albumin as a model protein. Because it is easily available, bovine serum albumin (BSA) has commonly been used. The protein is well characterized from the physicochemical and biochemical point of view [17]. In its native state, it has a molar mass of 66.4 kg/mol, and possesses one or fewer sulfhydryl groups and 17 disulfide bonds. Upon ligation with moderate amounts of SDS, about 0.4 g SDS per gram of protein is bound to both the nonreduced and the reduced protein; excess of SDS leads to about 0.9 g/g for the nonreduced and about 1.4 g/g for the reduced protein, while at large excess even higher amounts of surfactant can be bound [11, 12, 13]. In the study reported here, we use our comprehensive knowledge of the BSA–SDS system to derive the thermodynamic parameters of complex formation and to compare the results with those reported for other denaturants.

As has been shown by detailed hydrodynamic and spectroscopic investigations, the ligation of proteins with surfactants provides information on the binding and denaturation capacity of a given surfactant, on the one hand, and protein stability, on the other. Previous studies clearly established that neither potent surfactants, such as SDS, nor acid denaturation lead to a complete breakdown of the native structure; to accomplish maximal protein randomization (unfolding, complete loss of helicity) strong denaturants, such as 6 M guanidinium chloride (GdmCl), are required [8, 13]. However, ionic surfactants such as SDS are unique in so far as millimolar concentrations suffice for protein denaturation, while molar concentrations of the chaotropic reagents GdmCl or urea are required for the breakdown of the native structure [5].

Protein stability is an important feature in connection with our understanding of protein structure and function, especially in the context of biotechnological applications. In the present study, we apply the thermodynamics developed for chaotropic and thermal protein unfolding [18, 19, 20, 21] to surfactant-induced unfolding reactions. Though the theoretical background for these types of denaturing conditions seems to be established, application of the classical approaches to surfactants needs several additional aspects to be considered. The primary goal of our analysis was the determination of the Gibbs energy change, ΔG, for protein unfolding. As is well known, this quantity represents a suitable tool for comparing net stabilization energies of the protein structure [5, 18, 19, 20, 21, 22, 23, 24, 25, 26, 27, 28, 29, 30, 31, 32, 33, 34].

Thermodynamic analysis of protein denaturation

Approaches to follow protein unfolding

Aiming at the conformational stability of a protein means comparing the native, folded (F) conformation of a protein with the conformation of the denatured, unfolded (U) state. The F $\leftrightarrow$ U transition is considered to be a reversible process; irreversible covalent changes are not taken into account. After an examination of the reversibility of the reaction, this allows the determination of the equilibrium constant, K, and the free-energy change, ΔG, for the F $\leftrightarrow$ U transition. Usually the difference in free energy between the native and the denatured states, $\Delta G(H_2O)$, with the value of ΔG at 25 °C in the absence of a denaturant as a reference state, is considered to be representative to describe the conformational stability of a given protein [5, 18, 19, 20, 21, 22, 23, 24, 25, 26, 27, 28, 29, 30, 31, 32, 33, 34]. For the analysis of equilibrium transitions, in general, hydrodynamic or spectroscopic measurements are the most suitable means. For example, analytical ultracentrifugation (AUC), UV absorption spectroscopy, fluorescence and circular dichroism (CD) are widely used to follow the unfolding process of proteins. A necessary prerequisite for a chosen technique is the magnitude of the respective signals, independent of different sample size and time requirements. Commonly, the magnitude of the signal and the signal-to-noise ratio can be modulated by suitable registration conditions (appropriate choice of the wavelength, signal amplifications, etc.). The different techniques allow molecular characteristics to be analyzed. While AUC reveals changes in mass and/or size, UV absorption and fluorescence mainly reflect alterations of the tertiary structure of a given protein (environmental changes of tryptophan and tyrosine residues); far-UV CD responds to changes in protein secondary structure (changes in the content of α-helix, β-structure and random coil).

Transition curves may exhibit pretransition, transition and posttransition regions. The interpretation of thermodynamic reactions requires equilibrium conditions and complete reversibility for the unfolding reaction. Depending on the activation energy of the reaction, the $F \leftrightarrow U$ transition may take time; for example, at 25 °C, equilibrium may only be reached hours after adding the denaturant. Proteins with free sulfhydryl groups (such as reduced BSA) involve special precautions to keep the cystein residues in their reduced state; the simultaneous presence of sulfhydryls and disulfides may lead to significant irreversible disulfide exchange reactions. For experimental details, particularly with respect to thermal, GdmCl- or urea-induced denaturation, and for a discussion of pitfalls see Refs. [19, 20].

Analysis of unfolding curves

The simplest interpretation of unfolding transitions rests on the assumption of a two-state folding mechanism ($F \leftrightarrow U$ reaction). To a good approximation, this is observed for small, monomeric, globular proteins. However, if the unfolding curve exhibits more than one step, the unfolding mechanism is more complex and requires deconvolution procedures to be quantified. Multidomain or oligomeric proteins show appreciable amounts of intermediates [18, 19, 20, 21].

In the case of two-state folding mechanisms, it is assumed that only the folded and unfolded conformations are present:

$$f_F + f_U = 1 \ , \qquad (1)$$

with f_F and f_U as protein fractions in the folded and unfolded state. The experimental parameter, y, at any point along the denaturation profile is given by

$$y = y_F f_F + y_U f_U \ , \qquad (2)$$

with y_F and y_U as y values characteristic of the folded and unfolded conformations.

Combining Eqs. (1) and (2) with the definition of the equilibrium constant, K, and the correlation of the equilibrium constant with the free-energy change, ΔG, yields

$$K = f_U/f_F = (y_F - y)/(y - y_U) \qquad (3)$$

and

$$\Delta G = -RT \ln K = -RT \ln[(y_F - y)/(y - y_U)] \ , \qquad (4)$$

where R is the gas constant and T is the absolute temperature. The values in the transition range, y_F and y_U, are obtained by extrapolating the pretransition and posttransition ranges, and values of K from a series of data points near the point of inflection (midpoint) of the denaturation curves.

Experimental results for numerous proteins have shown that ΔG varies linearly with the denaturant concentration, at least in the concentration range near the midpoint. Therefore, it is generally assumed that the free-energy change in the absence of denaturant, $\Delta G(H_2O)$, can be determined without significant error by extrapolating the experimental data points to zero denaturant concentration:

$$\Delta G = \Delta G(H_2O) - m c_{denaturant} \ . \qquad (5)$$

Here, the cooperativity parameter, m, is a measure of the dependence of ΔG on the denaturant concentration, $c_{denaturant}$. The results of other (nonlinear) extrapolation methods do not differ significantly from the linear extrapolation [19, 20].

The determination of $\Delta G(H_2O)$ from reversible thermal unfolding curves requires the extrapolation of ΔG from the data points near the midpoint of the thermal transition to the standard temperature 25 °C.

The enthalpy change, ΔH, may be derived from the van't Hoff equation:

$$d(\ln K)/d(1/T) = -\Delta H/R \ . \qquad (6)$$

For protein unfolding transitions, owing to significant differences in the heat capacities, C_p, of the unfolded and folded states, $C_p(U)$ and $C_p(F)$, the van't Hoff plots are nonlinear so ΔH may be temperature dependent:

$$\mathrm{d}\Delta H/\mathrm{d}T = \Delta C_p = C_p(\mathrm{U}) - C_p(\mathrm{F}) \; . \tag{7}$$

At the inflection point of the thermal unfolding curve, T_m, the free energy vanishes:

$$\Delta G(T_\mathrm{m}) = 0 = \Delta H_\mathrm{m} - T_\mathrm{m}\Delta S_\mathrm{m} \; . \tag{8}$$

ΔG at any temperature, $\Delta G(T)$, can be derived, for example, by plotting ΔG versus T; the determination of ΔC_p is less straightforward and usually requires calorimetric experiments [19].

Establishing stability differences

To compare the differences in the conformational stabilities for a given set of proteins under suitable denaturation conditions, easily accessible parameters, such as midpoints of denaturant concentrations and thermal unfolding curves, $c_{1/2}$ and T_m, can be applied. This holds because these quantities do not depend significantly on the unfolding mechanism. Quantities such as the cooperativity parameter m reveal deviations from simple two-state behavior and may, therefore, be used as indicators of multistate behavior. With these caveats in mind, it is obvious from Eq. (5) that estimates of the Gibbs free energy of stabilization may be afflicted with relatively large errors.

Though the measurement of spectral or other changes of unfolding transitions provides a powerful tool for determining protein stability, evaluations rest on the validity of a series of assumptions regarding the folding/unfolding mechanism: folding model (two-state or multistate behavior), reaching equilibrium conditions, influence (neglect) of pretransition and posttransition ranges, choice of the extrapolation methods (i.e. fixing of the F and U reference states), reduction state of protein (integrity of sulfhydryl groups and disulfide bonds), etc. This should be borne in mind before changes in spectral or hydrodynamic parameters are attributed to changes in thermodynamic properties, and before final statements on stability differences are made.

Materials and methods

Samples and solutions

BSA and SDS were obtained from Serva, Heidelberg, GdmCl from Merck, Darmstadt, and all the other reagents used in this study were of analytical grade. The protein and the additives were dissolved in 0.05 M sodium phosphate buffer pH 7.0 (ionic strength $I \cong 0.1$ mol/l). For binding and denaturation studies, a wide range of concentrations was selected: 0–4 g/l SDS, 0–6 M GdmCl. Stock solutions of BSA and the corresponding additives were used to volumetrically prepare the protein–additive solutions. The pH of the resultant solutions was controlled after mixing the components; GdmCl concentrations were determined refractometrically. The solutions were incubated for at least 12 h at 25 °C, to guarantee sufficient binding/denaturation. All investigations were performed at 25 °C.

Experimental techniques

AUC was performed using a Beckman model E ultracentrifuge equipped with a high-sensitivity UV scanner, a multiplexer system, and a 10-inch recorder. Runs were performed in a six-hole rotor (AnG), using 12-mm double-sector cells (charcoal-filled epon). Molar masses, M, and molar mass weight averages, M_w, were determined by high-speed sedimentation equilibria (HSSE), using the meniscus-depletion technique [35].

Recording of absorption, fluorescence and CD spectra was performed using a Perkin-Elmer Lambda 5 spectrophotometer, a Jobin Yvon-Spex Fluoromax-2 spectrofluorometer and a JASCO J-500A spectropolarimeter, respectively.

As auxiliary tools, density measurements with a Paar digital density meter and determinations of the refractive index with a Zeiss Optron refractometer were used.

Hydrodynamic and spectroscopic data were evaluated and interpreted as described elsewhere ([36, 37, 38, 39, 40, 41, 42, 43, 44] and references therein). In making use of these techniques, the investigation of protein–surfactant complexes and of micellar surfactant systems requires consideration of several adaptations and precautions [11, 13, 45]. In context with the present study, the following aspects have to be considered:

1. For the analysis of BSA–SDS complexes and SDS-free proteins labeling is not required, since proteins usually contain aromatic chromophores (tryptophan, tyrosine, phenylalanine); for investigations of nonaromatic micelles, such as SDS, labeling with (fluorescent) dyes is recommended.
2. To allow a straightforward interpretation, AUC experiments of BSA–SDS complexes require SDS micelles and other macrosolutes in sample and reference cells to be absent. Except for detailed binding studies, this may be achieved by measurements of BSA–SDS complexes just below the critical micelle concentration (cmc). To avoid erroneous (skew) baselines, generally the reference cell should be devoid of surfactant. To check the quality of the data points near the meniscus and bottom, a critical inspection of the primary data (e.g. in terms of A versus r plots, Fig. 1a) is indispensable.
3. To avoid misinterpretations due to heterogeneity, AUC runs at different experimental conditions (different rotor speeds, loading concentrations, scanning wavelengths) are required. Nonlinear curves in $\log A$ versus r^2 curves (Fig. 1b) disclose either heterogeneity and/or nonideality of the system under investigation.
4. For a quantitative comparison of the effects of different surfactants, discrimination between the effects caused by environmental changes of aromatics (tryptophan, tyrosine), unfolding events and disturbances by quenching has to be taken into account. Fluorescence quenching has only to be considered if aromatic surfactants are used; in the case of SDS, quenching is negligible.
5. The problem of thermodynamically defining the native and denatured states is by no means trivial, because two-state or multistate behavior is influenced by numerous factors, including the nature of the proteins and denaturants as well as the sensitivity of the experimental techniques applied for the detection of intermediates. Solvent conditions such as ionic strength, pH, Hofmeister effects, and the presence of ligands may be of utmost importance. For the given reasons, it is advisable to use more than one experimental technique for the thermodynamic analysis. As a first approximation, the evaluations made use of the two-state model.
6. Transition curves may be obtained from spectroscopic investigations at different wavelengths. For the spectral

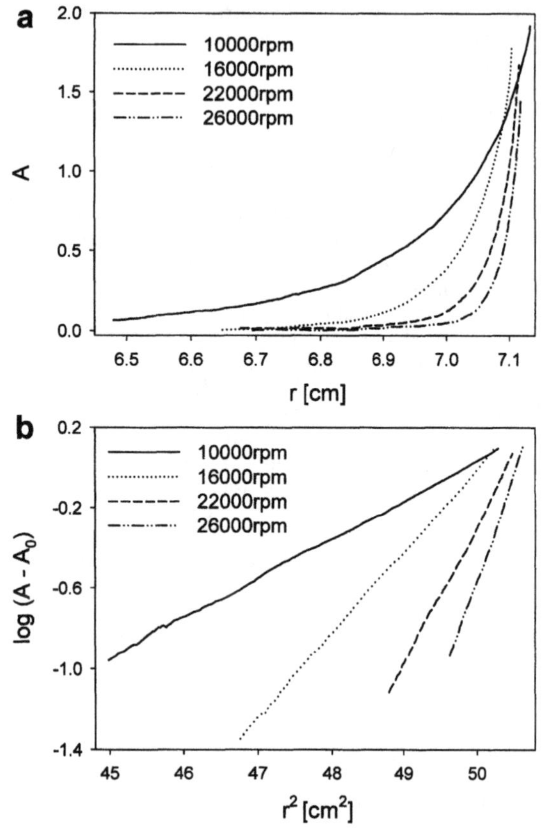

Fig. 1a,b High-speed sedimentation equilibria (*HSSE*) of native, surfactant-free bovine serum albumin (*BSA*) (0.66 g/l *BSA*) at different rotor speeds: **a** original data in A versus r format; **b** log (A–A_0) versus r^2 plot, with slopes proportional to molar mass, M, buoyancy, $(1-\bar{v}\rho)$, and squared angular velocity, ω^2. A is the experimentally observed absorbance of the sample, A_0 the absorbance of the assumed baseline, and r the radial distance of a selected mass point from the center of rotation. Observed masses taken from a representative scan, recorded at $\lambda = 280$ nm and amplification factor (*AF*) of 1: 91.5 kg/mol at 10,000 rpm, 75.3 kg/mol at 16,000 rpm, 67.2 kg/mol at 22,000 rpm, and 71.6 kg/mol at 26,000 rpm. Note that native BSA always contains a certain number of cross-links (dimers, trimers and higher oligomers), which leads to enhanced molar mass weight averages at low rotor speed (10,000 rpm) [11, 13]

analysis of protein unfolding, usually wavelengths are used which provide maximum effects of average proteins: 287 nm in the case of UV absorption of simple proteins, 320 nm for fluorescence emission and 238 nm for far-UV CD measurements [18, 19, 20, 21]; for experimental reasons (noise of signals) sometimes wavelengths close to the mentioned values have to be chosen, for example, 220 nm for CD studies. The wavelengths chosen do not correspond to the maxima in the corresponding spectral curves (as may be taken, for example, from the UV difference absorption spectrum in the inset to Fig. 5). Use of the exact positions of the maxima of the difference spectra (not of the original curves) of individual proteins might improve the results slightly, particularly with respect to the data points at low additive concentration, but it generally results in similar ΔG values (Table 1, footnote b).

Results and discussion

The investigation of the complexes between BSA and surfactants or chaotropic reagents at various additive concentrations by AUC and spectroscopy yields information both at the global and at the local level of molecular structure. For the thermodynamic analysis of the binding and denaturation behavior, only a limited number of parameters is considered appropriate. Molar masses from sedimentation equilibria were used to determine binding isotherms. On the other hand, spectroscopic measurements at selected wavelengths were used to derive transition curves and equilibrium constants to finally calculate free-energy changes ΔG.

For the following studies, primarily nonreduced BSA was used to avoid side reactions caused by reducing agents, as well as SH–SS exchange reactions.

Analytical ultracentrifugation

Numerous investigations of native, surfactant-free BSA and BSA–SDS complexes were performed at constant BSA concentration (0.66 g/l$\triangleq$10 μM) and varying amounts of SDS (0–3 g/l) under a variety of experimental conditions regarding rotor speed, etc. Typical data, obtained for the native protein by HSSE at different rotor speeds and scanning conditions (scanning wavelengths, amplification factors of the absorbances obtained by electronic magnification of the respective signals), are illustrated in Fig. 1. Applying different rotor speeds allows particle separations and molar mass weight averages to be determined. The latter indicate that the results at very low rotor speed (10,000 rpm) are influenced by the unavoidable presence of cross-links; the numerical values are in agreement with previous experiments [11, 13]. At very high rotor speed, masses may be affected by erroneous data points obtained near the cell bottom, again suggesting that experiments need to be performed under different experimental conditions, for example, different rotor speeds.

For some of the HSSE experiments, the accuracy was rather low, particularly when surfactants were present [11, 13, 45]. Therefore, for the following evaluations, at least as a first approximation, averages obtained for all the experimental conditions (rotor speed, scanning conditions) were applied. The results obtained for $M_w(1-\bar{v}\rho)$. $(1-\bar{v}\rho)$, and M_w are illustrated in Fig. 2. The values for the buoyancy term, $(1-\bar{v}\rho)$, obtained for BSA at different degrees of saturation with SDS (Fig. 2b) differ by a factor of about 2, revealing the pronounced influence of the partial specific volume, $\bar{v}$, on the molar mass estimates. As expected, the perturbations introduced by including the data obtained at low rotor speeds affect the calculated averages and lead to a rather

Table 1 Free-energy changes, ΔG, for probing the conformational stability of bovine serum albumin upon addition of sodium dodecyl sulfate (*SDS*) at 25 °C, as obtained by various techniques and the assumption of two-state or three-state transitions. For comparison some values obtained from other types of denaturation and a few values from the literature are also added

Technique	Approach	Additive	ΔG (kJ/mol)[a]	Reference
Calorimetry		None	−30	[26]
Hydrogen exchange		None	−32.6 to −39	[26]
Calorimetry		SDS	−24.8	[5, 22]
UV Absorption	A at 287 nm	SDS	−17.7 (−3.7)	This study
Fluorescence	λ_{exc} = 280 nm, λ_{em} = 320 nm	SDS	−17.9 (−0.5)[b]	This study
Fluorescence	λ_{exc} = 295 nm, λ_{em} = 320 nm	SDS	−13.5 (−0.2)	This study
Circular dichroism	$[\Theta]_{MRW}$ = 220 nm	SDS	−11.6 (−1.4)	This study
UV Absorption	A at 287 nm	Guanidinium chloride	−29.4 (−4.7)	This study
Fluorescence	λ_{exc} = 280 nm, λ_{em} = 320 nm	Guanidinium chloride	−24.7 (−4.2)	This study
Fluorescence	λ_{exc} = 295 nm, λ_{em} = 320 nm	Guanidinium chloride	−27.8 (−4.5)	This study
Circular dichroism	$[\Theta]_{MRW}$ = 220 nm	Guanidinium chloride	−27.4 (−4.2)	This study
Circular dichroism	$[\Theta]_{MRW}$ = 220 nm	Heat	−29.6 (−12.5)[c]	This study

[a] The values given for the spectroscopy techniques refer to evaluations based on a three-state model. For comparison the values based on a two-state model are also shown (in *parentheses*), which prove that this model is obviously unrealistic. The ΔG values given for the three-state models represent the sum of the two transitions

[b] Evaluation of the SDS-induced denaturation at other emission wavelengths yields similar ΔG values: −18.9, −18.3 and −18.3 kJ/mol at λ_{em} = 329, 339 and 350 nm, respectively

[c] Evaluation of the thermal denaturation allows derivation of the enthalpy change, ΔH = −380 kJ/mol, and entropy change, ΔS = −1.2 kJ/(mol K), neglecting their temperature dependence

pronounced scatter of the mass estimates (Fig. 2c), irrespective of the number of data used for averaging. Though the results found at 10,000 rpm seem to be rather smooth (Fig. 3a), the absolute values of the masses (about 80–160 kg/mol) reveal the significant influence of oligomers present in commercial BSA. The estimates at 16,000 rpm are much more reliable: they show a continuous increase of M_w with increasing surfactant binding (Fig. 3b). Correspondingly, the binding isotherms, derived from the experimentally observed masses of the protein–surfactant complexes and the mass of the native protein (calculated from its amino acid composition), yield erroneous results for the masses obtained by HSSE at 10,000 rpm (Fig. 4a), but render a relatively realistic binding behavior at 16,000 rpm (Fig. 4b) or higher rotor speeds. The amount of binding, δ, (0.3–0.4 g SDS per gram of BSA at low SDS concentrations and 0.8–0.9 g/g at excess of SDS) are in full accord with previous results [11, 12, 13].

After having established the progress of SDS binding to BSA by mass estimates with AUC, the previously mentioned spectroscopic techniques were applied to monitor the surfactant-induced denaturation of the protein.

Spectroscopy

As shown in Fig. 5, the UV absorption of native BSA and its SDS complex exhibits only marginal changes, except for a moderate blueshift of the maximum at about 280 nm and a minute decrease in absorbance. As a consequence of these small changes, both the transition curves (A versus c_{SDS}) and the ln$K$ versus c_{SDS} plots show a certain degree of ambiguity (even at 287 nm, where the maximum difference is observed; Fig. 5, inset), in particular regarding the linear approximations that are required to distinguish the two-state or multistate character of the SDS-induced transitions (Fig. 6). Clearly, the results obtained for ΔG exhibit the same deficiency. Thus, no clear-cut decision regarding a definite transition state can be given. Evidently, the two-state model fails to describe the transition in a quantitative fashion; fitting the data to a three-state model provides a better fit. The accuracy of the calculated ΔG values may be compared to the data obtained for other denaturants.

The formation of BSA–SDS complexes results in pronounced changes in fluorescence emission for two typical excitation wavelengths (tyrosine and tryptophan, λ_{exc} = 280 nm; or tryptophan alone, λ_{exc} = 295 nm). Both modes of excitation yield a prominent decrease in fluorescence intensity and remarkable blueshifts of their maxima (Fig. 7). Again, the transition curves monitored at 320 nm (F versus c_{SDS}) and the ln$K$ versus c_{SDS} plots reveal uncertainties in attempts to confirm the two-state model and to quantify ΔG (Figs. 8, 9).

In the case of CD spectra in the far-UV, SDS binding to BSA causes distinct changes which may be ascribed to a substantial decrease in helicity (67 → 44%) (Fig. 10). By exploiting these spectral changes to generate transition curves ($[\Theta]_{MRW}$ versus c_{SDS}) and their correspond-

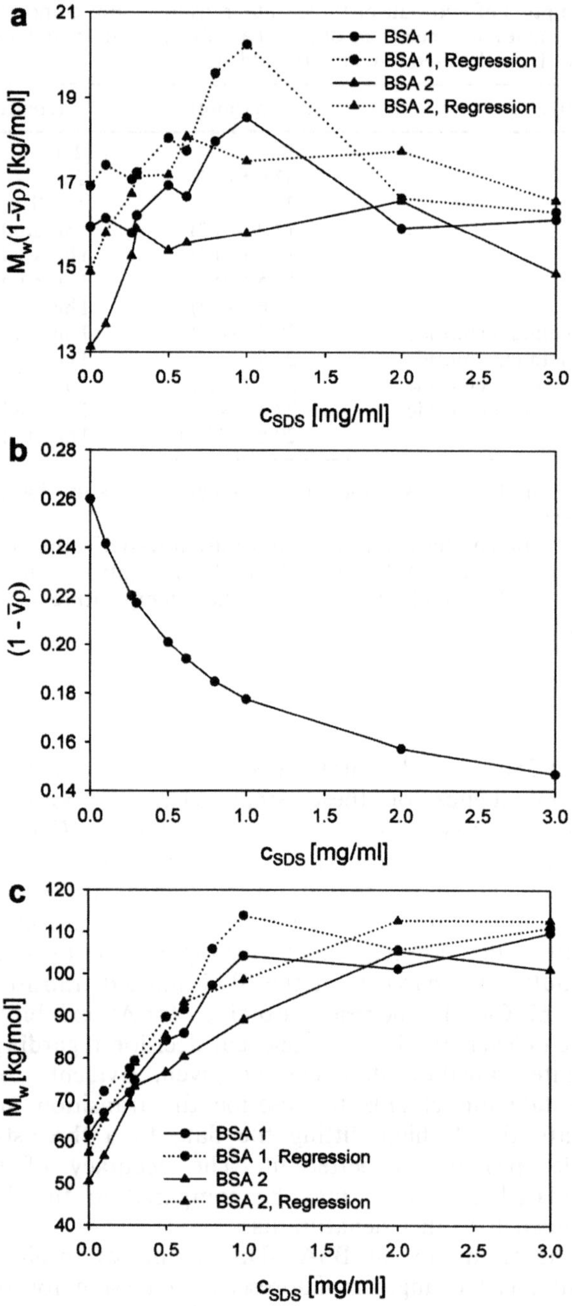

Fig. 2a–c HSSE of BSA–sodium dodecyl sulfate (*SDS*) complexes (0.66 g/l BSA, 0–3.0 g/l SDS) of two selected samples. **a** $M_w(1-\bar{v}\rho)$ plots. The data represent values averaged over all the HSSE conditions used (rpm, λ, AF) of the experimentally observed quantity "molar mass weight average, M_w, times buoyancy $(1-\bar{v}\rho)$". The indicated regressions represent the approximation of the A versus r profile by exponentials [$A(r) = A_0 + A_1\exp(br)$, where A_0, A_1 and b are the floating parameters of the exponential equation]. The results are strongly influenced by the assumed values for $(1-\bar{v}\rho)$. **b** $(1-\bar{v}\rho)$. The values for $\bar{v}$ of the protein–surfactant complex were calculated according to the method in Refs [47, 48, 49]; the values for ρ of the solvent were obtained by densimetric measurements. **c** M_w, corrected for buoyancy

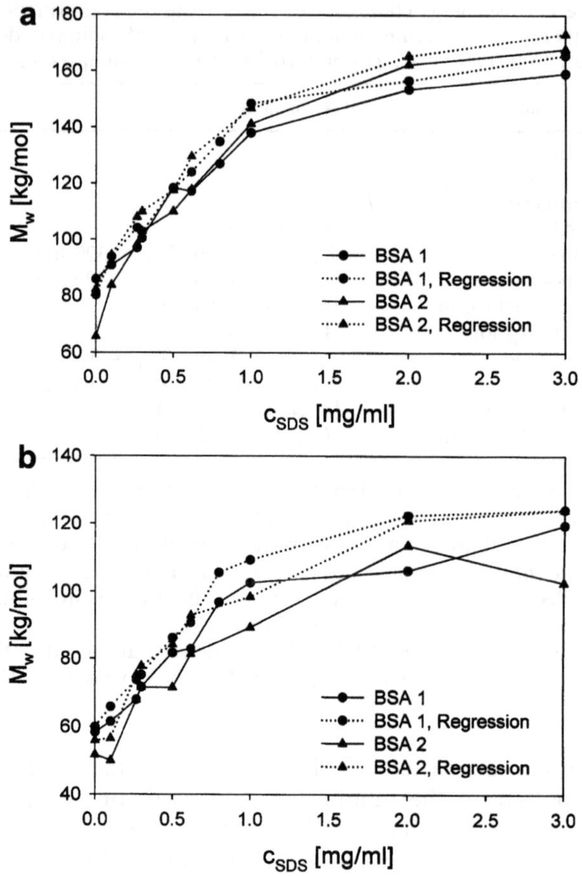

Fig. 3a,b HSSE of BSA–SDS complexes (0.66 g/l BSA, 0–3.0 g/l SDS) of two selected samples at two rotor speeds. M_w, obtained at **a** 10,000 and **b** 16,000 rpm. The data represent averages after recording at different λ (230, 280 or 295 nm) and AF (1 or 4)

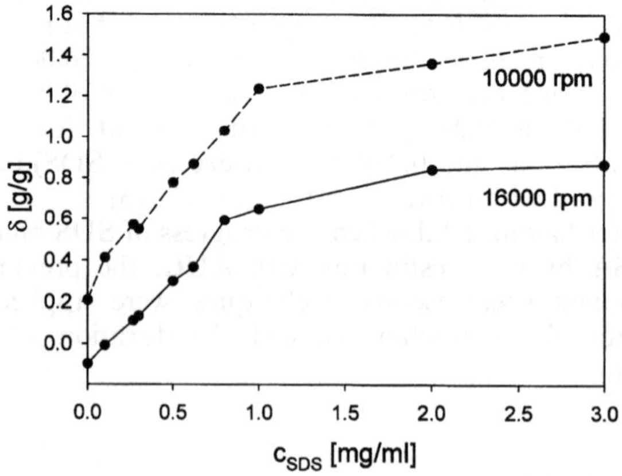

Fig. 4 Binding isotherms of BSA at various SDS concentrations, obtained from the averaged values of Fig. 3. Binding ratio, δ, obtained by HSSE at 10,000 or 16,000 rpm

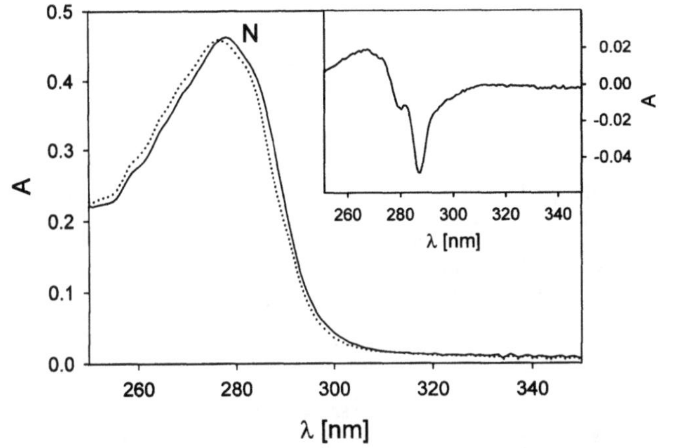

Fig. 5 UV absorption spectra of native BSA (0.66 g/l; *N*, *solid line*) and its complex at excess of SDS (4.0 g/l SDS). *Inset*: difference spectrum; the maximum effect is observed at 287 nm

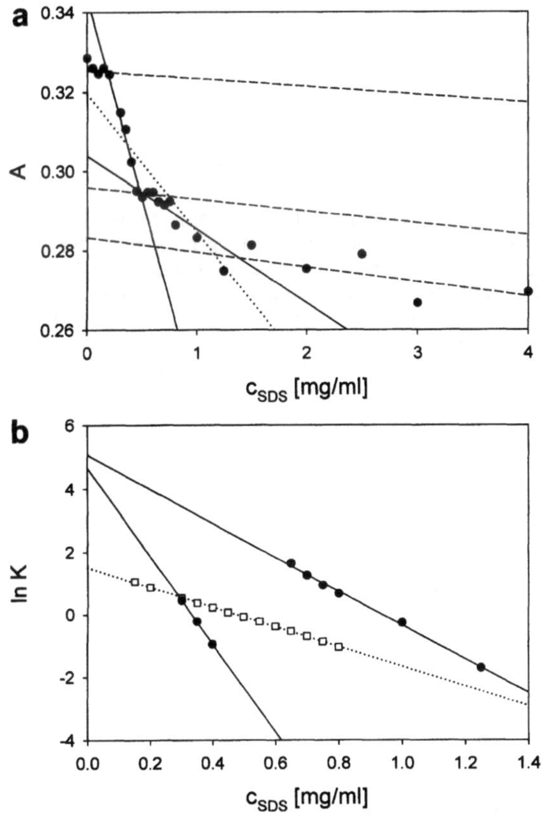

Fig. 6a,b Conformational stability of BSA in the presence of SDS, as monitored by UV absorption spectroscopy (0.66 g/l BSA, 0–4.0 g/l SDS). **a** Transition curve at 287 nm, obtained from the spectra given in Fig. 5. **b** Calculation of transition energies, ΔG, from the logarithm of the equilibrium constant, *K*, for two-state and three-state transitions. The *lines* depict the transitions for the two-state (·····) or three-state evaluation (—) and the pretransition and posttransition regions (- - - -); according to Eq. (4) the ordinate intercepts are proportional to ΔG

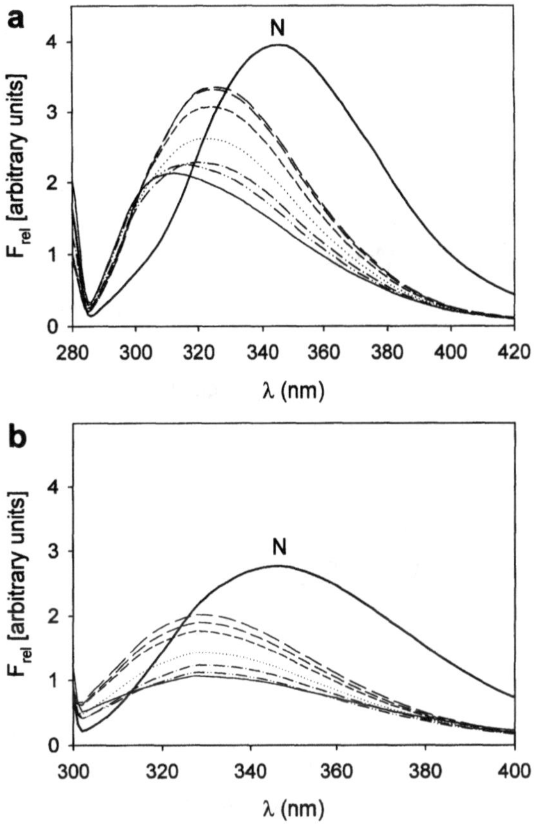

Fig. 7a, b Fluorescence emission spectra of native BSA (0.66 g/l; *N*, *bold solid line*) and BSA–SDS complexes (0.1, 0.2, 0.3, 0.4, 0.6, 1.0 and 4.0 g/l SDS). The SDS concentration increases from *top* to *bottom*: $\lambda_{exc} = 280$ nm (**a**) or 295 nm (**b**)

ing ln*K* versus c_{SDS} plots, the results point to a three-state rather than a two-state mechanism (Fig. 11), in accordance with the absorption and fluorescence spectra.

Conclusions

In the present study the relationship between surfactant binding and protein denaturation was studied using hydrodynamic and spectroscopic techniques. Sedimentation equilibria in the analytical ultracentrifuge were used to quantify the binding stoichiometry of surfactants; parallel measurements of UV–Vis absorption, fluorescence and CD spectra were applied to monitor denaturing effects. The complex formation of the anionic surfactant SDS with nonreduced BSA was used as a model reaction in the attempt to obtain thermodynamic quantities, such as the equilibrium constant and the Gibbs free-energy change of the surfactant-induced denaturation.

On the basis of HSSE experiments, the molar masses of the protein–surfactant complexes at varying surfactant concentration were interpreted in terms of binding curves, indicative of two separate binding steps at low

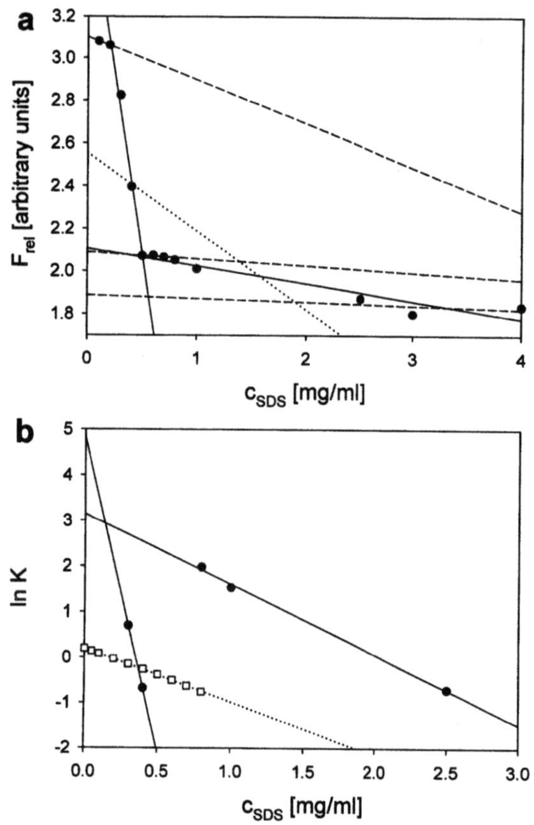

Fig. 8a,b Conformational stability of BSA in the presence of SDS, as monitored by fluorescence emission spectroscopy (0.66 g/l BSA, 0–4.0 g/l SDS). **a** Transition curve at 320 nm for $\lambda_{exc}=280$ nm, obtained from the spectra in Fig. 7a. **b** Calculation of transition energies, ΔG, from the logarithm of the equilibrium constant, K, for two-state and three-state transitions. The *lines* depict the transitions for the two-state (·····) or three-state evaluation (—) and the pretransition and posttransition regions (- - - -)

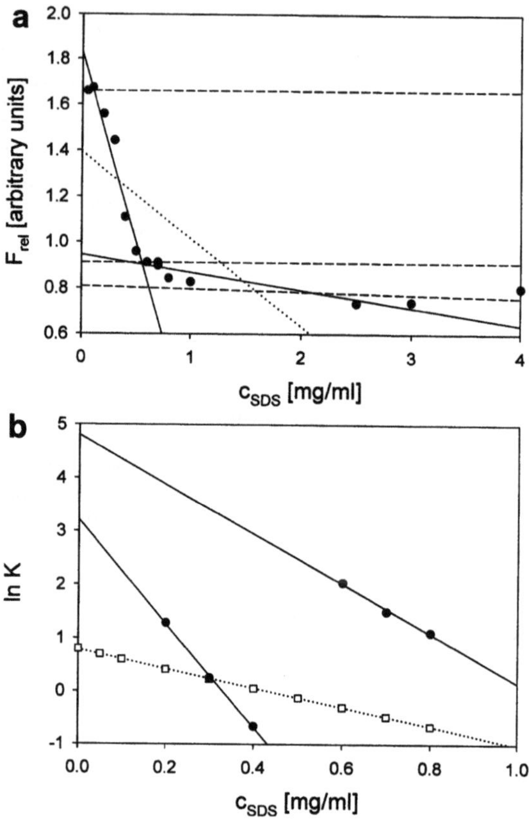

Fig. 9a, b Conformational stability of BSA in the presence of SDS, as monitored by fluorescence emission spectroscopy (0.66 g/l BSA, 0–4.0 g/l SDS). **a** Transition curve at 320 nm for $\lambda_{exc}=295$ nm, obtained from the spectra in Fig. 7b. **b** Calculation of transition energies, ΔG, from the logarithm of the equilibrium constant, K, for two-state and three-state transitions. The *lines* depict the transitions for the two-state (·····) or three-state evaluation (—) and the pretransition and posttransition regions (- - - -)

and high surfactant levels (0.4 and 0.9 g SDS per gram of BSA). The denaturation curves obtained by the spectroscopic methods could not be quantitatively described by the F↔U two-state model for protein unfolding. By making use of the three-state model, i.e. assuming one intermediate on the unfolding pathway to be significantly populated, the fit of the data was efficiently improved. The comparison of the free-energy changes, attributable to the SDS-induced partial denaturation of BSA, gives clear evidence that the corresponding Gibbs free energy is lower by a factor of 2 compared to the values observed for the protein at high GdmCl concentration or after heat treatment (Table 1). This result is in agreement with the insignificant denaturing effect of SDS, compared to the large effects of chaotropic reagents.

The comparison between binding and transition curves reveals obvious similarities. At SDS concentrations corresponding to the binding of up to 0.4 g SDS/g protein, only peripheral surfactant binding to the protein takes place, in accordance with the aforementioned

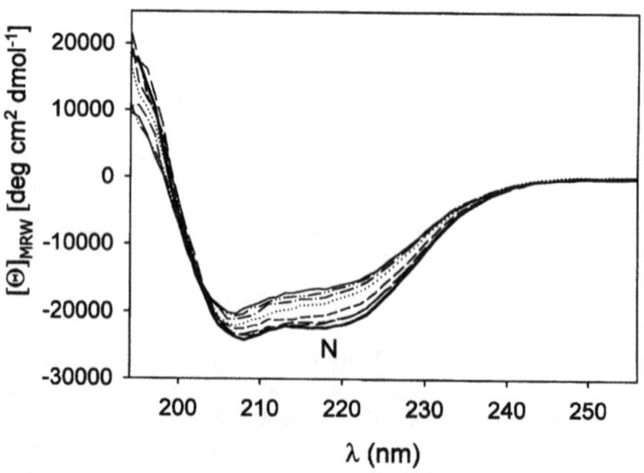

Fig. 10 Far-UV circular dichroism (*CD*) spectra of native BSA (0.66 g/l; *N*, *bold solid line*) and BSA–SDS complexes (0.1, 0.2, 0.3, 0.4, 0.6, 1.0 and 4.0 g/l SDS). The SDS concentration increases from *bottom* to *top*

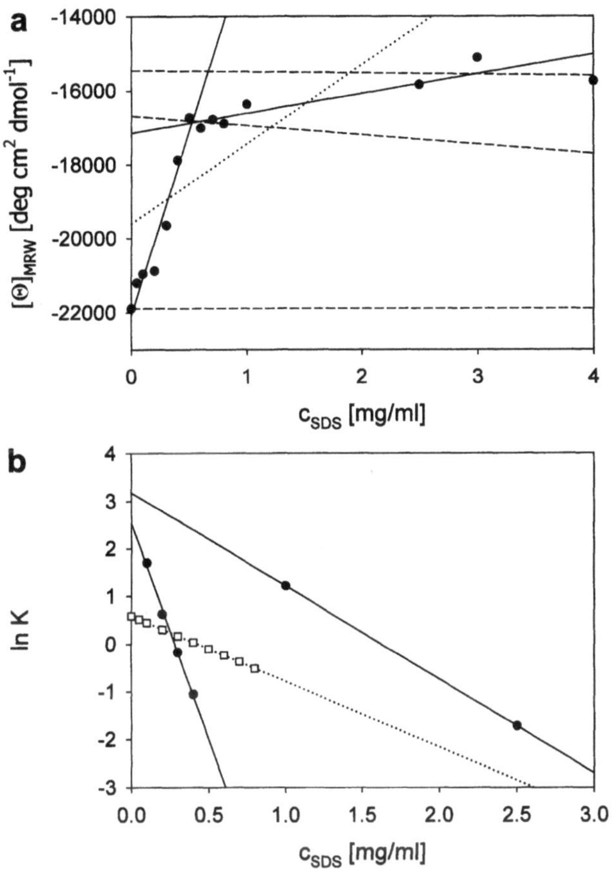

Fig. 11a, b Conformational stability of BSA in the presence of SDS, as monitored by far-UV CD spectroscopy (0.66 g/l BSA, 0–4.0 g/l SDS). **a** Transition curve at $[\Theta]_{MRW} = 220$ nm, obtained from the spectra in Fig. 10. **b** Calculation of transition energies, ΔG, from the logarithm of the equilibrium constant, K, for two-state and three-state transitions. The *lines* depict the transitions for the two-state (·····) or three-state evaluation (—) and the pretransition and posttransition regions (- - - -)

rather weak denaturing effect of moderate amounts of SDS. By contrast, at elevated surfactant concentrations, binding and denaturation occur in a cooperative fashion.

At excess free SDS, even the final steps of protein unfolding can be followed by conventional spectroscopic techniques, whereas monitoring the binding effects in the ultracentrifuge may be impaired by the occurrence of SDS micelles [11, 13, 45]. Further problems may arise from an incorrect choice of the surfactant concentrations applied and from incomplete unfolding of the protein as a consequence of nonreduced cross-links. Despite these disturbances, excessive unfolding of nonreduced BSA at high SDS concentration allows a maximum binding of about 0.9 g SDS/g protein to be determined with the analytical ultracentrifuge; the partial denaturation is accompanied by a free-energy change of −12 to −18 kJ/mol. Though a number of uncertain assumptions and limitations of the accuracy of the data do not allow a rigorous analysis of the transition curves, the results clearly indicate a lower degree of unfolding in the presence of SDS compared to the denaturation effect of strong chaotropic denaturants or elevated temperature. The extrapolation of the GdmCl data to zero denaturant concentration, heat denaturation, and the indirect calorimetric determination yield ΔG values of the order of −25 to −30 kJ/mol (Table 1), in agreement with structural data obtained from spectroscopic measurements [8, 13]. Maximum randomization of proteins can only be accomplished in the presence of strong chaotropic denaturants. In contrast, even at high concentrations of synthetic ionic surfactants, appreciable amounts of secondary structure remain unaffected, allowing "reconstructive denaturation", i.e. an increase in helicity in proteins with low helix content in their native state, and a decrease in helicity in proteins with a high helix content [46].

References

1. Tanford C (1980) The hydrophobic effect: formation of micelles and biological membranes, 2nd edn. Wiley, New York
2. Israelachvili JN (1991) Intermolecular and surface forces, 2nd edn. Academic, London
3. Jones MN (1992) Chem Soc Rev 21:127
4. Jones MN, Chapman D (1995) Micelles, monolayers, and biomembranes. Wiley-Liss, New York
5. Jones MN (1988) In: Jones MN (ed) Biochemical thermodynamics, 2nd edn. Elsevier, Amsterdam, pp 182–240
6. Durchschlag H, Christl P, Jaenicke R (1991) Prog Colloid Polym Sci 86:41
7. Durchschlag H, Binder S, Christl P, Jaenicke R (1994) Jorn Com Esp Deterg 25:407; Anexo 26
8. Durchschlag H, Weber R, Jaenicke R (1996) In: Proceedings of the 4th world surfactants congress, vol 1. CESIO, Barcelona, pp 519–534
9. Durchschlag H, Jaenicke R (1997) Chimica Oggi – Chem Today 15, No 9/10:15–24
10. Durchschlag H, Tiefenbach K-J, Jaenicke R (1997) Jorn Com Esp Deterg 27:185; Anexo 35
11. Tiefenbach K-J, Durchschlag H, Jaenicke R (1997) Prog Colloid Polym Sci 107:102
12. Durchschlag H, Kuchenmüller B, Tiefenbach K-J, Jaenicke R (1998) Jorn Com Esp Deterg 28:353
13. Durchschlag H, Tiefenbach K-J, Weber R, Kuchenmüller B, Jaenicke R (2000) Colloid Polym Sci 278:312
14. Durchschlag H Tiefenbach K-J, Fischer A, Weber R, Jaenicke R (2000) In: Proceedings of the 5th world surfactants congress, vol 1. CESIO, Florence, pp 634–646
15. Durchschlag H, Tiefenbach K-J, Gebauer S, Jaenicke R (2001) J Mol Struct 563–564:449
16. Durchschlag H, Fischer A, Tiefenbach K-J, Jaenicke R (2002) Jorn Com Esp Deterg 32:225
17. Peters T Jr (1985) Adv Protein Chem 37:161
18. Pace CN (1986) Methods Enzymol 131:266

19. Pace CN, Shirley BA, Thomson JA (1989) In: Creighton TE (ed) Protein structure: a practical approach. IRL, Oxford, pp 311–330
20. Shirley BA (1992) In: Ahern TJ, Manning MC (eds) Stability of protein pharmaceuticals, part A: chemical and physical pathways of protein degradation. Plenum, New York, pp 167–194
21. Shirley BA (1995) Methods Mol Biol 40:177
22. Tipping E, Jones MN, Skinner HA (1974) J Chem Soc Faraday Trans I 70:1306
23. Pfeil W (1986) In: Hinz H-J (ed) Thermodynamic data for biochemistry and biotechnology. Springer, Berlin Heidelberg New York, pp 349–376
24. Pfeil W (1988) In: Jones MN (ed) Biochemical thermodynamics, 2nd edn. Elsevier, Amsterdam, pp 53–99
25. Pfeil W (1996) In: Holtzhauer M (ed) Methoden in der Proteinanalytik. Springer, Berlin Heidelberg New York, pp 276–310
26. Pfeil W (1998) Protein stability and folding: a collection of thermodynamic data. Springer, Berlin Heidelberg New York
27. Pfeil W (2001) Protein stability and folding, supplement 1: a collection of thermodynamic data. Springer, Berlin Heidelberg New York
28. Gromiha MM, An J, Kono H, Oobatake M, Uedaira H, Sarai A (1999) Nucleic Acids Res 27:286
29. (a) Gromiha MM, An J, Kono H, Oobatake M, Uedaira H, Prabakaran P, Sarai A (2000) Nucleic Acids Res 28:283; (b) http://www.rtc.riken.go.jp/jouhou/Protherm/protherm.html
30. Greene RF Jr, Pace CN (1974) J Biol Chem 249:5388
31. Santoro MM, Bolen DW (1988) Biochemistry 27:8063
32. Bolen DW, Santoro MM (1988) Biochemistry 27:8069
33. Gupta R, Ahmad F (1999) Biochemistry 38:2471
34. Pace CN, Shaw KL (2000) Proteins Struct Funct Genet Suppl 4:1
35. Yphantis DA (1964) Biochemistry 3:297
36. Chervenka CH (1973) A manual of methods for the analytical ultracentrifuge. Spinco Division of Beckman Instruments, Palo Alto
37. Durchschlag H, Jaenicke R (1983) Int J Biol Macromol 5:143
38. Schmid FX (1997) In: Creighton TE (ed) Protein structure: a practical approach, 2nd edn. IRL, Oxford, pp 261–297
39. Shirley BA (ed) (1995) Methods in molecular biology, vol 40: protein stability and folding: theory and practice. Humana, Totowa, N.J.
40. Demchenko AP (1986) Ultraviolet spectroscopy of proteins. Springer, Berlin Heidelberg New York
41. Lakowicz JR (1999) Principles of fluorescence spectroscopy, 2nd edn. Kluwer/Plenum, New York
42. Fasman GD (ed) (1996) Circular dichroism and the conformational analysis of biomolecules. Plenum, New York
43. Holtzhauer M (ed) (1996) Methoden in der Proteinanalytik. Springer, Berlin Heidelberg New York
44. Winter R, Noll F (1998) Methoden der Biophysikalischen Chemie. Teubner, Stuttgart
45. Tiefenbach K-J, Durchschlag H, Jaenicke R (1999) Prog Colloid Polym Sci 113:135
46. Jirgensons B (1976) Biochim Biophys Acta 434:58
47. Durchschlag H, Zipper P (1994) Prog Colloid Polym Sci 94:20
48. Durchschlag H, Zipper P (1995) Jorn Com Esp Deterg 26:275
49. Durchschlag H (2002) In Hinz H-J (ed): Landolt–Börnstein new series VII/2A. Springer, Berlin Heidelberg New York (in press)

Progr Colloid Polym Sci (2003) 122: 141–148
DOI 10.1007/978-3-540-36114-5

José L. López-Fontán
Pablo Martínez-Landeira
Cibrán Santamarina
Juan M. Ruso
Gerardo Prieto
Félix Sarmiento

The surfactant characteristics of short-chain lecithins analyzed through lecithin–lecithin and lecithin–biopolymer interactions

J.L. López-Fontán · P. Martínez-Landeira
C. Santamarina · J.M. Ruso · G. Prieto
F. Sarmiento (✉)
Group of Biophysics and Interfaces,
Department of Applied Physics,
Faculty of Physics, University of Santiago
de Compostela, 15782 Santiago
de Compostela, Spain
e-mail: fsarmi@uscmail.usc.es
Tel.: +34-981-563100
Fax: +34-981-520676

Abstract The micellar behaviour of the binary mixed systems dihexanoylphosphatidylcholine (diC_6PC)/diheptanoylphosphatidylcholine (diC_7PC), diC_7PC/dioctanoylphosphatidylcholine (diC_8PC) and diC_6PC/ diC_8PC has been studied. Critical micelle concentrations of the mixtures were quantitatively estimated from surface tension measurements versus solution composition plots. The micellar composition in the micelles was determined by the Motomura thermodynamic model. The nature and the strength of the interaction between two surfactants in the mixture were estimated by calculating the values of their β_{12}^{M} parameter from Holland and Rubingh's treatment. DiC_6PC and diC_7PC mix ideally with an interaction parameter near zero. DiC_7PC/diC_8PC and diC_6PC/diC_8PC show nonideality for about 75 and 40% diC_8PC in the mixture, respectively. The nonideality has been attributed to differences in the acyl chain length, the size of the micelles formed by pure components and the composition of mixed micelles. Zeta potential and microcalorimetric measurements were used to investigate the effect of diC_8PC on human serum albumin. The results revealed that the interaction is hydrophobically driven and depends on lecithin concentration.

Keywords Mixed micelles · Lecithin–lecithin interaction · Lecithin–protein interaction · Surface tension · Thermodynamic models

Introduction

Mixed surfactants producing mixed micelles are of great interest for practical use and for scientific research. In recent years many studies on mixed micelles have been reported by a number of workers [1, 2, 3] and important theoretical models have been proposed and used [4, 5, 6, 7, 8, 9, 10, 11]. In general, mixtures of nonionic surfactants show ideal behaviour, while other combinations exhibit nonideality through attractive or antagonistic interactions between the components. Among the various types of surfactants, those of the zwitterionic type are of particular interest because they often exhibit significant synergism (i.e., favorable interactions) when mixed with ionic surfactants, while comparably little synergism is observed when mixed with nonionic surfactants [12, 13, 14]. The nature and factors determining these interactions have not been sufficiently assessed, so investigations of these matters are of research interest.

Zwitterionic surfactants exhibit unique properties in aqueous solution, such as pH dependence of the critical micelle concentration (cmc), foam stability, lowering of the Krafft point by addition of salts, good solubility in water and reduced sensitivity to salts and temperature [15, 16, 17, 18]. Short-chain lecithins are very good model of zwitterionic molecules. From a structural point of view, these lecithins are composed by two acyl chains, each with 4–8 carbon atoms, one glycerol molecule, a phosphate group and a choline residue (Scheme 1).

As observed, the molecules are zwitterionics owing to the negative charge attached on the phosphate group and the positive charge on the nitrogen atom of choline. Then, the properties of solutions of these amphiphiles depend, on the one hand, on structural parameters of the amphiphile itself, i.e., on its chemical structure and, on the other hand, on the pH of the medium which can change the charge of the head group towards positive or negative values.

The structures and thermodynamic parameters of pure one-component short-chain dialkylphosphatidylcholine (diC_nPC, with $n = 4$–8) micelles have been well studied and characterized by different physical techniques [19, 20, 21, 22, 23, 24]. Hauser [25] has recently reviewed the physicochemical properties of short-chain phosphatidylcholines, emphasizing their important detergent-like behaviour. An important application of this property is the use of short-chain lecithins in the solubilization of biological membranes and the reconstitution of membrane proteins into simple, well-defined membrane systems.

Our most recent studies [26, 27, 28] deal with the influence of pH and acyl chain length on the micellization of diC_5PC, diC_6PC, diC_7PC and diC_8PC and the interaction of diC_8PC with human serum albumin (HSA) [29]. Since the properties of diC_6PC, diC_7PC and diC_8PC are well determined, the mixed diC_6PC/ diC_7PC, diC_6PC/diC_8PC and diC_7PC/diC_8PC are ideal model systems for studying mixed micelles. Some thermodynamic modelling on the diC_6PC/ diC_7PC system has been performed [30, 31] to predict the micelle size and the composition of mixed micelles at different mixing ratios.

In this work, we extended the previously mentioned studies to the mixtures diC_6PC/diC_7PC, diC_7PC/diC_8PC and diC_6PC/diC_8PC, and determined their cmcs by the surface tension technique. The results are analysed in light of the treatments of Holland and Rubingh [6] and Motomura et al. [7]. With the model of Motomura et al., the composition of each component in the mixture can be obtained. The model of Holland and Rubingh permits us to obtain the β_{12}^{M} parameter as a measurement of the interactions and nonideality of the mixtures. The lecithin–lecithin interaction in mixed micelles was compared with that of the lecithin–protein interaction using the data obtained in a previous study on diC_8PC and HSA [29]. The study of the HSA–diC_8PC interaction is of interest in relation to the widespread occurrence of phosphatidylcholines in biological organisms, whereas their short-chain precursors and breakdown products are present in anabolic and catabolic processes, respectively.

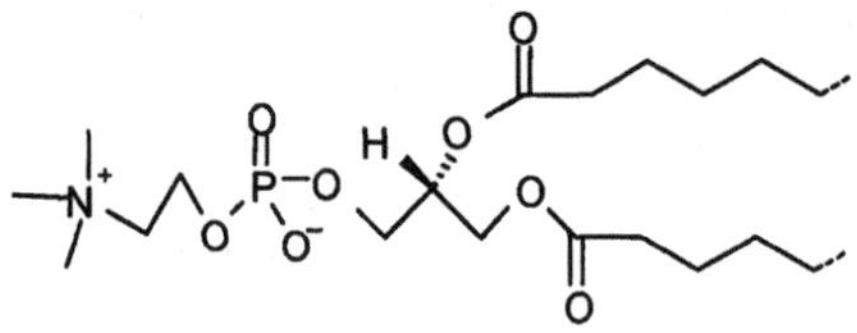

Experimental

Materials

The lecithins diC_6PC, diC_7PC and diC_8PC were purchased from Avanti Polar Lipids (Birmingham, AL) in powder form with purity greater than 99%, stored at −20 °C and used without any further purification. The buffer system used was phosphate, pH 7.4. Water was from a Milli-Q reverse-osmosis purification system.

HSA (fraction V A 1653) was obtained from Sigma Chemical Company and was used as supplied. The phosphate buffer systems, pH 7.4, used in the HSA–diC_8PC interaction were at an ionic strength of 0.188 and 0.0188, respectively.

Surface tension measurements

Surface tension was measured by the Wilhelmy plate technique using a Krüss K12 surface tension apparatus, equipped with a processor to acquire the data automatically. The equipment was connected to a circulating water bath to keep the temperature constant to 298.15 ± 0.01 K. Phospholipid solutions were prepared by diluting with buffer and determining the concentration by mass with a precision of ±0.00001 g. The instrument was calibrated against the Milli-Q water and measurements were taken until the surface tension was constant at least for two periods of 20 min each. The reproducibility of the surface tension was ±0.01 mN m^{-1}. The usual precautions were taken to ensure cleanliness.

Electrophoretic mobility measurements

The electrophoretic mobilities, u, of HSA plus diC_8PC were measured using a Malvern Instruments, Zetasizer 3000. ζ potentials were calculated from the Henry equation [32, 33]:

$$\zeta = \frac{3\eta u}{2\varepsilon_0\varepsilon_r f(\kappa a)} , \quad (1)$$

where the permittivity of a vacuum, ε_0, the relative permittivity, ε_r, and the viscosity of water, η, were taken as 8.854×10^{-12} J^{-1} C^2 m^{-1}, 78.5 and 8.904×10^{-4} N m^{-2} s, respectively. To calculate the Henry factors, $f(\kappa a)$, the Debye length, $1/\kappa$, and the measured complex radii, a, from dynamic light scattering were used [29].

Microcalorimetry

Enthalpy measurements were made at 298.15 K using an LKB-Produkter 10700 twin-cell batch microcalorimeter system [34]. The instrument was used on the 30-μV range, where the mean sensitivity of the detectors in the heat sinks of the two vessels was 14.66 ± 0.32 W V^{-1}. The sample cell was charged with 2 g buffered HSA, concentration 0.25% w/v, and 2 g diC_8PC of the required concentration in a range up to approximately 0.3 mM. The reference cell was charged with 2 g diC_8PC solution of identical concentration to that in the sample cell and 2 g buffer solution. On mixing, the enthalpies of dilution of the phospholipid solution cancel and the enthalpy of dilution of the HSA was measured in a separate experiment and was used to correct the data.

Results and discussion

Thermodynamic model

A plot of surface tension, γ, against total molar concentration, c_{12}^0, for the diC$_7$PC/diC$_8$PC binary system is shown in Fig. 1. Similar plots (not shown) were obtained for the other mixtures.

The cmcs, c_{12}^M, were calculated from the intersection of the best straight lines through the two branches of the γ/c_{12}^0 curves and the values are given in Table 1. The c_{12}^M values are lower than the cmc of component 1 and are near the cmc of pure component 2 for a mixture composition higher than 1:1. This suggests the strong influence of the second component in the micellization process.

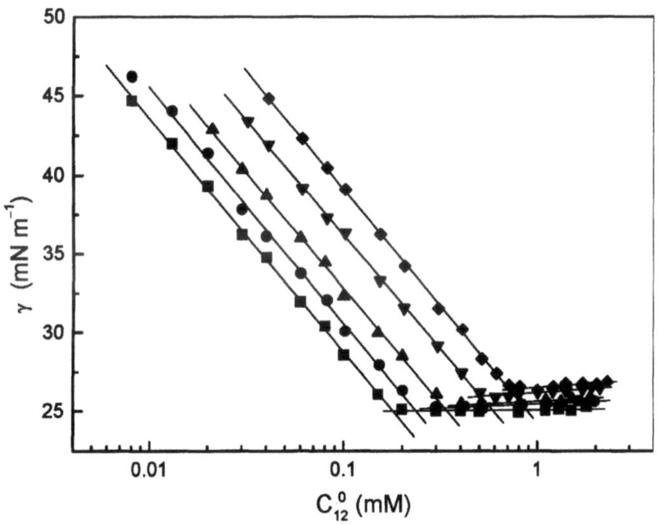

Fig. 1 Surface tensions, γ, as a function of molar concentration, c_{12}^0, for the binary system diheptanoylphosphatidylcholine (diC$_7$PC)–dioctanoylphosphatidylcholine (diC$_8$PC) at 298.15 K and pH 7.4: 10% diC$_8$PC (*diamonds*); 20% diC$_8$PC (*down triangles*); 40% diC$_8$PC (*up triangles*); 60% diC$_8$PC (*circles*); 80% diC$_8$PC (*squares*)

The composition of the monomeric and micellar phases of these systems was evaluated by the treatment of Motomura et al. [7], which is based on excess thermodynamic quantities. This method of estimating the equilibrium distribution of components in a mixed system has been shown to give values in good agreement with those derived from direct measurements by gel permeation chromatography [35]. The surface tension from an aqueous solution of a two-component surfactant mixture as a function of temperature, T, and pressure, p, is written as

$$d\gamma = -s^E dT + v^E dp - \Gamma_1^E d\mu_1 - \Gamma_2^E d\mu_2, \tag{2}$$

where s^E and v^E are the excess entropy and volume per unit interfacial area defined with respect to the two dividing planes, making the excess number of moles of water and air zero. μ_i is the chemical potential of the components and Γ_i^E represents their surface densities.

By assuming the aqueous solutions to be ideal, and substituting the chemical potentials expressed as a function of the total composition, $c_{12}^0 = c_1^0 + c_2^0$, of the mole fraction of surfactant 2 in the aqueous solution, x_2, Eq. (2) can be written

$$d\gamma = -\Delta s dT + \Delta v dp - \left(\frac{RT\Gamma_{12}^E}{c_{12}^0}\right) dc_{12}^0 - \left(\frac{RT\Gamma_{12}^E}{x_1 x_2}\right)\left(x_2^E - x_2\right) dx_2 \ . \tag{3}$$

Δy is the thermodynamic quantity of adsorption of surfactant per unit surface area defined by [31]

$$\Delta y = y^E - \left(\Gamma_1^E y_1 + \Gamma_2^E y_2\right) \ , \tag{4}$$

$$\Gamma_{12}^E = \Gamma_1^E + \Gamma_2^E \tag{5}$$

is the total surface density of surfactants and

$$x_2^E = \frac{\Gamma_2^E}{\Gamma_1^E + \Gamma_2^E} \tag{6}$$

is the mole fraction of surfactant 2 in the surface separation.

Table 1 Critical micelle concentrations, c_{12}^M, of the mixed systems dihexanoylphosphatidylcholine (diC$_6$PC)–diheptanoylphosphatidylcholine (diC$_7$PC), diC$_7$PC–dioctanoylphosphatidylcholine (diC$_8$PC) and diC$_6$PC– diC$_8$PC in aqueous solution (pH 7.4) at 298.15 K

Molar fraction diC$_7$PC	c_{12}^M diC$_6$PC/diC$_7$PC (mM)	Molar fraction diC$_8$PC	c_{12}^M diC$_7$PC/diC$_8$PC (mM)	Molar fraction diC$_8$PC	c_{12}^M diC$_6$PC/diC$_8$PC (mM)
0.00	11.0	0.00	1.3	0.00	11.0
0.14	5.2	0.10	0.72	0.01	6.0
0.22	4.6	0.20	0.51	0.05	2.2
0.41	2.8	0.40	0.33	0.10	1.2
0.45	2.4	0.60	0.24	0.40	0.33
0.65	1.9	0.80	0.22	0.78	0.22
1.00	1.3	1.00	0.21	1.00	0.21

The micelle formation is described by the analogue of Eq. (2) by using the excess thermodynamic quantity y^M of mixed micelles, which is defined with respect to the dividing surface, making the number of moles of water zero:

$$-s^M dT + v^M dT - N_1^M d\mu_1 - N_2^M d\mu_2 = 0 \ , \quad (7)$$

where N_i^M is the number of molecules of surfactant i in the mixed micelle. By introducing the total excess number of molecules, N_{12}^M and the mole fraction, x_2^M of surfactant 2, respectively, by

$$N_{12}^M = N_1^M + N_2^M \quad (8)$$

and

$$x_2^M = \frac{N_2^M}{N_1^M + N_2^M} \ , \quad (9)$$

x_2^M is evaluated by

$$x_2^M = x_2 - \left(\frac{x_1 x_2}{c_{12}^M}\right)\left(\frac{\partial c_{12}^M}{\partial x_2}\right)_{T,p} \ , \quad (10)$$

where c_{12}^M is the cmc.

Diagrams of c_{12}^M versus composition for the diC_6PC/diC_7PC, diC_7PC/diC_8PC and diC_6PC/diC_8PC systems are shown in Figs. 2, 3 and 4, respectively. In these curves line a represents the relationship between c_{12}^M and the mole fraction of surfactant 2 (diC_7PC, diC_8PC and diC_8PC, respectively) in the solution, x_2, line b the variation of c_{12}^M with the mole fraction of surfactant 2 in micelles, x_2^M calculated by Eq. (10) and line c the variation of c_{12}^M with x_2^M of ideal mixing. As a criterion for ideal mixing we used the equation proposed by Aratono et al. [11]:

$$c_{12}^M = c_1^M + (c_2^M - c_1^M)x_2^M \ . \quad (11)$$

The analysis of Fig. 2 suggests that diC_6PC and diC_7PC mix practically as the ideal form in all range of compositions. Slight differences exists when the percentage of diC_7PC in the mixture is bigger than

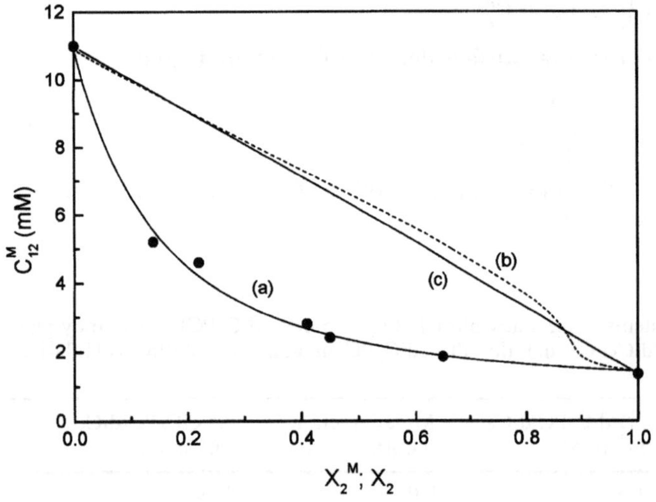

Fig. 2 Phase diagram of micelle formation of the dihexanoylphosphatidylcholine (diC_6PC)—diC_7PC system. *Line a* represents the relationship between c_{12}^M and the mole fraction of diC_7PC in the solution, x_2, *line b* represents the variation of c_{12}^M with the mole fraction of diC_7PC in micelles, x_2^M calculated by Eq. (9) and *line c* the variation of c_{12}^M with x_2^M of ideal mixing

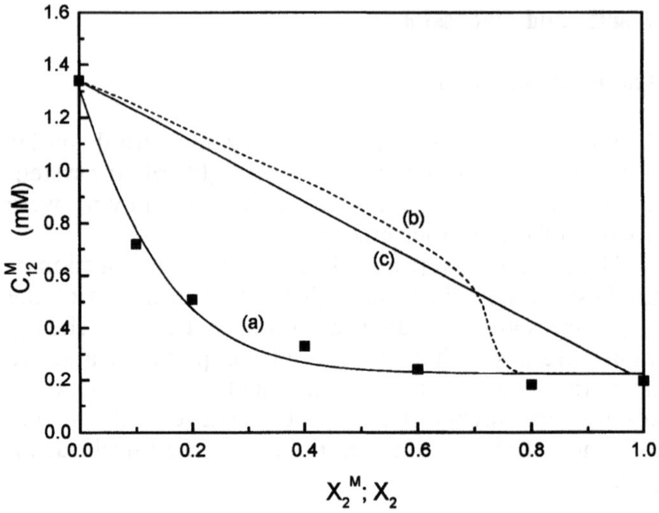

Fig. 3 Phase diagram of micelle formation of the diC_7PC– diC_8PC system. *Line a* represents the relationship between c_{12}^M and the mole fraction of diC_8PC in the solution, x_2, *line b* represents the variation of c_{12}^M with the mole fraction of diC_8PC in micelles, x_2^M calculated by Eq. (9) and *line c* the variation of c_{12}^M with x_2^M of ideal mixing

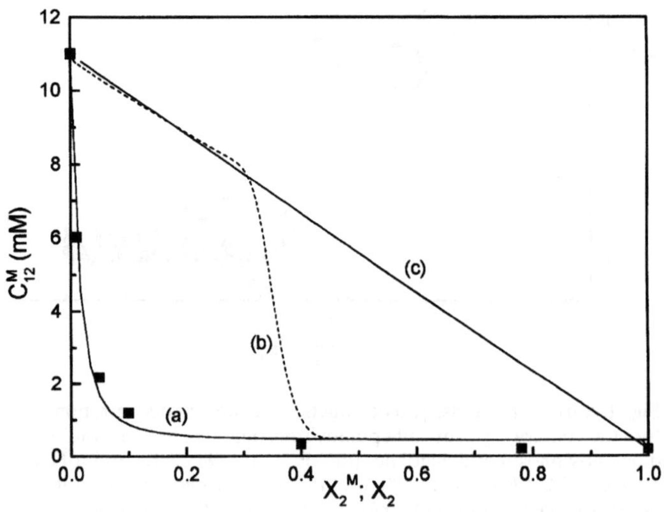

Fig. 4 Phase diagram of micelle formation of the diC_6PC– diC_8PC system. *Line a* represents the relationship between c_{12}^M and the mole fraction of diC_8PC in the solution, x_2, *line b* represents the variation of c_{12}^M with the mole fraction of diC_8PC in micelles, x_2^M calculated by Eq. (9) and *line c* the variation of c_{12}^M with x_2^M of ideal mixing

approximately 90%. This behaviour could be explained by the influence that diC_7PC has on the size of the mixed micelle. As generally known, diC_7PC forms rodlike micelles in solution, while diC_6PC forms spherical micelles [22, 23, 24]. An increase in the percentage of diC_7PC in the mixture must cause the micelles to become larger. Lin and coworkers [30, 31] have elaborated a thermodynamic theory to predict the microscopic change in the formation and growth of the diC_6PC/diC_7PC micelles. Their predictions, corroborated by dynamic light scattering experiments, show that the mixed diC_6PC/diC_7PC micelles are still rodlike micelles as the ratio of diC_7PC to diC_6PC increases. Our observations confirm this fact, although the thermodynamic model applied by us is different.

DiC_7PC/diC_8PC and diC_6PC/diC_8PC systems have more complexity. The number of carbon atoms in the acyl chain determines the variety of the shapes and sizes of the aggregates formed in aqueous solution: from micelles (spherical and nonspherical) to vesicles and lyotropic liquid-crystalline phases [36, 37, 38, 39]. DiC_8PC represents the transitions between the straightforward micellar structures that in aqueous solution are formed by phosphatidylcholines with acyl chains of fewer than eight carbon atoms and vesicles formed when this number is bigger than eight [22, 23, 24]. This suggests that the influence of diC_8PC becomes more dramatic in the nonideality of the mixtures that contain this component than, for example, diC_7PC mixed with diC_6PC. This can be seen in Figs. 3 and 4. When comparing these figures, note that the difference in the hydrophobicity between diC_6PC and diC_8PC, which is bigger than between diC_7PC and diC_8PC, influences the observed behaviour.

Lecithin–lecithin interaction

The nature and strength of the interaction between two surfactants in the mixture can be determined by calculating the values of their β_{12}^{M} parameter [6, 40]. Following the simplified approach proposed by Holland and Rubingh [6], the monomer concentration of each component in the micelle mixture, c_i^M, is given by

$$c_i^{M} = \alpha_i c_{12}^{M} \ , \tag{12}$$

where α_i is the mole fraction of the ith component in the total mixed solute.

For nonideal binary mixtures of surfactants, the activity coefficients are given by

$$f_i = \frac{\alpha_i c_{12}^{M}}{x_i^{M} c_i^{M}} \ . \tag{13}$$

Using a simple regular solution approximation, the activity coefficients can be expressed as functions of the mole fractions of each of the components in the mixed micelle, x_i^M and an appropriate interaction parameter, β_{12}^{M}, that is related to the net interaction between lecithin molecules in the mixed micelles. The regular solution approximation for the activity coefficients in the mixed micelles gives

$$f_1 = \exp \beta_{12}^{M}\left(1 - x_1^{M}\right)^2 \tag{14}$$

and

$$f_2 = \exp \beta_{12}^{M}\left(X_1^{M}\right)^2 \ . \tag{15}$$

β_{12}^{M} can be readily determined when the mixed cmc for the binary system is known, by interactively solving for x_i^M at the cmc using a relationship such as

$$\left(X_1^{M}\right)^2 \ln\left(\frac{\alpha_1 c_{12}^{M}}{x_1^{M} c_1^{M}}\right) = \left(1 - x_1^{M}\right)^2 \ln\left(\frac{\alpha_2 c_{12}^{M}}{\left(1 - x_1^{M}\right) c_2^{M}}\right) \tag{16}$$

obtained from Eqs. (13), (14) and (15). β_{12}^{M} can then be directly obtained by combining Eq. (13) with Eq. (14) to give

$$\beta_{12}^{M} = \frac{\ln\left(\alpha_1 c_{12}^{M} / x_1^{M} c_1^{M}\right)}{\left(1 - x_1^{M}\right)^2} \ . \tag{17}$$

Negative β_{12}^{M} values indicates attractive interaction and positive, repulsive interactions.

Values of β_{12}^{M} were obtained for the mixtures diC_6PC/diC_7PC, diC_6PC/diC_8PC and diC_7PC/diC_8PC under study using the computational method proposed by Holland and Rubingh [6] which is based on Eq. (16) and using cmc data of pure components.

The values obtained are shown in Fig. 5 as a function of molar fractions of component 2. The mixture of

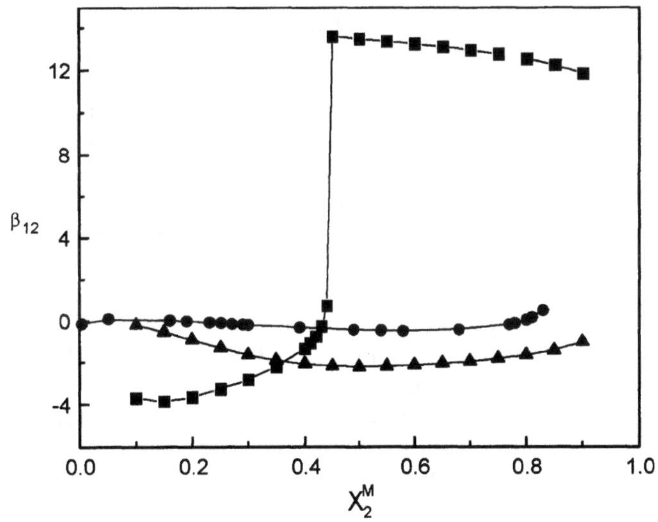

Fig. 5 Variation of molecular interaction parameter, β_{12}^{M}, against the mole fraction of component 2 in the solution, x_2, for the binary systems diC_6PC–diC_7PC (*circles*), diC_7PC–diC_8PC (*triangles*) and diC_6PC–diC_8PC (*squares*)

diC_6PC with diC_7PC does not show any significant interaction, as already discussed. The β_{12}^M are near zero (ideal mixing) over practically all the concentration range and confirm the results shown in Fig. 2. Values of β_{12}^M of about zero have been reported by Hines et al. [12] for mixtures of the zwitterionic C_{12} betaine and the nonionic C_{12} maltoside. Rosen and Zhu [41] obtained small values of β_{12}^M for the system $C_{12}BMG/C_{12}E_8$, which is a mixture of a zwitterionic similar to C_{12} betaine and the nonionic octaethylene glycol monodecylether. At pH 7.4 the lecithins studied are electrically neutral (electrophoretic mobility measurements [42] show an isoelectric point for synthetic lecithins of about 6) so there are no electrostatic interactions and hence the nonideality must be due to the influence of the acyl chain length of pure diC_6PC, diC_7PC and diC_8PC. DiC_8PC has more influence on the nonideality than diC_7PC as can be see by comparing the corresponding lines in Fig. 5. The interactions between molecules in the diC_6PC/diC_8PC system change dramatically at a ratio of approximately 1.5:1, from attractive ($\beta_{12}^M \approx -3$) to strongly repulsive ($\beta_{12}^M \approx 12$). Repulsive interaction in mixed micelles in aqueous media is unusual and for hydrocarbon-chain surfactants has been reported only in mixtures of long-chain carboxylates and long-chain alkylbenzensulfonates [43]. It is more commonly found in mixtures of anionic fluorocarbon-chain and anionic hydrocarbon-chain surfactants [1].

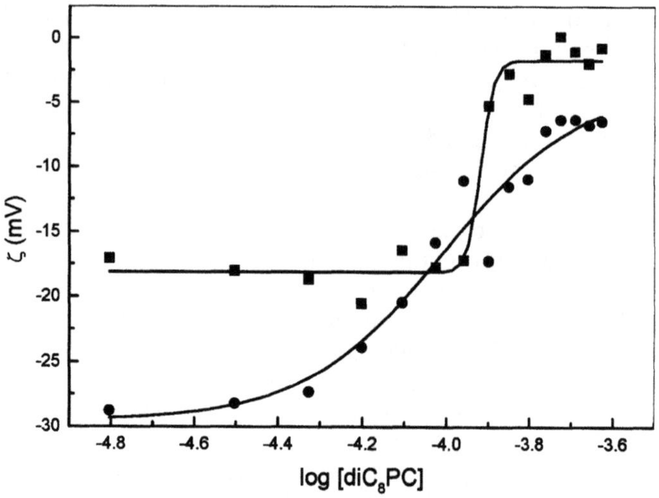

Fig. 6 ζ-potentials of human serum albumin (*HSA*) in aqueous solution at pH 7.4 in presence of diC_8PC. Ionic strength: 0.0188 M (*circles*), 0.188 M (*squares*)

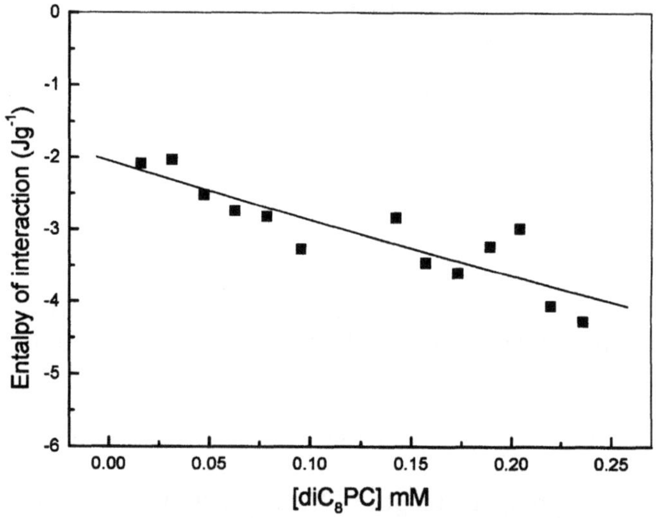

Fig. 7 Enthalpy of interaction of HSA in aqueous solution (pH 7.4, ionic strength 0.188 M) with diC_8PC at 25 °C

Lecithin–protein interaction

The HSA–diC_8PC interaction can be followed from measurements of the ζ potentials and by microcalorimetry.

The ζ potential of HSA + diC_8PC at pH 7.4 and two ionic strengths is shown in Fig. 6. As can be seen these ζ potentials are negative, which is consistent with previously reported isoelectric points of 4.2 [44], 4.9 [45], and 4.7–4.9 [46] for HSA and become less negative with increasing phospholipid concentration, smoothly at low ionic strength and more sharply at high ionic strength. The decrease in the negative ζ potential of HSA on addition of diC_8PC must arise from hydrophobic interactions of the acyl chains of the phospholipid and a change in the structure of the double layer.

The enthalpy of interaction between HSA and diC_8PC at pH 7.4 and ionic strength 0.188 M is shown in Fig. 7 as a function of the final diC_8PC concentration after mixing. Note that these data predominantly relate to diC_8PC concentrations below the cmc. The data were corrected for the enthalpy of dilution of the HSA, which was found to be exothermic (-1.04 ± 0.02 J g^{-1}), and for a final HSA concentration of 0.125% w/v. The enthalpy of dilution of diC_8PC was cancelled by mixing of diC_8PC and buffer in the reference cell. The exothermic effect of the HSA + diC_8PC interaction increases with phospholipid concentration. A possible explanation of this fact is the dissociation of dimers of HSA to monomers that become incorporated into HSA–diC_8PC complexes [29]. If we write the process occurring at low diC_8PC concentrations as an interaction between HSA dimers and $x$$diC_8PC$ molecules as

$$HSA_2 + x\,diC_8PC \leftrightarrow HSA_2diC_8PC_x \qquad (18)$$

and at high diC_8PC concentration an interaction between HSA monomers and the dimer–diC_8PC complexes which requires the partial dissociation of n dimers HSA_2, thus

$$nHSA_2 \leftrightarrow (n-1)HSA_2 + 2HSA \qquad (19)$$

$$(n-1)HSA_2diC_8PC_x+2HSA \leftrightarrow (n-1)HSA_{(2n/n-1)}diC_8PC_x \quad (20)$$

then at high diC_8PC concentrations the overall process may be written

$$nHSA_2 + (n-1)diC_8PC_x \leftrightarrow (n-1)HSA_{(2n/n-1)}diC_8PC_x. \quad (21)$$

If the dissociation of dimeric HSA to monomers is exothermic as the dilution data shows (see previously), then the increasing exothermicity observed at higher diC_8PC concentrations arises from dissociation of dimers to monomers that are incorporated into complexes.

Conclusions

Mixtures of short-chain lecithins are strongly influenced by the higher acyl chain component and the size and shape that the lecithin molecules form. DiC_6PC and diC_7PC mix ideally, with an interaction parameter near zero. DiC_7PC/diC_8PC and diC_6PC/diC_8PC show nonideality for approximately 75 and 40% diC_8PC, respectively. This fact suggests the prevalence of diC_8PC in the micellar behaviour of the mixture related with the type of aggregate that forms in solution.

The interaction between diC_8PC and HSA, followed from measurements of the ζ potential of HSA plus diC_8PC, shows a decrease in the negative ζ potential of HSA on addition of diC_8PC which must arise from hydrophobic interactions of the acyl chains of the phospholipid and a change in the structure of the double layer. The study of the interaction between diC_8PC and HSA, followed by microcalorimetry, shows an exothermic interaction at the pH studied (7.4). A possible explanation for the increasing exothermicity with increasing diC_8PC concentrations is the exothermic dissociation of HSA dimers to monomers that are incorporated into the HSA–diC_8PC complexes.

References

1. Scamehorn JF (ed) (1986) Phenomena in mixed surfactant systems. ACS symposium series 311. American Chemical Society, Washington, DC
2. Holland PM, Rubingh DN (eds) (1992) Mixed surfactant systems. ACS symposium series 501. American Chemical Society, Washington, DC
3. Ogino K, Abe M (eds) (1993) Mixed surfactant systems. Surfactant science series, vol 46. Dekker, New York
4. Clint J (1975) J Chem Soc Faraday Trans I 171:1372
5. Rubingh DN (1979) In: Mittal KL (ed) Solution chemistry of surfactants, vol 1. Plenum, New York
6. Holland PM, Rubingh DN (1983) J Phys Chem 87:1984
7. Motomura K, Matsukiyo M, Aratono M (1984) Colloid Polym Sci 262:948
8. Puvvada S, Blankschtein D (1992) J Phys Chem 96:5567
9. Sarmonia C, Puvvada S, Blankschtein D (1992) Langmuir 8:2690
10. Motomura K, Aratono M (1993) In: Ogino K, Abe M (eds) Mixed surfactant systems. Surfactant science series, vol 46. Dekker, New York
11. Aratono M, Villeneuve M, Takiue T, Ikeda N, Iyota H (1998) J Colloid Interface Sci 161:171
12. Hines JD, Thomas RK, Garrett PR, Rennie GK, Penfold J (1997) J Phys Chem 101:9215
13. Hines JD, Thomas RK, Garrett PR, Rennie GK, Penfold J (1998) J Phys Chem 102:8834
14. Mulqueen M, Blankschtein D (2000) Langmuir 16:7640
15. Tsujiik K, Mino J (1978) J Phys Chem 82:1610
16. Ikeda S, Tsunoda M, Maeda H (1979) J Colloid Interface Sci 70:448
17. Bluestein BR, Hilton CL (eds) (1982) Amphoteric surfactants. Surfactant science series, vol 12. Dekker, New York
18. Iwasaki T, Ogawa M, Esumi K, Meguro K (1991) Langmuir 7:30
19. Tausk RMJ, Karmiggelt J, Oudshoorn C, Overbeek JTG (1974) Biophys Chem 1:175
20. Tausk RMJ, Esch J van, Karmiggelt J, Voordouw G, Overbeek JTG (1974) Biophys Chem 1:184
21. Tausk RMJ, Oudshoorn C, Overbeek JTG (1974) Biophys Chem 2:53
22. Lin LT, Chen SH, Gabriel NE, Roberts MF (1986) J Am Chem Soc 108:3499
23. Lin LT, Chen SH, Gabriel NE, Roberts MF (1987) J Phys Chem 91:406
24 Lin LT, Tseng MY, Chen SH, Roberts MF (1990) J Am Chem Soc 94:7239
25. Hauser H (2000) Biochim Biophys Acta 1508:164
26. Martinez-Landeira P, Prieto G, Ruso JM, Sarmiento F (2002) Colloids Surf A 203:67
27. Martínez-Landeira P, Besada L, Seoane L, Ruso JM, Prieto G, Sarmiento F (2002) Mol Phys 100:1633
28. Martínez-Landeira P, Ruso JM, Prieto G, Sarmiento F (2002) J Chem Eng Data 47:1017
29. Martínez-Landeira P, Ruso JM, Prieto G, Sarmiento F, Jones MN (2002) Langmuir 18:3300
30. Lin T, Hu Y, Liu WJ, Samseth J, Mortensen K (1996) Prog Colloid Polym Sci 100:19
31. Lin TL, Hu Y, Liu WJ (1997) Langmuir 13:1422
32. Henry DC (1931) Proc R Soc Lond Ser A 133:106
33. Hunter RJ (1981) Zeta potential in colloid science. Academic, London, chap. 3
34. Wadsö I (1968) Acta Chem Scand 22:92
35. Attwood D, Patel HK (1989) J Colloid Interface Sci 129:222
36. Missel PJ, Mazer, NA, Benedek GB, Young CY, Carey MC (1980) J Phys Chem 84:1044
37. Hayter JB, Penfold P (1983) Colloid Polym Sci 261:1022
38. Israelachvili JN (1995) Intermolecular and surface forces. Academic, New York
39. Jones MN, Chapman D (1995) Micelles, monolayers and biomembranes. Wiley-Liss, New York

40. Rosen MJ (1989) Surfactant and interfacial phenomena, 2nd edn. Wiley, New York
41. Rosen MJ, Zhu BH (1984) J Colloid Interface Sci 99:427
42. Phillips MC, Chapman D (1968) Biochim Biophys Acta 163:301
43. Rosen MJ, Zhu ZH (1989) J Colloid Interface Sci 133:473
44. Taboada P, Mosquera V, Ruso JM, Sarmiento F, Jones MN (2000) Langmuir 17:6795
45. Houska M, Brynda E (1997) J Colloid Interface Sci 188:243
46. Bundschuh I, Jacklemeyer I, Luneberg E, Bentzel C, Petzodt R, Stotle H (1992) Eur J Clin Biochem 30:651

Progr Colloid Polym Sci (2003) 122: 149–156
DOI 10.1007/978-3-540-36114-5

Kazuyuki Miyazawa
Francoise M. Winnik

Isothermal titration calorimetry and fluorescence spectroscopy studies of the interactions between surfactants and a phosphorylcholine-based polybetaine

K. Miyazawa · F.M. Winnik (✉)
Department of Chemistry and Faculty of Pharmacy, Université de Montréal, CP 6128 Succursale Centre Ville, Montréal, QC H3C 3J7, Canada
e-mail: francoise.winnik@umontreal.ca
Tel.: +1-514-3436123;
Fax: +1-514-3432362

K. Miyazawa
Shiseido Basic Research Center, 2-2-1 Hayabuchi, Tsuzuki, Yokohama 224 8558, Japan

Abstract The interactions between a polybetaine and anionic, cationic, zwitterionic and neutral surfactants were studied by a fluorescence experiment with pyrene as a probe, isothermal titration calorimetry, and ^{1}H NMR spectroscopy. The polybetaine was a phosphorylcholine-based polymer (PNIPAM-PC) consisting of equimolar amounts of *N*-isopropylacrylamide and N-phosphorylcholine-*N'*-ethylenedioxybis (ethyl)acrylamide. Strong association took place between PNIPAM-PC and the anionic surfactants sodium *n*-dodecyl sulfate (SDS) and sodium *n*-hexadecyl sulfate (SHS) via a cooperative mechanism driven by electrostatic interactions between the surfactant headgroup and the trimethylammonium group of the PC moiety linked to the polymer. The onset of binding between PNIPAM-PC and SDS or SHS takes place for surfactant concentrations of 2.0 and 0.028 mmol l^{-1}, respectively, which are lower than their respective critical micelle concentrations (8.3 and 0.058 mmol l^{-1}). No interactions were detected between PNIPAM-PC and zwitterionic surfactants bearing either a phosphorylcholine headgroup (*n*-hexadecanoyl lysophosphocholine) or a dimethyl-3-ammonio-1-propanesulfonate group, cationic surfactants bearing a trimethylammonium headgroup, or neutral surfactants bearing a hepta(ethyleneglycol) headgroup. The mechanism of binding is discussed in terms of surfactant charge and chain length and is compared to the association of surfactants to polyampholytes and polyelectrolytes.

Keywords Polymer–surfactant interactions · Thermodynamics of micellization · Electrostatic binding · Pyrene fluorescence · Microcalorimetry

Introduction

Mixtures of water-soluble polymers and surfactants have found many applications in industrial fluids, such as those employed in enhanced oil recovery [1], and in consumer products formulations, such as cosmetics [2] and drug delivery systems [3]. In parallel with product development investigations, systematic fundamental studies of various combinations of surfactants and macromolecules have been carried out. There are numerous studies on systems consisting of surfactants and either nonionic polymers, polyelectrolytes, or polyampholytes, but data on the interactions of polybetaines, which are polyampholytes carrying oppositely charged groups on the same monomer unit, are rather scarce. We prepared recently phosphorylcholine-based polybetaines, which may present applications in cosmetics and pharmaceutical formulations [4]. It was important to assess the interactions of these polymers with various surfactants in order to lead formulation chemists in their product design. The results of this fundamental study are reported here.

As little is known on polybetaine/surfactant systems, it is useful to review briefly the trends that have emerged from studies of various polymer/surfactant systems. The strength of the interactions between nonionic polymers and surfactants depends both on the chemical nature of the polymer and on the charge of the surfactant headgroup [5]. Whereas neutral polymer/negatively charged surfactant interactions are usually strong, neutral polymers interact weakly, or not at all, with cationic or neutral surfactants [6]. When interactions occur, they begin at a "specific" surfactant concentration, usually called the critical aggregation concentration (cac or T_1), which is below the surfactant critical micelle concentration (cmc) [5].

Interactions between polyelectrolytes and surfactants of opposite charge are strong, owing to the Coulombic attraction between surfactant headgroups and polymer-linked charges. The cac is 2–3 orders of magnitude lower than the cmc [7]. Mixing polyelectrolytes and oppositely charged surfactants often results in the precipitation of a polymer/surfactant complex of stoichiometric composition. In contrast, polyelectrolytes and surfactants of the same charge have very low affinity to each other, owing to electrostatic repulsion effects. Nonetheless, charged surfactant binding to polyelectrolyte of the same charge may occur if the polyelectrolyte carries hydrophobic groups, so that attractive lipophilic interactions overcome the electrostatic repulsion [8, 9].

There have been a number of studies on the interactions between anionic surfactants and natural polyampholytes, such as gelatin [10, 11], collagen [12], and lysozyme [13]. These systems exhibit an increase in solution viscosity with increasing surfactant concentration. Studies by small-angle neutron scattering [11] and pulsed-gradient spin-echo NMR spectroscopy [14] suggest that this increase can be attributed to the formation of a network. An early study of systems of charged surfactants and synthetic polyampholytes consisting of cationic and anionic groups linked to different monomer units suggested that the interactions trigger a reduction of the polymer chain dimension and that the properties of the resulting polyampholyte/surfactant complex depend greatly on the surfactant structure [15]. Partially charged synthetic polyampholytes were shown to interact with cationic or anionic surfactants, leading to soluble or insoluble systems depending on polyampholyte composition and concentration [16, 17, 18]. Mixed systems consisting of fatty acids and polyampholytes with alternating cationic and anionic monomer units were shown to assemble in the form of particles ranging in size from 3 to 5 nm [19]. Mamytbekov et al. [20] have investigated the interactions of anionic surfactants with monochloroacetic acid–alkylated poly(2-methyl-5-vinylpyridine) by potentiometric titrations and spectrophotometry. However, the interactions of surfactants with typical polybetaines, such as polycarboxybetaines, polysulfobetaines, polyphosphobetaines, poly[(*N*-3-sulfopropyl)-*N*-methacryloylethyl-*N*,*N*-dimethylammonium betaines, and poly(2-methacryloyloxyethylphosphorylcholines) [20], have not been investigated.

The polybetaine under scrutiny in this report is a copolymer of *N*-isopropylacrylamide (NIPAM) and *N*-phosphorylcholine-*N'*-ethylenedioxybis(ethyl)acrylamide (PNIPAM-PC, Fig. 1), where the neutral and zwitterionic comonomers are distributed at random along the chain in approximately equimolar amounts [4]. The polymer was shown to form aggregates in water, approximately 170 nm in diameter, presumably as a result of interpolymeric and intrapolymeric ion-pair formation among the phosphorylcholine residues [22]. We report here a study by fluorescence spectroscopy, isothermal tiration calorimetry, and ^{1}H NMR spectroscopy of the interactions of PNIPAM-PC with surfactants possessing anionic, cationic, zwitterionic, or neutral headgroups linked to octadecyl or dodecyl chains (Fig. 1).

Experimental

Materials

Water was deionized using a Milli-Q water purification system (Millipore). Sodium *n*-hexadecyl sulfate (SHS) was obtained from Lancaster Chemicals. *N*-*n*-Hexadecyl-*N*,*N*,*N*-trimethylammonium chloride (HTAC) was purchased from Tokyo Kasei Chemicals. Sodium *N*-*n*-hexadecyl-*N*,*N*-dimethyl-3-ammonio-1-propanesulfonate, sodium *n*-dodecyl sulfate (SDS), *N*-*n*-dodecyl-*N*,*N*,*N*-trimethylammonium bromide (DTAB), and sodium *N*-*n*-dodecyl-*N*,*N*-dimethyl-3-ammonio-1-propanesulfonate (DDAMPS) were obtained from Sigma Chemicals Corporation. The nonionic surfactants heptaethyleneglycol mono-*n*-hexadecyl ether ($C_{16}H_{33}EO_7$, Nikkol BC-4SY) and heptaethyleneglycol mono-*n*-dodecyl ether ($C_{12}H_{25}EO_7$, Nikkol BL-7SY) were purchased from Nikko Chemicals. 1-Hexadecanoyl-2-hydroxy-*sn*-glycero-3-phosphocholine (hexadecanoyl lysophosphocholine HL-PC) was purchased from Avanti Chemicals. The PNIPAM-PC (M_n 20,400, M_w 35,500, M_w/M_n 1.71, NIPAM 49.6 mol %, NPC 50.4 mol %), was prepared as reported previously [4].

Measurements

Fluorescence measurements were carried out with a Fluorolog Tau-3 spectrometer (Jobin-Yvon Horiba) equipped with a GRAMS/32 (Galactic) data analysis system. Temperature control of the samples was achieved using a water-jacketed cell holder connected to a Neslab circulating bath. All measurements were carried out at 25 °C. The slits were set at 0.5 mm (excitation) and 0.25 mm (emission). The excitation wavelength was 334 nm. The excitation spectra were monitored at 376 nm. The I_1/I_3 ratio was calculated as the ratio of the intensity at 376 nm to that at 386 nm. The cmc or cac values were taken as the surfactant concentration corresponding to the midpoint of the transition of a plot of I_1/I_3 versus surfactant concentration. Samples for analysis were prepared by allowing PNIPAM-PC (polymer concentration 0.1 g l^{-1}) to dissolve in pyrene-saturated water previously filtered to remove pyrene microcrystals. Small aliquots of surfactant solutions in pyrene-saturated water were added to the polymer

Fig. 1 Chemical structures of the copolymer of *N*-isopropylacrylamide and *N*-phosphorylcholine-*N'*-ethylenedioxybis(ethyl) acrylamide (*PNIPAM-PC*) and of the surfactants used in this study

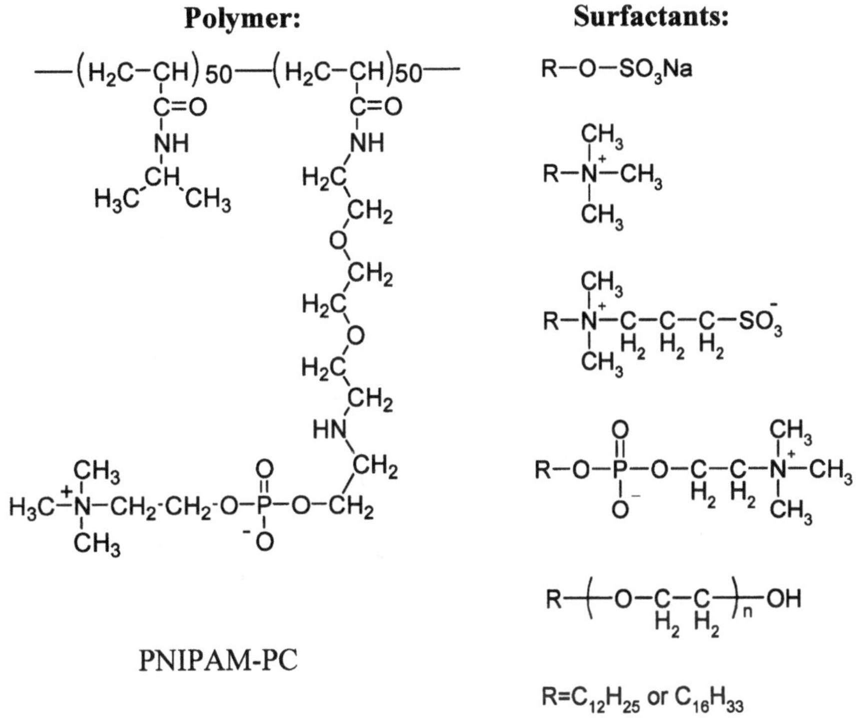

solution in amounts such that the polymer concentration remained constant within 10%. Mixed solutions were kept in the dark at room temperature for at least 12 h prior to measurements.

Isothermal titration calorimetry (ITC) measurements were performed with a VP-ITC titration microcalorimeter (MicroCal). The microcalorimeter consists of a reference cell and a sample cell both of 1.43 ml volume. The sample cell is connected to a 250-μl syringe. The tip of the syringe is fitted with a blade-type stirrer rotating at 400 rpm that ensures continuous mixing. Titrations were carried out by consecutive 5-μl injections of an aqueous surfactant solution from the syringe into the sample cell filled with water or an aqueous PNIPAM-PC solution (1.0 g l^{-1}). A total of 58 consecutive injections were performed. The delay between two consecutive injections was 300 s. All measurements were carried out at 25 °C. The data were analyzed using the MicroCal ORIGIN software. The surfactant cmc was taken as the midpoint of the transition of the surfactant concentration-dependent heat of dilution curve determined for each surfactant. The heat of micellization was taken as the difference between the heat of surfactant dilution in water before and after the transition of the curve Q versus surfactant concentration corresponding to the cmc.

^{1}H NMR spectra were recorded at room temperature using a Bruker ARX-400 spectrometer operating at a frequency of 400 MHz. Solutions for analysis were prepared by adding aliquots of surfactant (SDS or DTAC) solutions in D_2O to a solution of PNIPAM-PC in D_2O (4.0 g l^{-1}).

Results and discussion

Fluorescence probe studies

A polymer-induced shift to lower concentration of cmc of a surfactant is diagnostic of the occurrence of polymer/surfactant interactions [5]. This shift is easily detectable by fluorescence probe experiments. The method involves the use of pyrene, a fluorescent dye which exhibits different fluorescence characteristics depending on the polarity of the medium [23]. The intensity, I_1, of the (0,0) band of the pyrene emission increases with increasing polarity of the probe environment, while the intensity, I_3, of the (0,2) band is not affected, so the ratio I_1/I_3 reflects the micropolarity sensed by the probe. In solutions of surfactant concentration below the cmc, pyrene resides in water and its emission spectrum reports a polar environment ($I_1/I_3 \sim 1.80$). In surfactant solutions of concentration higher than the cmc, pyrene partitions preferentially into the micellar interior, where it experiences a nonpolar environment, and, consequently, the I_1/I_3 ratio takes a lower value, characteristic of the polarity within the micelle. This technique was used to assess the effect of the charge of the surfactant headgroup on its interactions with PNIPAM-PC using a series of hexadecyl derivatives. The cmc of each surfactant was measured in the presence of PNIPAM-PC (0.1 g l^{-1}, 0.19 mmol l^{-1} PC units) and this value was compared to the cmc of the surfactant in water, determined under identical conditions.

The results of the probe experiment obtained for the PNIPAM-PC/HTAC system are presented in Fig. 2, where we plot the variations of the ratio I_1/I_3 as a function of HTAC concentration in the presence of PNIPAM-PC and in water. In both cases the ratio remains constant (1.82 ± 0.2) up to an HTAC concentration of around 0.95 mmol l^{-1} and thereafter it decreases sharply to reach a constant value (1.35 ± 0.02). We note that the midpoint

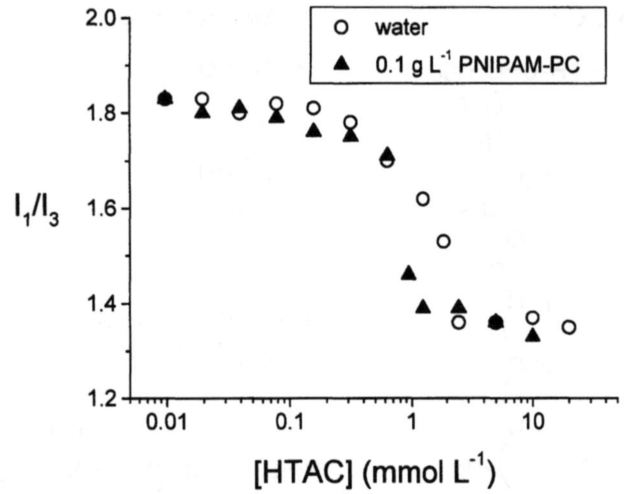

Fig. 2 Plot of the changes of I_1/I_3 for the emission of pyrene in aqueous PNIPAM-PC (0.1 g l^{-1}) and in water as a function of *N-n*-hexadecyl-*N,N,N*-trimethylammonium chloride (*HTAC*) concentration (logarithmic scale); temperature 25 °C, λ_{exc} 334 nm

of the transition is slightly lower for the mixed polymer/surfactant system, compared to the surfactant solution in water. However, the two curves nearly coincide, indicating that the polymer does not interfere significantly with the micellization of HTAC, or that the effect is not large enough to be detected by fluorescence probe techniques. Similar results were obtained with the PNIPAM-PC/neutral surfactant [$C_{16}H_{33}(OCH_2CH_2)_7OH$] system, as well as the systems of PNIPAM-PC and the two zwitterionic surfactants DDAMPS and HL-PC. It is only in the case of the PNIPAM-PC/SHS that we observed a significant effect of the polymer on the micellization of the surfactant: the midpoint of the transition in the SHS-concentration-dependent I_1/I_3 curve decreased by nearly 1 order of magnitude with respect to the cmc of the surfactant. Data gathered for all the surfactants are presented in Table 1.

Isothermal titration calorimetry

The fluorescence probe experiments described in the previous section allow us to conclude that only anionic surfactants are able to interact with PNIPAM-PC, presumably via ionic attraction between the negatively charged surfactant headgroups and the trimethylammonium group of the phosphorylcholine units linked to the polymer. To gain further insight into the mechanism of binding of anionic surfactants to PNIPAM-PC we carried out measurements by ITC, a titration technique giving access to the thermodynamic parameters associated with the binding process [24]. In a typical ITC experiment, aliquots of a micellar surfactant solution are titrated at constant temperature into a polymer solution or into water. After each injection, heat is released or absorbed, as a result of the various processes occurring in the solution. The enthalpy per injection is determined as a function of surfactant concentration. The enthalpogram recorded over the entire surfactant concentration range is then analyzed in terms of the following concentrations, referring to the total amount of added surfactant to the polymer solution:

1. The concentration signaling the onset of binding (cac).
2. The concentration, c_2, corresponding to the saturation of polymer with bound surfactant.
3. The concentration for which free micelles start to coexist in solution with polymer-bound surfactant (cmc) [25].

A titration of the surfactant into water needs to be carried out under strictly identical conditions, in order to

Table 1 Spectroscopic data for pyrene in an aqueous solution of the copolymer of *N*-isopropylacrylamide and *N*-phosphorylcholine-*N'*-ethylenedioxybis(ethyl)acrylamide (*PNIPAM-PC*) and critical micellar concentrations (*cmc*) of the surfactants in water and in an aqueous PNIPAM-PC solution. Surfactants: sodium *n*-hexadecyl sulfate (*SHS*), *N-n*-hexadecyl-*N,N,N*-trimethylammonium chloride (*HTAC*), sodium *N-n*-hexadecyl-*N,N*-dimethyl-3-ammonio-1-propanesulfonate (*HDAMPS*), heptaethyleneglycol mono-*n*-hexadecyl ether ($C_{16}H_{33}EO_7$), 1-hexadecanoyl-2-hydroxy-*sn*-glycero-3-phosphocholine (*HL-PC*)

Surfactant	cmc (mol l^{-1})		I_1/I_3 (initial)	I_1/I_3 (final)[a]	cac (mol l^{-1})
	Reported	Measured			
SHS	5.8×10^{-4}[c,d]	5.8×10^{-4}	1.81	1.22	2.8×10^{-5}
HTAC	1.3×10^{-3}[e]	1.9×10^{-3}	1.81	1.32 (1.33)	1.9×10^{-3}[b]
HDAMPS	1.7 to 2.1×10^{-6}[f]	4.4×10^{-5}	1.81	1.36 (1.36)	3.9×10^{-5}[b]
$C_{16}H_{33}EO_7$	7.0×10^{-6}[g]	5.2×10^{-5}	1.81	1.41 (1.41)	5.2×10^{-5}[b]
HL-PC		2.5×10^{-5}	1.82	1.30 (1.30)	2.7×10^{-5}[b]

[a] The values in parentheses were determined for solutions of the surfactants in water (no added polymer)
[b] The cmc values recorded for solutions of the surfactants in an aqueous PNIPAM-PC solution; polymer concentration 0.1 g l^{-1}
[c] Measured at 40 °C
[d] Ref. [29]
[e] Ref. [30]
[f] Ref. [31]
[g] Ref. [32]

determine the micellization parameters of the surfactant and to compare these with the thermodynamics of surfactant binding to the polymer.

Experimental titration curves obtained for an SDS solution (80 mmol l^{-1}) in water and in an aqueous PNIPAM-PC solution (1.0 g l^{-1}, 1.9 mmol l^{-1} PC units) are shown in Fig. 3. For the titration in water (Fig. 3a) each injection produces an endothermic signal which increases then decreases with the number of injections. The dilution enthalpy of the surfactant as a function of surfactant concentration is determined by integration of each peak. The resulting enthalpogram (Fig. 3b) presents a transition point (around 8.3 mmol l^{-1}) corresponding to the cmc of SDS, in agreement with previous studies. The heat absorbed upon injection of the surfactant into water before reaching the cmc is the sum of several contributions: the heat of dilution of the micelles, the enthalpy of demicellization, and the heat of dilution of individual surfactant molecules. In the 8.3–13 mmol l^{-1} concentration range the enthalpy decreases sharply, signaling that the cmc has been reached in the cell and, as more micellar solution is added, the enthalpy remains nearly constant. The micelles no longer disintegrate, and the only heat recorded is caused by the dilution of the micelles.

Injection of aqueous SDS into a PNIPAM-PC solution initially produces an endothermic signal (SDS concentration below 2 mmol l^{-1}), then the signal becomes more and more exothermic, passes through a minimum, (SDS concentration around 3.5 mmol l^{-1}), before finally merging with the corresponding values measured in the absence of polymer for an SDS concentration above 12 mmol l^{-1} (Fig. 3c, d). We note that the SDS concentration for which the ITC signal becomes exothermic (around 2 mmol l^{-1}) corresponds to the solution concentration of polymer-linked PC units (1.9 mmol l^{-1}). The onset of surfactant binding to the polymer, which should be observed as a small maximum of the dilution enthalpy [26], can hardly be detected in the enthalpogram. The cooperative formation of bound SDS aggregates takes place almost immediately after initial surfactant/polymer binding, as expected for an electrostatically driven process. Note that the SDS concentration for which the dilution curves in water and in aqueous polymer solution merge is slightly higher than the cmc of SDS, implying that there may exist a surfactant concentration range for which binding to the polymer and micellization are in competition.

Titrations of neutral, zwitterionic, and cationic surfactants into PNIPAM-PC solutions were carried out as well. Titration curves obtained upon addition of micellar solutions of the zwitterionic surfactant DDAMPS and of the neutral surfactant $C_{12}H_{25}(OCH_2CH_2)_7OH$ into water or into a PNIPAM-PC solution (1.0 g l^{-1}) are presented in Fig. 4. For each system, the enthalpograms recorded in the presence and in the absence of polymer superimpose. Neither the cmcs nor the enthalpies of micellization of the surfactants are affected by the polymer (Table 2), confirming the conclusions drawn from the fluorescence probe experiments.

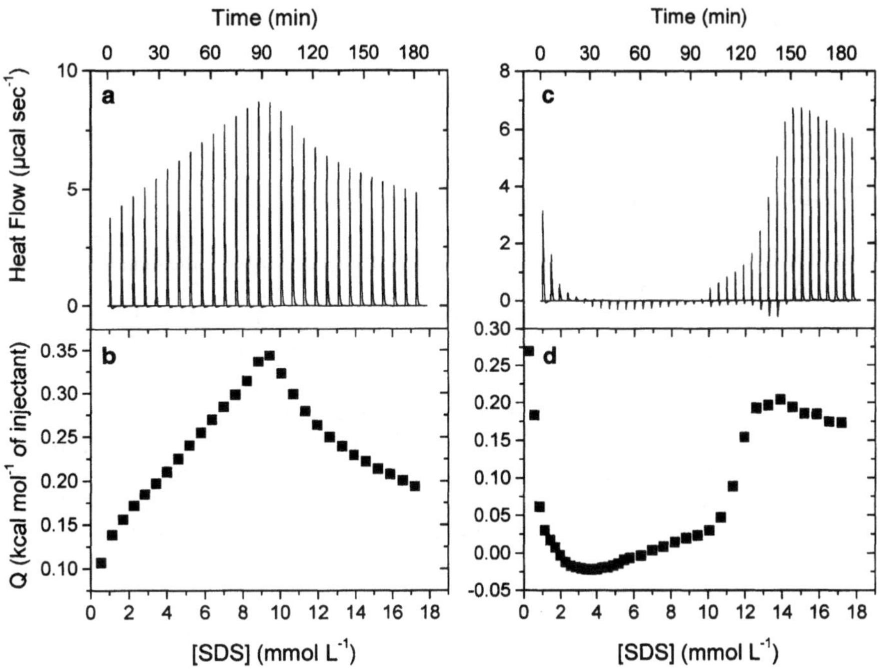

Fig. 3 **a** Calorimetric trace of the titration of sodium dodecyl sulfate (*SDS*) micelles in water at 25 °C; **b** reaction enthalpy, *Q*, observed upon dilution of 80 mmol l^{-1} SDS into water at 25 °C; **c** calorimetric trace of the titration of SDS micelles into an aqueous PNIPAM-PC solution (1.0 g $^{-1}$) at 25 °C; **d** *Q* observed upon dilution of 80 mmol l^{-1} SDS into an aqueous PNIPAM-PC solution (1.0 g l^{-1}) at 25 °C. The *x*-axis is the concentration of the surfactant in the cell; note that the scales of the *y*-axes are different in **a**, **c** and **b**, **d**, respectively

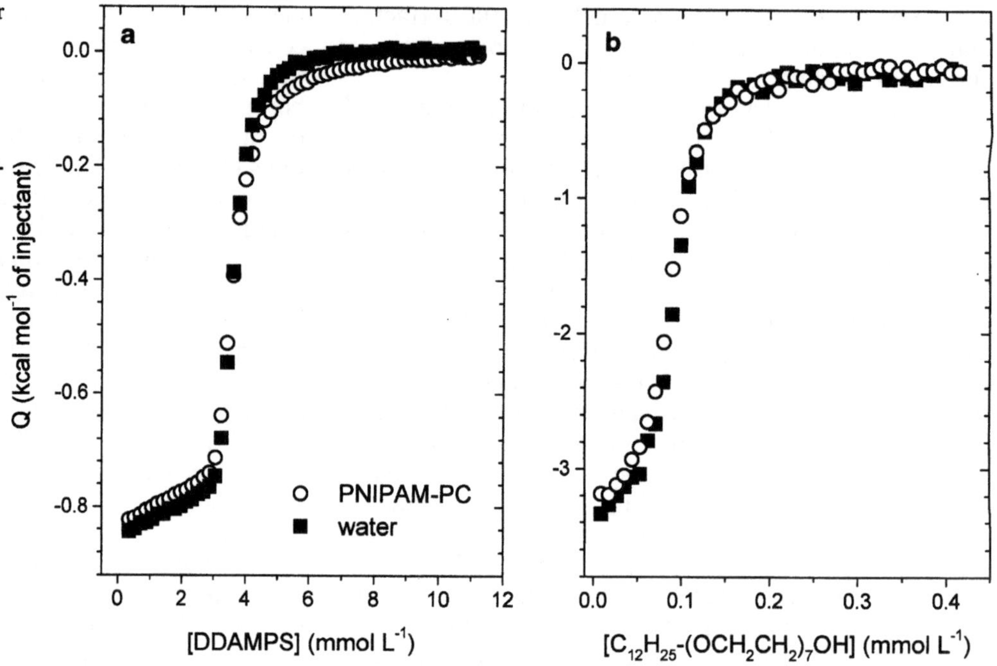

Fig. 4 Dilution enthalpograms for **a** sodium *N-n*-dodecyl-*N,N*-dimethyl-3-ammonio-1-propanesulfonate (*DDAMPS*) and **b** $C_{12}H_{25}(OCH_2CH_2)_7OH$ in water and in an aqueous PNIPAM-PC solution (1.0 g l^{-1}) at 25 °C. The *x*-axis represents the concentration of the surfactant in the cell; note that the scales of the *y*-axes are different in **a** and **b**

Table 2 Micellization parameters of various surfactants in water at 25 °C. Surfactants: sodium *n*-dodecyl sulfate (*SDS*), *N-n*-dodecyl-*N,N,N*-trimethylammonium bromide (*DTAB*), odium *N-n*-dodecyl-*N,N*-dimethyl-3-ammonio-1-propanesulfonate (*DDAMPS*), heptaethyleneglycol mono-*n*-dodecyl ether ($C_{12}H_{25}EO_7$)

Surfactant	cmc (mol l^{-1})		ΔH_m (kJ mol^{-1})
	Measured	Reported	
SDS	8.2×10^{-3}	8.1×10^{-3}[a]	−0.65
DTAB	1.6×10^{-3}	1.6×10^{-2}[b]	−1.86
DDAMPS	3.5×10^{-3}	–	3.13
$C_{12}H_{25}EO_7$	8.6×10^{-5}	8.0×10^{-5}[c]	

[a] Ref. [33]
[b] Ref. [34]
[c] Ref. [35]

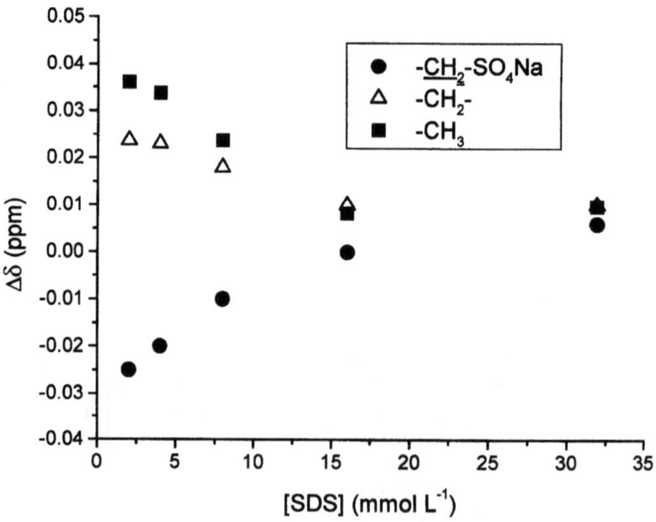

Fig. 5 Plots of the changes, Δδ, in the proton chemical shifts (ppm) of the signals due to the terminal methyl group, the methylene chain, and the α-methylene group of SDS as a function of surfactant concentration for SDS in water and in an aqueous PINAM-PC solution (4.0 g l^{-1}); Δδ = δ(PNIPAM−PC/SDS)−δ(SDS)

^{1}H NMR measurements

Further insight into the mechanism of binding of SDS to PNIPAM-PC was obtained by measuring the polymer-induced changes in the chemical shifts of the surfactant proton resonances. The underlying assumption of the approach is that the chemical shift values, δ, of the protons of the surfactant hydrophobic chain depend on the aggregation state of the surfactant and that the chemical exchange between all aggregation states is rapid on the NMR scale [27]. In this study, we examined the changes of the chemical shifts of the SDS signals due to the resonances of the α-methylene group (−CH_2-SO_4^-), the methylene groups of the alkyl tail [−$(CH_2)_{10}$−CH_2−SO_4^-], and the terminal methyl group [H_3C−$(CH_2)_{11}$−SO_4^-] in D_2O solutions of constant PNIPAM-PC concentration (4.0 g l^{-1}, 7.6 mmol l^{-1} PC units) and increasing amounts of SDS, ranging from 2.5 to 32 mmol l^{-1}. The chemical shifts were compared to the corresponding signals in solutions of SDS of identical concentration, but without added polymer. The differences in the chemical shifts Δδ = δ(PNIPAM−PC/SDS solution)−δ(SDS solution) for the terminal methyl group, the methylene chain, and the α-methylene group are plotted in Fig. 5 as a function of SDS concentration. The largest differences are observed in the premicellar range. The chemical shift of the headgroup of SDS experiences a significant upfield shift in the presence of the polymer, while the chemical shifts of the methylene chain and terminal methyl groups exhibit a downfield shift, compared to the values recorded in the absence of polymer. It can be inferred from these results that the anionic headgroup is involved in the interactions driven primarily by Coulombic forces. An identical set of measurements was performed on the PNIPAM-PC/DTAB system. The chemical shifts of the surfactant protons were unaffected by the presence of polymer, and this for surfactant concentrations ranging from 2 to 30 mmol l^{-1}.

It has been shown in previous ^{1}H NMR studies of polymer/surfactant interactions that line shapes also provide information about the structure of the polymer/surfactant complex. For example, line broadening of the surfactant proton signals suggests a restriction of the surfactant motion as a result of its interactions with a polymer [27, 28]. While such effects may take place in the case of PNIPAM-PC/SDS, the surfactant proton signals, in the low surfactant concentration range, were too weak for accurate assessment of the effect.

Conclusions

The main results obtained in this study by fluorescence spectroscopy, ITC and NMR spectroscopy of the interactions of surfactants with a phosphorylcholine-bearing polybetaine can be summarized as follows.

1. The charge of the surfactant headgroup is the major determinant that controls the interactions: only anionic surfactants associate with PNIPAM-PC. There is no binding between PNIPAM-PC and zwitterionic, cationic, or neutral surfactants, at least in the premicellar concentration domain. The experimental methods employed here did not allow us to assess the interactions of surfactants micelles with the polymer for surfactant concentrations above the cmc. Fluorescence measurements with a pyrene labeled PNIPAM-PC have been carried out in order to elucidate this point.
2. For systems where interactions occur, PNIPAM-PC/SDS and PNIPAM-PC/SHS, the surfactant concentration at which binding first takes place is very low. It was not possible to determine it accurately by ITC. Both ITC and fluorescence probe measurements yielded cacs significantly lower than the cmc. The enthalpograms and the curves of I_1/I_3 versus surfactant concentration obtained in the fluorescence probe experiments point to a cooperative binding mechanism.
3. The first step in the association between PNIPAM-PC and anionic surfactants occurs via electrostatic attraction between the surfactant headgroups and the polymer-bound trimethylammonium groups. As more and more surfactant molecules bind to the polymer, the PC/PC intrapolymeric and interpolymeric ion pairs that exist in surfactant-free polymer solutions are disrupted and the large polymer aggregates disintegrate and are replaced by free polymer chains decorated with SDS clusters.
4. None of the mixed systems investigated underwent macroscopic phase separation, a fact which may be attributable to the presence of the neutral comonomer (NIPAM), which may act as a solubilizing agent.

At this stage it is interesting to compare the effect of surfactants on the behavior of polyelectrolytes (with one type of charge) and of polyampholytes to that of the polybetaines used here. Addition of an ionic surfactant to a solution of oppositely charged polyelectrolyte results in the precipitation of a stoichiometric polyelectrolyte/surfactant complex, even at relatively low polymer concentration. This was not observed in the case of PNIPAM-PC/SDS or PNIPAM-PC/SHS. With polyampholytes, interactions occur with both anionic and cationic surfactants [16]. Partially charged polyampholytes are often insoluble in water and association with either anionic or cationic surfactants results in solubilization via neutralization of one type of charge. The remaining nonneutralized charges confer polyelectrolyte character to the polyampholyte, ensuring its solubility. The polybetaines described here act as weak polycations in their interactions with surfactants, but their solution properties are modulated also by the occurrence of ion pairing amongst the polymer-bound PC moities.

Acknowledgements This work was supported in part by a grant of the Natural Sciences and Engineering Research Council of Canada to F.M.W. K.M. thanks Shiseido Co. Ltd. for giving him the opportunity to study in the Université de Montréal and for financial support.

References

1. Desai NN, Shah DO (1981) Polym Prepr Am Chem Soc Div Polym Chem 22:39
2. Goddard ED (1999) In: Goddard ED, Gruber JV (eds) Principles of polymer science and technology in cosmetics and personal care. Dekker, New York, pp 181–216
3. Saltzman, WM (2001) Drug delivery, engineering principles for drug therapy. Oxford University Press, Oxford, pp 199–314

4. Miyazawa K, Winnik FM (2002) Macromolecules 35:2440
5. Goddard ED (1993) In: Goddard ED, Ananthapadmanabhan KP (eds) Interactions of surfactants with polymers and proteins. CRC, Boca Raton, pp 123–170
6. Saito S, Anghel DF (1998) In: Kwak JTC (ed) Polymer–surfactant systems. Surfactant science series, vol 77. Dekker, New York, pp 357–408
7. Goddard ED (1993) In: Goddard ED, Ananthapadmanabhan KP (eds) Interactions of surfactants with polymers and proteins. CRC, Boca Raton, pp 171–203
8. Winnik FM, Regismond S, Goddard ED (1996) Colloids Surf A 106:243
9. Anthony O, Zana R (1996) Langmuir 12:667
10. Whitesides TH, Miller DD (1994) Langmuir 10:2899
11. Cosgrove T, White SJ, Zarbakhash A, Heenan RK, Howe AM (1995) Langmuir 11:744
12. Henriquez M, Lissi E, Abuin E, Ciferri A (1994) Macromolecules 27:6834
13. Murakami K (1999) Langmuir 15:4270
14. Griffith PC, Stilbs P, Howe AM, Whitesides TH (1996) Langmuir 12:5302
15. Bekturov EA, Kudaibergebov SE, Kanapyanova GS (1984) Polym Bull 11:551
16. Harrison IM, Candau F, Zana R (1999) Colloid Polym Sci 277:48
17. Berkturov EA, Kudaibergenov SE, Kanapyanova GS (1991) Makromol Chem Rapid Commun 12:37
18. Kudaibergenov (2002) Polyampholytes: synthesis, characterization and application. Kluwer/Plenum, New York, pp 96–101
19. Thünemann AF, Sander K, Jaeger W, Dimova R (2002) Langmuir 18:5099
20. Mamytbekov GK, Abilov ZA, Shekeeva BZ, Musabekov KB, Bekturov EA (1996) Kazakh Izvest Minister Nauk-Akad Nauk Resp Kazakh Ser Khim 6:85 (CAN 127:347921)
21. Kudaibergenov (2002) Polyampholytes: synthesis, characterization and application. Kluwer/Plenum, New York, pp 137–152
22. Miyazawa K, Winnik FM (2002) Macromolecules 35:9536
23. Winnik FM, Regismond STA (1998) In: Kwak JTC (ed) Polymer–surfactant systems. Surfactant science series, vol 77. Dekker, New York, pp 267–316
24. Blandamer MJ (1998) In: Ladbury JE, Chowdry BZ (eds) Biocalorimetry: applications of calorimetry in the biological sciences. Wiley, New York, pp 5–26
25. Olofsson G, Wang G (1998) In: Kwak JTC (ed) Polymer–surfactant systems. Surfactant science series, vol 77. Dekker, New York, pp 317–356
26. Li Y, Ghoreishi SM, Warr J, Bloor DM, Holzwarth JF, Wyn-Jones E (1999) Langmuir 15: 6326
27. Stilbs P (1998) In: Kwak JTC (ed) Polymer–surfactant systems. Surfactant science series, vol 77. Dekker, New York, pp 239–316
28. Chen A, Wu D, Johnson CS Jr (1995) J Phys Chem 99:828
29. Evans HC (1956) J Chem Soc 576
30. Ralson AW, Eggenberg DN, Du Brow PL (1943) J Am Chem Soc 65: 686
31. van Os NM, Haak JR, Rupert LAM (1993) Physico-chemical properties of selected anionic, cationic and nonionic surfactants. Elsevier, Amsterdam, p 222
32. Stafford RE, Fanni T, Dennis EA (1989) Biochemistry 28: 5113
33. Lin W (1955) Bull Chem Soc Jpn 28:227
34. Scott AB, Tartar HV (1943) J Am Chem Soc 66:1083
35. van Os NM, Haak JR, Rupert LAM (1993) Physico-chemical properties of selected anionic, cationic and nonionic surfactants. Elsevier, Amsterdam, pp 216

Progr Colloid Polym Sci (2003) 122:157

AUTHOR TITLE INDEX

Progr Colloid Polym Sci (2003) 122: 158

KEY WORD INDEX

MIX
Papier aus verantwortungsvollen Quellen
Paper from responsible sources
FSC® C105338
FSC
www.fsc.org

If you have any concerns about our products,
you can contact us on
ProductSafety@springernature.com

In case Publisher is established outside the EU,
the EU authorized representative is:
Springer Nature Customer Service Center GmbH
Europaplatz 3, 69115 Heidelberg, Germany

Printed by Libri Plureos GmbH
in Hamburg, Germany